Engineering Design and Pro/ENGINEER

GUANGMING ZHANG

Director, Advanced Design and Manufacturing Laboratory

Department of Mechanical Engineering and the Institute for Systems Research

The University of Maryland @ College Park

COLLEGE HOUSE ENTERPRISES, LLC
Knoxville, TN

USE AT YOUR OWN RISK

The author and the publisher have expended every effort to avoid errors in preparing this textbook. However, we are human and errors are inevitable. Readers and users of this book are cautioned to verify the results and the designs suggested in the text. The author and the publisher cannot guarantee, and do not assume responsibility for any aspect of any development based on the information contained herein.

THANKS FOR PERMISSIONS

The author and the publisher thank the copyright and trademark owners providing permission for use of copyrighted materials and trademarks. We apologize for any errors or omissions in obtaining permissions. Where appropriate, we referenced similar treatments and gave credit for prior work. Errors or omissions in obtaining permissions or in giving proper references are not intentional. We will correct them at the earliest opportunity after the error or omission is brought to our attention. Please contact the publisher at the address given below.

The manuscript was prepared in Microsoft Word 97 using 12 point Times New Roman font. The book was printed from camera ready copy by Publishing and Printing Inc., Knoxville, TN.

College House Enterprises, LLC.
5713 Glen Cove Drive
Knoxville, TN 37919-8611, U. S. A.
Phone (423) 558 6111
FAX (423) 584 1766

ISBN 0-9655911-5-8

To my wife Jinyu
and my children Zumei and Haowei

who have always been with me especially at those difficult times. Their love and understanding have supported me in a journey for survival and success.

ABOUT THE AUTHOR

Guangming Zhang obtained a bachelor degree and a master degree both in mechanical engineering from Tianjin University, the Peoples Republic of China. He obtained a master degree and a Ph. D. degree in mechanical engineering from the University of Illinois at Urbana-Champaign. He is currently an associate professor in the Department of Mechanical Engineering and Director of the Advanced Design and Manufacturing Laboratory at the University of Maryland at College Park. He holds a joint appointment with the Institute for Systems Research.

Professor Zhang worked at the Northwest Medical Surgical Instruments Factory in China where he served as a principal engineer to design surgical instruments and dental equipment. He also taught at the Beijing Institute of Printing and received the National Award for Outstanding Teaching from the Press and Publication Administration of the Peoples Republic of China in 1987. In 1992 he was selected by his peers at the University of Maryland to receive the Outstanding Systems Engineering Faculty Award. He was the recipient of the E. Robert Kent Outstanding Teaching Award of the College of Engineering in 1993. He was a recipient of the 1992 Blackall Machine Tool & Gage Award of the American Society of Mechanical Engineers. In 1995 he received the Award of Commendation from the Society of Manufacturing Engineers, Region 3, for his outstanding service as the faculty advisor to the SME Student Chapter at College Park.

He has actively participated in the NSF sponsored ECSEL grant since 1990. He currently serves as the principal investigator for this grant at the University of Maryland, and coordinates the ECSEL sponsored projects on integration of design, on active learning and hands-on experiences, and on developing methods for team learning. He is a member of the editorial boards of the International Journal of Flexible Automation and Intelligent Manufacturing and the International Journal of Advanced Manufacturing Systems. He has written about 70 technical papers and holds one patent.

PREFACE

This book presents a comprehensive treatment of engineering design with a focus on solutions based on information technology. With capabilities of computers expanding at an unthinkable rate, the importance of using advanced computer-aided design (CAD) systems in engineering design must be emphasized. Pro/ENGINEER, a leading CAD system, is described to demonstrate the role of the computer in assisting engineers in the design process. The book is written as an introductory text for undergraduate students in engineering in all specialty areas (e.g., mechanical, aerospace, civil, electrical, chemical, industrial, materials, and fire protection engineering). The book should also be useful to those engaged in the product design.

The book is organized into six chapters. The first three chapters provide a fundamental coverage of engineering design. A practical approach that follows national and international standards related to engineering graphics, dimensioning, and tolerancing is stressed. Representative examples are provided to demonstrate industrial applications. A systematic description of the Pro/ENGINEER design system is presented in Chapter 4. The chapter begins with a description of feature-based solid modeling, and then the preparation engineering drawings is illustrated with many examples. The concept of virtual assembly is introduced, and the chapter ends with a discussion of the concept of parametric design. Chapter 5 describes the application of finite element analysis, and the application of numerically controlled machining is covered in Chapter 6. Both Chapters 5 and 6 demonstrate the integration of a CAD system, an engineering analysis tool, and a computer-aided manufacturing system under the Pro/ENGINEER design environment.

The material covered in this book is an outgrowth of several design and manufacturing courses taught by the author at the University of Maryland. Therefore, the author is indebted to many people, both students and faculty, who have made many contributions. The author especially thanks Mr.Lixun Qi, Mr. Yuqing Cao, Mr. Bing Chen, and Mr. Xin Che for their assistance in publishing this book. Special thanks go to Dr. David Anand Chairman of Mechanical Engineering, and Dr. Gary Rubloff, Director of the Institute for Systems Research, for their support and encouragement. The support from Mr. David Pettine at the Parametric Technology Corporation is appreciated and acknowledged.

And last, but not least, the author would like to extend his special thanks to Dr. William Destler, Dean of the A. James Clark School of Engineering, and Dr. Thomas Regan, Associate Dean and Director of the ECSEL Coalition. Their support of the concept of teaching Pro/ENGINEER to freshman students under the National Science Foundation sponsored ECSEL program has made the publication of this book a reality. Thanks are also extended to Dr. James W. Dally, President of the College House Enterprises. His support has been invaluable to the author, not only in the academic area, but also in many aspects of the author's life and career development.

Guangming Zhang
College Park, MD

CONTENTS

CHAPTER 3 DIMENSIONING ENGINEERING DRAWINGS

CHAPTER 4 Pro/ENGINEER DESIGN SYSTEM

CHAPTER 5 ENGINEERING ANALYSIS

CHAPTER 6 ENGINEERING APPLICATIONS

APPENDEXES

INDEX

CHAPTER 1

COMPUTER AIDED DESIGN (CAD)

1.1 Introduction

In the nineteenth century, new machinery and new manufacturing processes were introduced that so significantly affected society that historians refer to the period as the industrial revolution. The machinery and the processes not only markedly reduced the time and labor to produce products and materials, but they also provided a richer life style than previously existed. several innovations were introduced to improve the methods used in meeting the needs of society in daily life. As an example, the appearance of locomotives revolutionized the entire transportation system. Steam powered machinery replaced much of the physical power previously generated by humans or horses. Since the industrial revolution, we have worked continuously to invent new machines and to redesign older ones. The purpose is to reduce the time and effort of society in producing the goods and services consumed. We also seek to create new ways to improve living standards.

In the late 20th century, a second industrial revolution is taking place. Computers enable us to enhance our capabilities, and provide a means for greatly increasing our productivity. The emergence of information technology has revolutionized the ways we think, work, communicate, and live. In the engineering profession today, it is almost unthinkable to undertake an engineering project without the use of computers. Applications of computers and information technology are a central element in all engineering disciplines. Among these applications, computer aided design (CAD) and computer aided manufacturing (CAM) have already gained wide acceptance in industry because they enable the development teams to more quickly generate innovative product designs. CAD and CAM also enable major increases in productivity.

Today industries are adopting new technologies in their design and development organizations to reduce the time and the cost of product development. Development teams attempt to extend their product lines to address new markets. Simply decreasing the time for specific tasks such as component sizing is insufficient. To meet the more demanding requirements, they are upgrading their engineering skills, and employing very effective CAD systems that enhance the visualization and analytical capabilities of their engineers. As engineering managers redefine the product development process, developers of CAD systems are making sweeping changes in their products to support the emerging streamlined engineering processes.

Engineering managers in industry are now facing new challenges: they must select a CAD system from the commercially available products, ranging from more costly, highly integrated CAD systems to lower-priced CAD systems with better functionality but less integration. Upon the selection of one system or another, they have to retrain the design engineers, and/or find new personnel qualified to use the new CAD system.

Over the past decade, CAD systems have become more intelligent and are more focused on system integration such as assembly modeling and concurrent design. Understanding the fundamentals of modern CAD systems and imparting the skills necessary to efficiently operate one of these systems, represents a significant challenge in teaching CAD methods and techniques at universities. This book is written to support those students attempting to become efficient with modern CAD methods being used in industry by an effective development organization. The three objectives that are addressed in this book include:

1. Describe the fundamentals of engineering design.
2. Impart skills required in employing modern CAD programs
3. Demonstrate CAD programs by providing several real-life problems.

1.2 Concept of Computer-Aided Design

Computer aided design involves the creation and manipulation of drawings representing the design of some component, subsystem or complete assembly. The computer loaded with a CAD program assist the engineer in the design process. It is a techniquc in which the engineer and the computer together form a team with enhanced capabilities. The combination provides better results with higher productivity than the engineer or the computer would create individually. The computers' graphical capability and computing power allow the designers to fashion and test their ideas interactively in real time. In this regard, we emphasize two equally important factors, the person and the machine. The concept of CAD implies that the machine is not used to substitute for the person when the designer is most effective. On the other hand, the computer should be used in those tasks where the person is limited. Therefore, the machine and the person are complementary. In fact, good designs are rarely produced today without a multi-disciplinary approach. This includes an integrated development team and experienced personnel proficient on a modern CAD system.

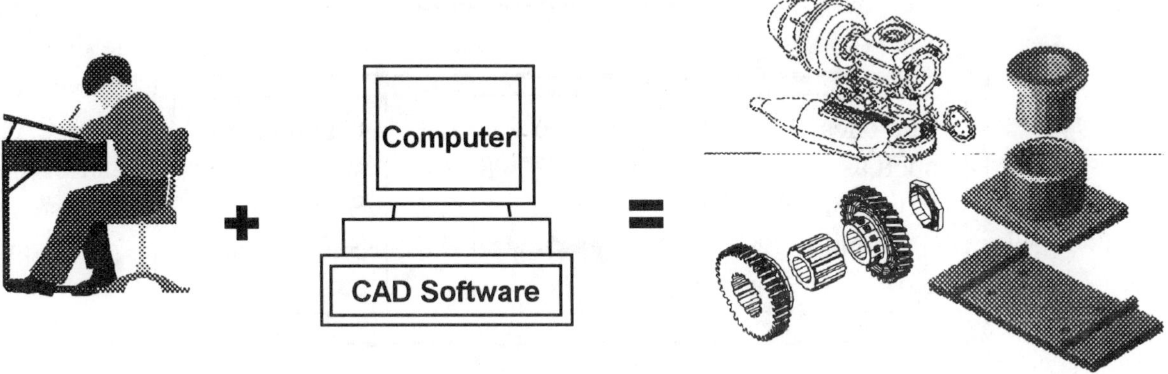

Figure 1-1 The person, machine and output with CAD.

The development process is a method for generating innovative solutions to well formulated problems leading to the production of new products or processes. Traditionally, the main purpose of engineering design is to produce a clear definition of a part or subsystem to be manufactured. The definition is usually made by producing engineering drawings of the component and/or assembly. These engineering drawings establish the physical configuration of the part or system. The uniqueness of modern computer-aided design is the creation of a solid model for the part (or solid models for the system) in the form of a geometric database. The drawings of the component and/or assembly are derived from this data base.

Major characteristics of CAD systems are presented below:

1. **A creative and innovative method of representing a design conceived by the developer who is often an engineer.** The judgement of an experienced designer is vital to the process. Moreover, this process must be controlled by the designer. This fact implies that the designer must have the flexibility to work on various parts of the assembly at any time and in any sequence. By displaying component designs side by side and by recalling and comparing with older designs a better component or assembly is produced. Computers with significant storage capability, rapid recall, and the ability to display complex three dimensional offer a superb information environments with which the designer can work.
2. **An effective method of communicating design information.** Technology and design are becoming broad and complex. As a result engineering developments rely on the work of teams much more than individual efforts. The design team shares experiences among team members, exchange information, and make assessments on an hourly basis. Computers provide a unique means for communication. Today the rapid advancement of information technology employing the Internet and the World Wide Web allows design engineers to work freely in "cyber space" gathering, exchanging, and utilizing information.

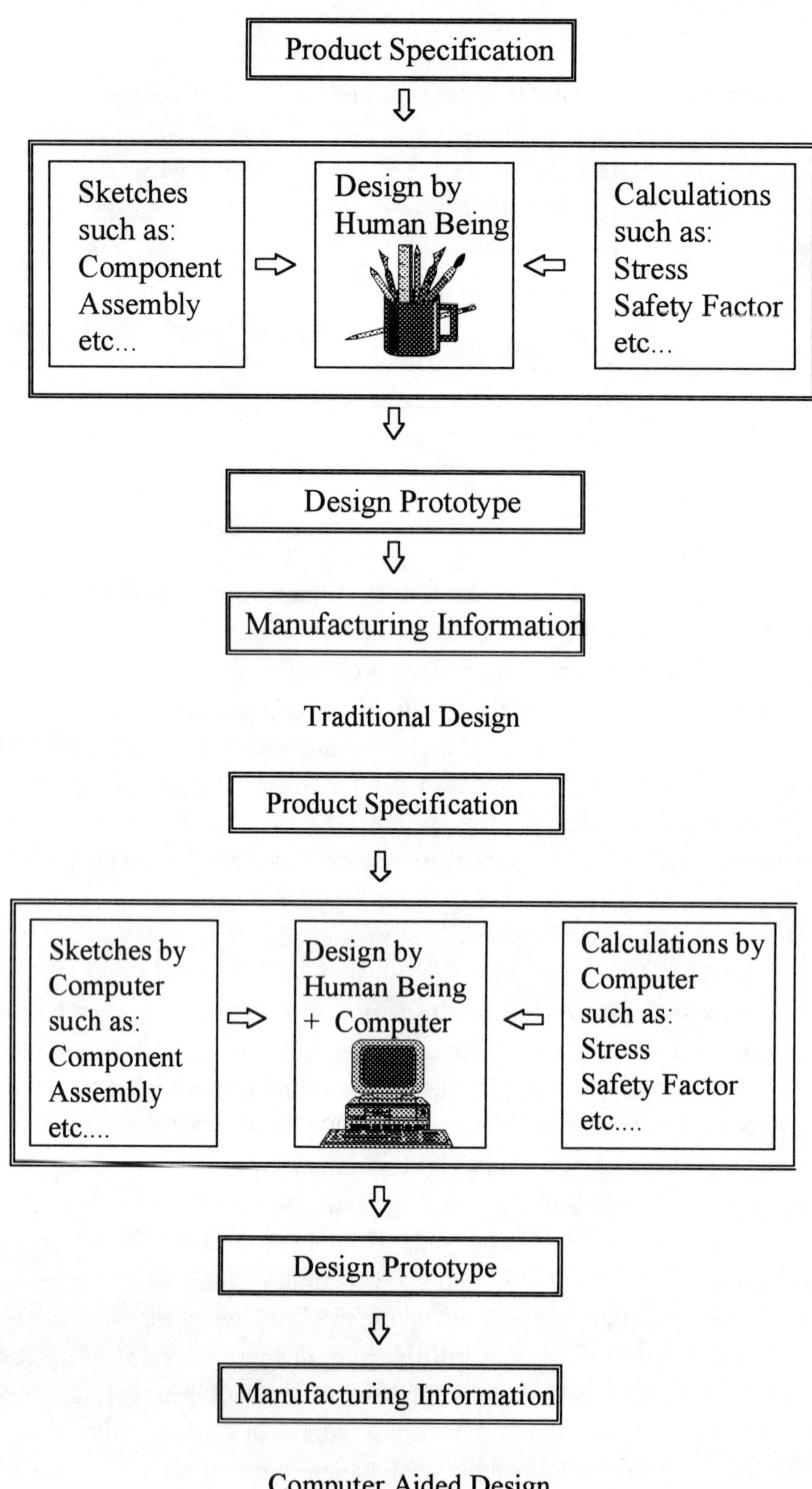

Figure 1-2 Traditional Design Process Compared to a Computer Aided Design Process

3. **Facilitate design modification and redesign.** The design process includes many steps to refine ideas and concepts. Creative and innovative designs come from thinking, experience, and hard work. Making modifications during the phase to design a component or assembly is an essential part of the process. The geometric database serves to store information that can be retrieved at any time during the design process. More important is the fact that the computer can be programmed to detect design errors which are systematically definable. For instance, the computer program can compute the torque capacity of a shaft storage, while the designer may be limited to his/her experience and judgment ascertain if the shaft is sized correctly. It is too much to expect the computer to automatic correct design errors. This is a task for the designer as he/she monitors errors and makes appropriate design changes to eliminate them.
4. **Execute long and complex engineering analyses.** A computer with the appropriate software is extremely capable of numerical analysis. Individuals find these same analyses time-consuming and tedious. In the modern CAD systems the analysis tools are included with the graphics tools. It is possible to conduct finite element analysis and predict performance in the design stage. Today it is common practice that as much as possible of the numerical analysis involved in the design is performed on the computer, leaving the designer free to make decisions based on the output from these analyses and his/her intuitive judgement.
5. **Couple computer aided design with computer-aided manufacturing.** Computer controlled machine tools are being employed by many companies as they seek to improve product quality, increase productivity and flexibility, and reduce the cost of the product realization process. The geometric database generated by CAD systems is tailored so that compatibility of the data between CAD systems and CAM systems is ensured.

These five characteristics clearly illustrate that there is a sharp division between the functions of designer and computer in a design process that incorporates a CAD system. The principle functions performed by the computer are:

- To relieve the designer from routine and repetitious tasks.
- To serve as an extension of the designer's memory.
- To provide a variety of communication tools for information gathering and dissemination.
- To expand the analytical capabilities of the designer.

The principle functions performed by the designer are:

- To make the fullest use of computer and information technology.
- To apply creativity, ingenuity and experience in the design process.
- To control the information distributed during the design process.
- To organize and manage design information.

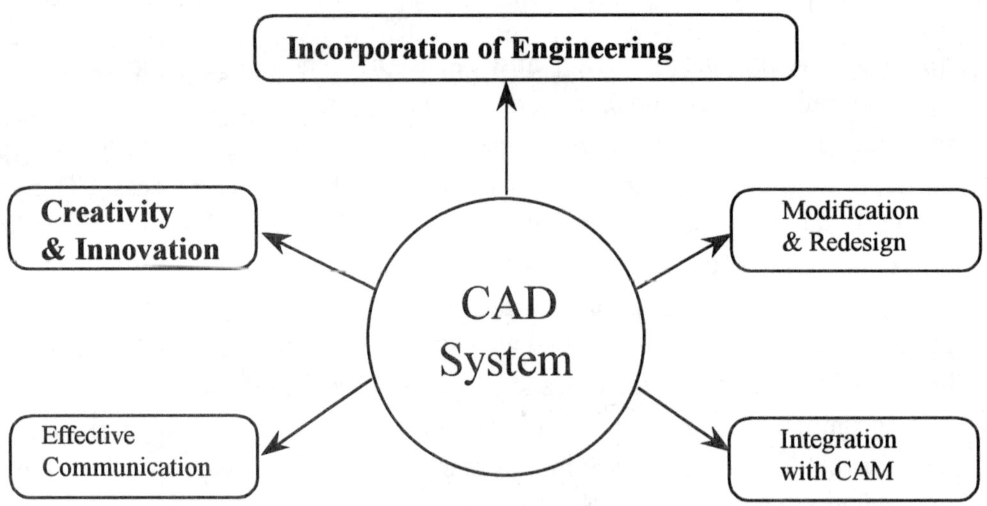

Figure 1-3 The Five Characteristics of a Modern CAD System.

1.3 Process of Product Realization

As science and technology advance, the needs of our society change accordingly. Demands for new products are increasing at an unforeseeable pace. To meet the new demands of the market, the product realization process must incorporate:

- **Customer-oriented design and production.** Managing a wide variety of options has become a business strategy. Being able to deliver many versions of a standard product targeted toward niche markets has become a critical business objective of many corporations. Currently the typical practice in developing a new product model, involves standard design for about 80% of the output and customized designs for the remaining 20%. The automobile industry serves as a unique example. The industry has to provide a wide variety of new car models to attract its customers. For example, the Honda Prelude is designed for the young generation. The Honda Accord is to attract the favor of middle class households. The Honda Civic is the choice of most first new car owners. The three Honda models share 40 - 60 % of the standardized parts. Because of the customized features included in the designs, the three car models change their appearances every year to attract new car buyers as illustrated in Fig. 1-3.
- **High quality and low cost.** The price of almost every product is increasing due to inflation; there is a continuous annual increase of labor costs and raw materials. On the other side of the price equation, the increase in household income often does not keep pace with the increase of product price. Consequently, a significant percentage of people are losing some purchasing power each year. This appears to be true for a segment of the population in the United States. Under the resulting financial pressure, people pay much more attention to reliability and quality of the product(s) that they intend to purchase. At the same time, people are looking at the price tags very carefully. They go shopping and make comparisons in price to make certain that they procure

products with value. Product price is a major competitive factor among companies who making similar products. This competition is global requiring companies to maintain a very tight control of their profit margins. Profit margins are enhanced by designing and manufacturing quality products at low cost. Driving production costs lower is the open secret adopted in the business strategy of every company competing in the global marketplace.

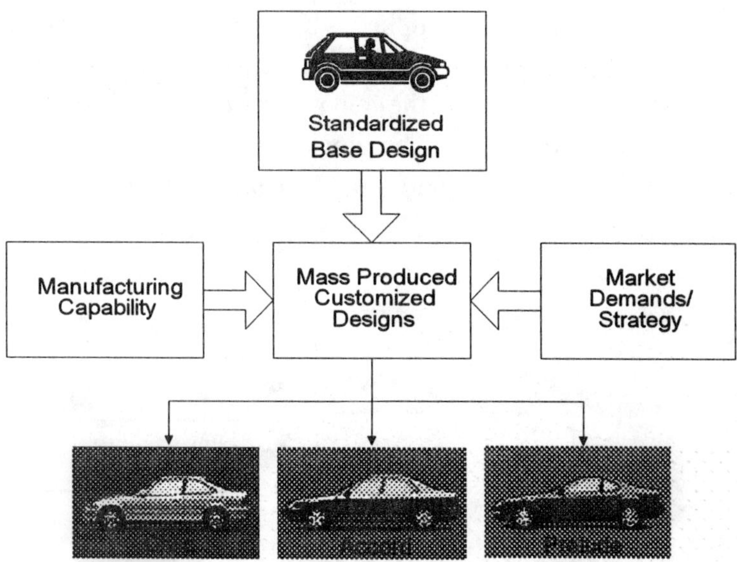

Figure 1-4 Customer oriented design and production

- **Globalization of marketing, suppliers, and competitors.** The availability of rapid transportation at low cost and the capability of quickly communicating with anyone in the world have opened the world market to all commodities and even high performance products. Today we do business on a world wide scale. When we design a new product, our customers are not limited to the domestic market. Asia and Latin America are two areas with a high population density, giving a probability for many companies to increase their base of customers. These geographic regions have offered attractive markets for decades. Many of the countries in these two regions are developing as capital moves in to provide the means for improving the industrial base and the infrastructure. More important is the fact that the low cost of labor and raw material in these areas offers great opportunities for companies from the more developed countries to establish their supply base. Competition for these desirable markets and supply bases is continuing. Doing business in the world market is an endless challenge to product designers. As time to market has become progressively shorter, and mass customization has become a reality, design engineers must accommodate more design changes in a shorter development cycle to capture and maintain market share.

- **Environmental protection.** With increasing population throughout the world, pollution problems are becoming more severe. Society in many of the developed countries is becoming much more concerned with preserving the environment. Any product produced in the U. S. must meet regulations set by either federal or local environmental protection agencies. If product development is viewed as planning for manufacturing, design engineers have the responsibility for selecting the raw materials, ensuring environmentally conscious manufacturing processes, and allowing for remanufacture or recycling of the aging products.

The product realization process is complex. It plays a critical role in ensuring the production of a reliable product that meets the needs of the customers, providing a profit for the stockholders of the company, and treating the environment in a friendly manner during production, usage and in end of life disposal. The block diagram presented in Fig. 1-5 depicts the central role of the product design process and the major factors that must be accommodated in the process.

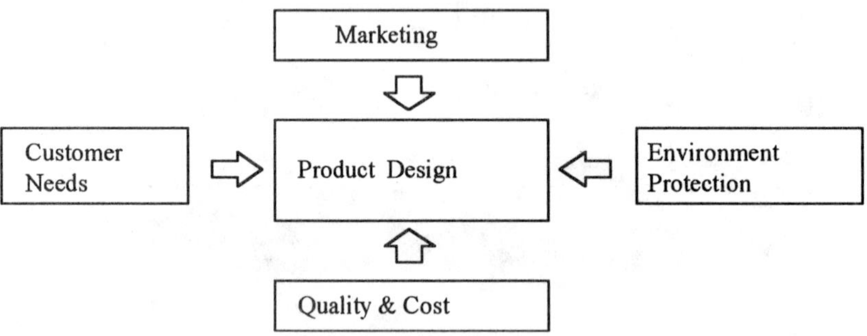

Figure 1-5 Product Design and Related Factors

1.4 Design Process

Design is a process used to create a product or process that satisfies a clearly defined set of requirements. The design problem has multiple solutions, and these solutions are constrained by the need to employ only available resources. In essentially all cases, the final design must be completed within a budget and on schedule. With the rapid advancement in technological knowledge and increased global pressure to be the first to market, a new concept known as concurrent (or simultaneous) engineering has been introduced. Concurrent engineering is incorporated into the design process to shorten the product development cycle and to enhance the profit potential of the product. Concurrent engineering integrates process design with product design reducing product lead time. It also provides a means for balancing the conflicting requirements between the design and manufacturing processes.

There have been a variety of definitions pertaining to the design process. Procedure have been advanced proposing design for manufacturing, design for assembly, design for profits, or design for X where X is some objective function that is to be achieved. However, the design process in the engineering is best illustrated in the four stage sequence illustrated in Fig. 1-6.

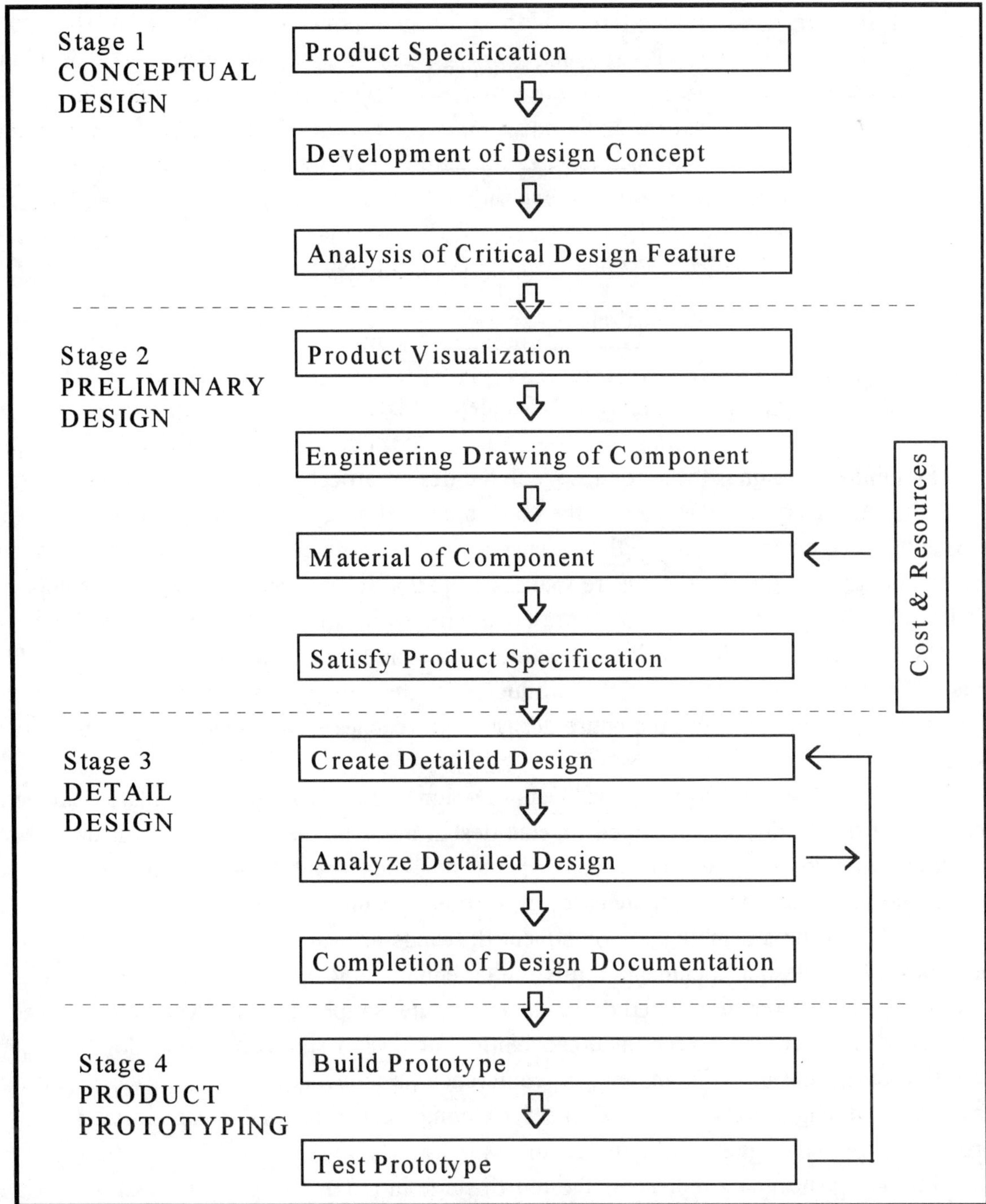

Figure 1-6 Essential Stages in the Design Process

The design process begins with the product specification, which is followed by the generation of design concepts, and the identification and analysis of critical product features. This first stage is known as "conceptual design," an involves the definition of objectives, identification of alternative approaches, and the evaluation of the merits of each approach in terms of both technical and financial feasibility. The flow of information in completing the conceptual design is illustrated in Fig. 1-7.

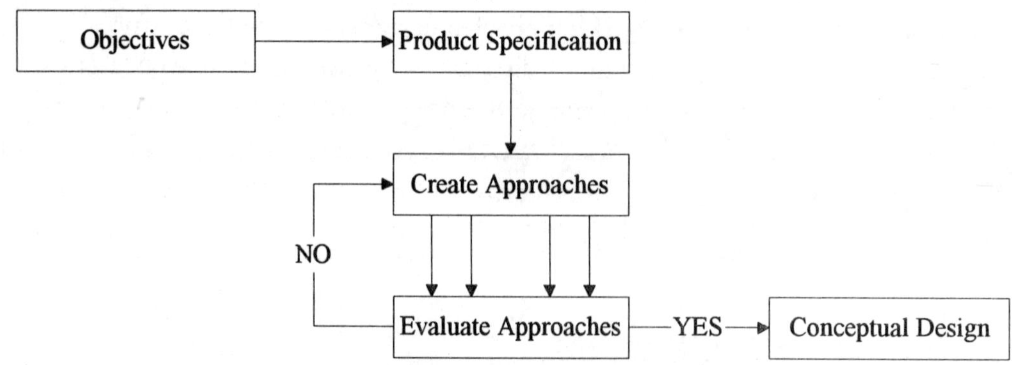

Figure 1-7 Conceptual Design

Preliminary design is the second stage in the design process. At this stage of the process, the structure necessary in implementing the conceptual design is visualized. Isometric drawings of the product are made to facilitate the visualization of the new product. Engineering drawings of the component are sketched to ensure spatial compatibility, and materials for each component are selected. Finally, analyses and tests are conducted to insure that the product will meet the product specifications. Information on resource availability and component cost is sought to estimate the cost of production of the product. In this stage, modifications to component designs are made routinely until the entire design team reaches an agreement to pursue detailed design.

The third stage in the process is "Detail Design." In this stage, a complete engineering description of the product is developed. Detail design involves the specification of dimensional and geometric tolerances for each component, analytical evaluations of performance are conducted at the system and component levels, cost and quality of the manufacturing processes are reviewed, and the assembly and component drawings are subjected to careful examination. Completion of the design documentation is the essential element in the detail design stage. It should be noted that computer-aided design and computer graphics, which has matured over the past two decades, provides exceptional capabilities for design and engineering documentation. When the detail design is complete, a conference is convened to "release" the design. Managerial and engineering personnel meet to conduct a thorough design review, and to approve the design by signing the final documents.

Product prototyping represents the fourth stage in the design process. After the design release, production of the prototype representing the first article is initiated. It is necessary to build scaled or full-size working models to verify performance, and demonstrate the operations required in manufacturing and assembling the final product. Very often the first prototype, often called the first article build, provides valuable information identifying design deficiencies that were not detected prior to he release of the design. Corrections to the design, or design modifications, must be made and verified prior to initiating production of the product. The time required to generate these working models of new or redesigned components is extremely important in controlling the time to bring a new product to market. A new technology, called rapid prototyping, has recently been introduced to reduce the time needed to fabricate prototypes of any component. The rapid prototyping equipment integrates laser and computer technologies to create a free form method of fabrication. Layers of a polymers are created by exposing a liquid to a focused laser beam which enables designer-directed construction of three dimensional models and prototypes. Testing the first article prototype represents a critical step in this stage of the design process. It determines if the design meets product specifications or if it is deficient. For example, the aerodynamic characteristics of an automobile can be evaluated by wind tunnel tests using the first article prototype as the wind tunnel model. The design process ends when the building and the testing of the prototypes verify performance and manufacturability.

1.5 The Manufacturing Process

Upon completion of the design process, we begin the manufacturing process. The first stage in the manufacturing process is process planning. This is the principal activity of manufacturing in the product realization process. Process planning includes determine the methods and processes to be used in producing each component, the sequence of each operation, determining tooling requirements, selecting production equipment and systems, and estimating costs of production based on these selections. If the product requires assembly, process planning includes determining an appropriate sequence of assembly steps and often the design of the assembly line or cell. The process plan must be developed within the limitations imposed by available processing equipment and the productivity capacity of the factory available for the production. Components or subassemblies that cannot be produced in-house must be purchased from external suppliers. In some cases, components that can be produced internally may be purchased from outside suppliers for economic reasons. At this stage, routing sheets are prepared to specify the details of the process plan.

The use of computers in manufacturing, or Computer-Aided Manufacturing (CAM), has grown in importance as companies seek to improve product quality, increase productivity, enhance flexibility, and reduce operational costs. Computer-aided process planning systems have been developed to automate process planning by employing computer hardware and software systems. Considerable effort has been expended in developing a link between the CAD and CAM systems. The basis of link between the two systems is the design data stored in memory in a digital form. Because of the shared database, numerical control (NC) codes can be generated directly from the CAD files with additional information on tool selection and machine settings.

After the planning process is completed the components that have been designed are manufactured, inspected, and tested if required and finally assembled. The complete product is tested after assembly. Let's summarize the design and manufacturing processes by referring to Fig. 1-8.

1. Engineering design is a sequential process. As illustrated in Fig. 1-8, the design process evolves from concept through realization. It is not a reversible process. Component cannot be assembled until they are machined, and the components cannot be machined until NC codes are created. The NC codes cannot be created without a dimensioned drawing of the component, and the drawing cannot be dimensioned without a set of requirements and a general notion of the shape and

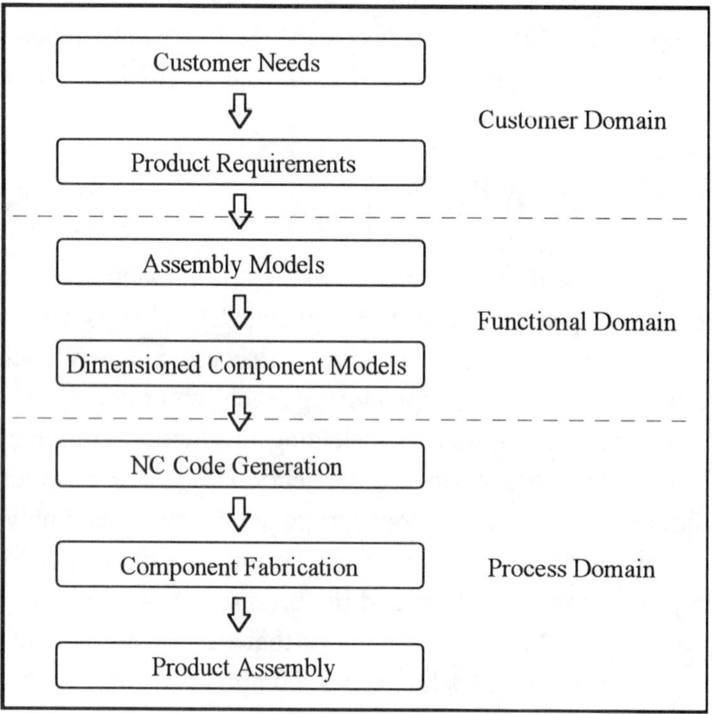

Figure 1-8 Sequential Process of Engineering Design

size of the component. These size and shape evolve in order to meet the needs of the customers that must be identified before the process can begin. All these factors indicate that there is an inherent, sequential order to the design process. Loops are possible, but it's not possible to reverse the process.

2. Design is an iterative process. This is due to the fact that designers are humans with bounded capabilities. Engineering designers cannot simultaneously consider every relevant aspect of a given design. More important recognize that the design process is a process of information accumulation. As shown in Fig. 1-9, when new ideas, information, and technologies become available, design modifications follow. In the competitive global market, companies seek an optimal design of their new products, which provides the best possible performance with the lowest cost. Design through iteration is essential in achieving these optimal designs. These considerations all clearly indicate that design is inherently an iterative process.

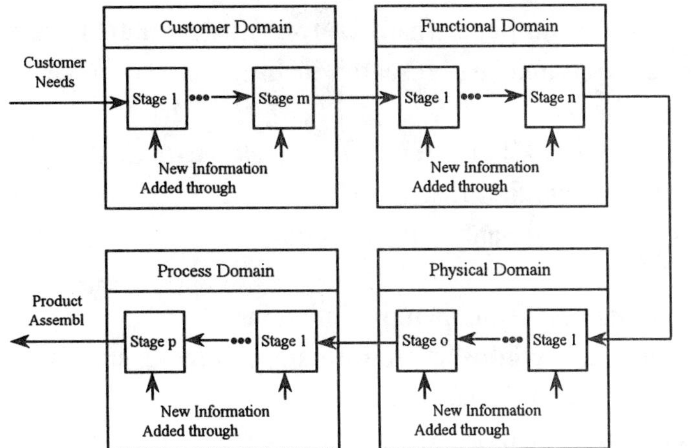

Figure 1-9 The Iterative Process of Engineering Design

1.6 CAD Systems

Computer-aided design is a graphical means of communication. It can be used to convey information much more accurately and rapidly than even a very well qualified person equipped with paper, and pens or pencils. The central elements of a CAD system are the computer hardware and software systems.

1.6.1 Computer Hardware Systems

All the machinery that constitutes a computer is known as hardware. The general configuration of a typical computer system is illustrated in Fig. 1-10. The hardware system can be divided into four main components, namely, input devices, output devices, a central processing unit, and memory.

- Input Devices. We are familiar with the keyboard and the mouse because we use them to type information into the computer or to choose commands. The keyboard and mouse are the two most useful input devices that permit users to communicate with computers. Other input devices are the card and paper tape readers, which are out-of-date devices. A digitizing tablet, is a device consisting of a flat operating surface under which is a fine wire grid that sense position and feeds coordinates into the computer. Other input devices include optical scanners, light pens, and joy-sticks.

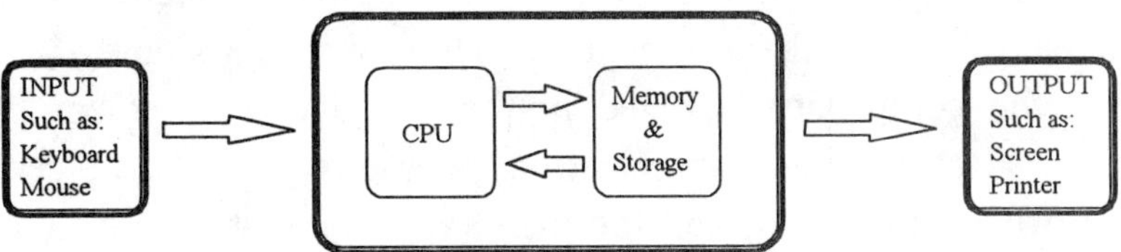

Figure 1-10 General Configuration of a Computer Hardware System

- Output Devices. The output from a computer is usually displayed on the screen of a monitor (soft copy) or on paper (hard copy). Laser printers are widely used as the most popular device in industry, research institutions and universities for generating hard copy.
- Central Processing Unit (CPU). The CPU is the heart of the computer. When we purchase a computer, we find that the CPU is the most expensive component. The CPU controls the performance of the computer. The CPU is performs three functions:
    1. Manipulating data from memory locations.
    2. Monitoring and controlling information flow from input and output (I/O) devices and within the computer
    3. Interacting with memory to carry out instructions stored in the memory.

  For many PCs, there is a label "Intel Inside", and "Pentium Processor", which characterizes the type of CPU which operates the computer.
- Memory and Storage. The computer uses primary and secondary memory systems to store the information that it processes. The primary high performance storage system is known as random access memory (RAM), which is contain on chips located in the computer. The secondary storage system is provided by a number of different less expensive but somewhat slower performing devices, which include hard drives, zip drives, floppy drives, tape drives and compact disks drives.

1.6.2 Computer Software Systems

During the past two decades, the development of computer software systems has revolutionized development of technology and information exchange. We are now living in an information age. The emergence of information technology gave the birth of the software industry. Its impact to the economy of our society is evidenced by the fact that Bill Gates, Chairman of Microsoft Inc., has become the richest person on the earth.

In the 1980s, CAD systems were developed using mainframe computers because of the significant requirement for memory capacity and CPU speed. With rapid advancement of micro electronics, the computer industry began a transformation by shifting from mainframe computers, to mini computers, to work stations, to personal computers (PCs). Software companies immediately responded to the transformation. The development of CAD systems operating in a PC environment, led by AutoDesk, began in 1980s. Today design engineers, managers, and corporate executives are amazed and pleased with the development of powerful Windows-based systems that enable CAD programs to be operated on common PCs. They believe that the Windows-based systems give design engineers much more performance for the investment than that offered by the proprietary or UNIX-based systems.

However, the complexity involved in the product-development process requires for system integration, automation and optimization. These requirements place very high demands for memory and processing speed, and the mainframe computer based CAD systems are still employed by the larger companies. On the other hand, PC-based CAD systems with more limited capabilities are employed by small businesses because of their lower costs. The field is developing rapidly and several midrange CAD systems have been marketed in recent years.

These programs include many of the features previously available only with the high-end systems. They are commercially available at reasonable prices and operate effectively on reasonably priced PCs.

We present brief introduction to several modern CAD systems in the paragraphs below:

1. **CATIA** is a computer-aided design and computer-aided manufacturing graphic system marketed by IBM. The system consists of a base module, which provides the basic interactive graphics functions, and several application modules to perform specific functions. These functions include drafting, 3-D design, sculptured surfaces design, solid geometry, library, NC part programming, and simulation tools, such as kinematics and robotics. Therefore, CATIA is characterized by integration, automation and optimization of not only individual tasks, but also all the product-development process as a whole. CATIA mainly runs on IBM computers, such as 308X, 303X, and 43XX under the UNIX operating system.

2. **I-DEAS** is a computer-aided design and computer-aided manufacturing graphic system marketed by SDRC. Data is managed and shared inherently, whereby engineering can be done concurrently. Products can be created, simulated, optimized, documented, built and tested all within the I-DEAS environment. Work done in one application can automatically update the master model, and team members notified of changes thereafter. Versions of I-DEAS exist for all major platforms, including SGI, Sun, and Windows NT. Regularly, it is found on UNIX workstations.

3. **Pro/ENGINEER** is a computer-aided design and manufacturing graphic system marketed by Parametric Technology Corporation in Waltham, MA. The Pro/Engineer library of programs and modules extends to many areas of mechanical and industrial design and manufacture. Individual modules may be added depending upon the applications desired. These may include: NC generation, piping, electronic schematics, finite element analysis, sheet metal processes, welding, die-making, casting, and many other add-ons. Pro/Engineer encompasses the idea of associative relationships. Features on a model are directly related to that model. Changing the base model will automatically update features added to it. Likewise, a completed part is directly related to assemblies to which it is linked. Part changes then flow through the design from original concept to completed assembly. Like most high solid modeling software, Pro/Engineer typically requires reasonable computing power for general use. Versions are available for all major platforms, including PCs and UNIX workstations.

4. **EUCLID** is a computer-aided design and computer-aided manufacturing graphic system marketed by MATRA DATAVISION in Andover, Mass. Its flagship CAD/CAM system is EUCLID Designer. The uniqueness of EUCLID is the system that has adopted object-oriented programming and database technology. The design data are stored in an object-oriented database, which enables the designer to assign relationships, "like a 3-D relational database", much more efficiently. This characteristic simplifies the assignment, modification, tracking, and control of the complex relationships that exist not only

between individual piece parts, but also among many parts within a complex assembly. The result is a more flexible and more efficient parametric design process, thus adding intelligence to parts designed with EUCLID Designer and increasing the efficiency of downstream applications. With EUCLID Designer, engineers can associate feature-based holes with NC tool paths.

5. **AutoCAD** is a computer-aided drafting and design system implemented on a personal computer. Although AutoCAD is used in engineering workstations such as Sun and SGI, the majority of AutoCAD implementation is still based on PCs. AutoCAD supports a large number of I/O devices, and is a low-cost yet effective solution to many CAD needs. AutoCAD supports 2-D drafting and 3-D wire-frame models. The system is designed primarily as a single-user CAD package. It has been used for all kinds of 2-D applications. A standard drawing interchange file format, DXF, is used which has become an industry standard file type. Users can develop application programs around the system, and a large number of commercial "add-ons" exist to enhance its capabilities. IGES preprocessors and postprocessors are also available to transfer data to and from other CAD systems.

6. **CADKEY**: BAYSTATE Technologies (Marlborough, Massachusetts) has recently acquired the mechanical product line for CADKEY from Cadkey, Inc. Primarily for PC's, its current versions are available for DOS and Windows 3.1/95/NT. It has minimal hardware requirements, and is generally most useful for standalone systems. CADKEY is a 2D and 3D mechanical design and drafting package, including 3D solids and surface modeling, and engineering analysis. Special features include model-to-drawing associativity, assembly design, drafting standards support, automatic hidden line removal, shading, mass properties, and stereolithography output for rapid prototyping applications. Designs may be output directly for machining and to other applications via CADKEY's direct DXF, DWG and IGES bi-directional data translators.

REFERENCES

1. F. L. Amirouche, Computer-aided Design and Manufacturing, Prentice Hall, Englewood Cliffs, New Jersey, 1993.
2. R. E. Barnhill, IEEE Computer-Graphics and Applications, 3(7), 9-16, 1983.
3. H. R. Buhl, Creative Engineering Design, Iowa State University Press, Ames, Iowa, 1960.
4. J. Dieter, Engineering Design, McGraw-Hill, New York, 1983.
5. J. H. Earle, Graphics for Engineers, AutoCAD Release 13, Addision-Wesley, Reading, Massachusetts, 1996.
6. J. Encarnacao, E. G. Schlechtendahl, Computer-aided Design: Fundamentals and System Architectures, Springer-Verlag, New York, 1983..
7. S. Fingers, J. R. Dixon, A review of research in mechanical engineering design, Part I: Descriptive, prescriptive and computer-based models of design processes, Research in Engineering Design, 1(1), 51-68, 1989.
8. S. Fingers, J. R. Dixon, A review of research in mechanical engineering design, Part II: Representations, analysis, and design for the life cycle, Research in Engineering Design, 1(2), 121-38, 1989.
9. J. D. Foley, A. Van Dam, Feiner S. and J. Hughes, Computer Graphics, Principles and Practice, 2nd edition, 1990.
10. P. Ingham P. CAD System in Mechanical and Production Engineering, London, 1989.
11. W. M. Newman, R. F. Sproull, Principles of Interactive Computer Graphics, New York, McGraw-Hill, 1979.
12. R. L. Norton, Machine Design: An Integrated Approach, Prentice Hall, Upper Saddle River, New Jersey, 1996.
13. A. A. G. Requicha and H. B. Voelcker, IEEE Computer Graphics and Applications, 2(2), 9-24, 1982.
14. N. P. Suh, The Principles of Design, Oxford University Press, New York, NY, 1990.
15. D. L. Taylor, Computer-aided Design, Reading MA: Addision-Wesley, 1982.

EXERCISES

1. Use your personal experience to describe the essential steps involved in engineering design. Make sure that the following information is presented.
   (1) What was the product you designed?
   (2) Was the design process a team effort or your own effort?
   (3) What was the design objective(s)? Did you or your team achieve the design objective(s) at the end of the project?
   (4) What was the successful experience gained? What was the unsuccessful lesson learned?

2. Based on your personal experience or your observation, describe the importance of documentation in engineering design. Why do computers and information technology play an important role in today's engineering design community?
3. Present the experience you have had in using CAD packages. Did you use AutoCAD? Did you use CADKEY? Did you use KeyCAD? Present the experience of having difficulties in presenting your design intent when using the CAD package(s).
4. Have you made any engineering drawings in your education? If yes, how did you prepare your engineering drawings? By hand? or Using computer?
5. How to make an assessment on cost when using a CAD system to prepare an engineering document? Compare the cost of using CAD with the cost required for making the identical document without using computer.
6. What is your expectation from taking a course in CAD systems?

CHAPTER 2

ENGINEERING GRAPHICS

2.1 Introduction

Engineering graphics is the primary medium used in developing and communicating design concepts. The solution to many engineering problems requires an integrated systems engineering approach, and engineering graphics plays a central role as the language for communication. Engineering graphics focuses on three-dimensional spatial visualization. It covers two major areas, namely: descriptive geometry and documentation drawings. Descriptive geometry is the study of points, lines, and surfaces, and serves as the basis to represent objects in the three-dimensional space. Drawings for documentation involves the preparation of working documents or blueprints required to execute the design and to guide operations in production.

2.2 Three-Dimensional (3-D) Spatial Visualization

Visualization is critical for success in engineering design. We all know that a picture is better than a thousand of words to aid in understanding. People learn new information most efficiently through their visual sense. It is recognized that the ability to visualize ideas is an extremely important talent that an engineer must develop if he or she is to function effectively.

Results from scientific research reveal that an individual acquires spatial visualization ability through three distinct stages of development. In the first stage, children learn topological spatial visualization. They are able to discern the topological relationship of an object relative to other objects (i.e., How close are the subjects to one another? What is the object's location within a group of objects?). In the second stage of spatial visualization development, projective representation is acquired. People are able to conceive the appearance of an object from different perspectives. In the final stage, a person learns to combine projective abilities with the concept of measurement. A sample problem to test your 3-D spatial ability is presented in Fig. 2-1. We hope your visualization skills are sufficient to select the correct object and pass the test.

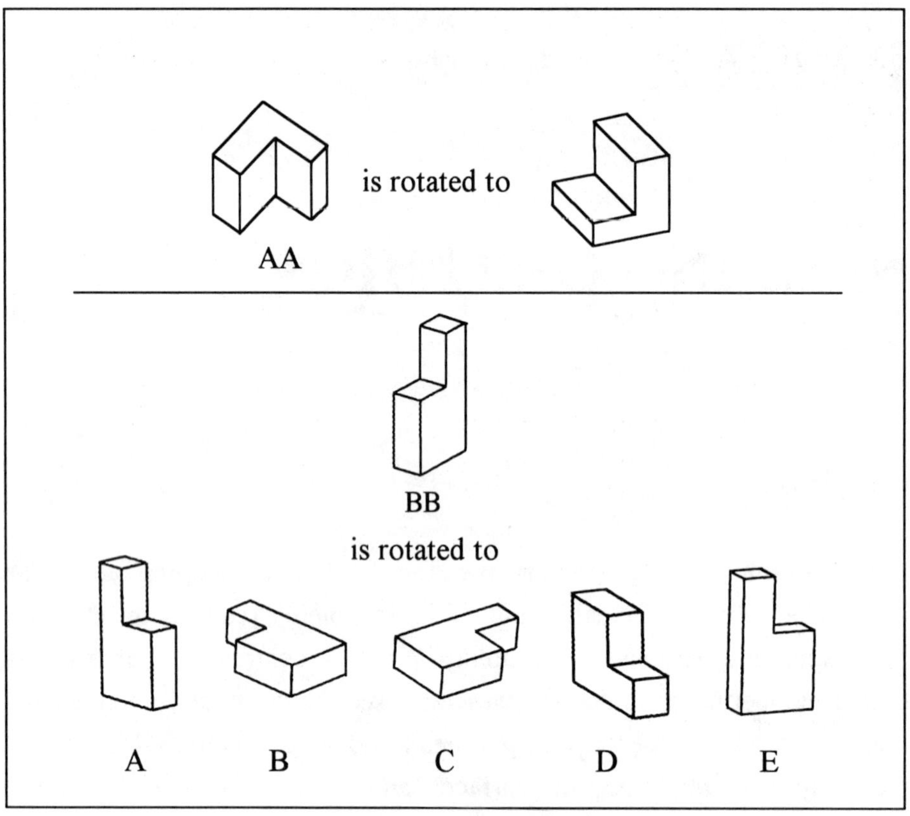

Figure 2-1 A Test for 3-D Spatial Ability

2.3 Product Design and Engineering Graphics

To illustrate the critical role engineering graphics plays in the process of product design, we describe a case study involving the design of a seesaw commonly found in a playground. School administrators in a local area requested that seesaws be installed for children attending kindergarten. A project sponsored by the school administrators was organized. A group of five high school students formed a design team. They designed a seesaw, manufactured the required components, assembled them, installed it at the playground, and finally tested it. The flow chart shown in Fig. 2-2 outlines the three phases in the product design process, including design, manufacturing, and assembly. In each phase, specific tasks were planned. The team followed this plan, and seesaw development progressed form one phase to the next. A comparison is made among the sketch of the seesaw drawn when phase 1 was initiated and the assembly drawing of the seesaw drawn at the conclusion of phase 1.

From this example, it is clear that the application of engineering graphics in the design process included the following steps:

(1) Sketch the appearance of the product and the assembly model in an isometric format.
(2) Refine the design of each component based on analytical evaluations.
(3) Model the assembly by defining the geometrical and functional relations among the components.
(4) Finalize the design through engineering evaluation.
(5) Prepare the documentation drawings

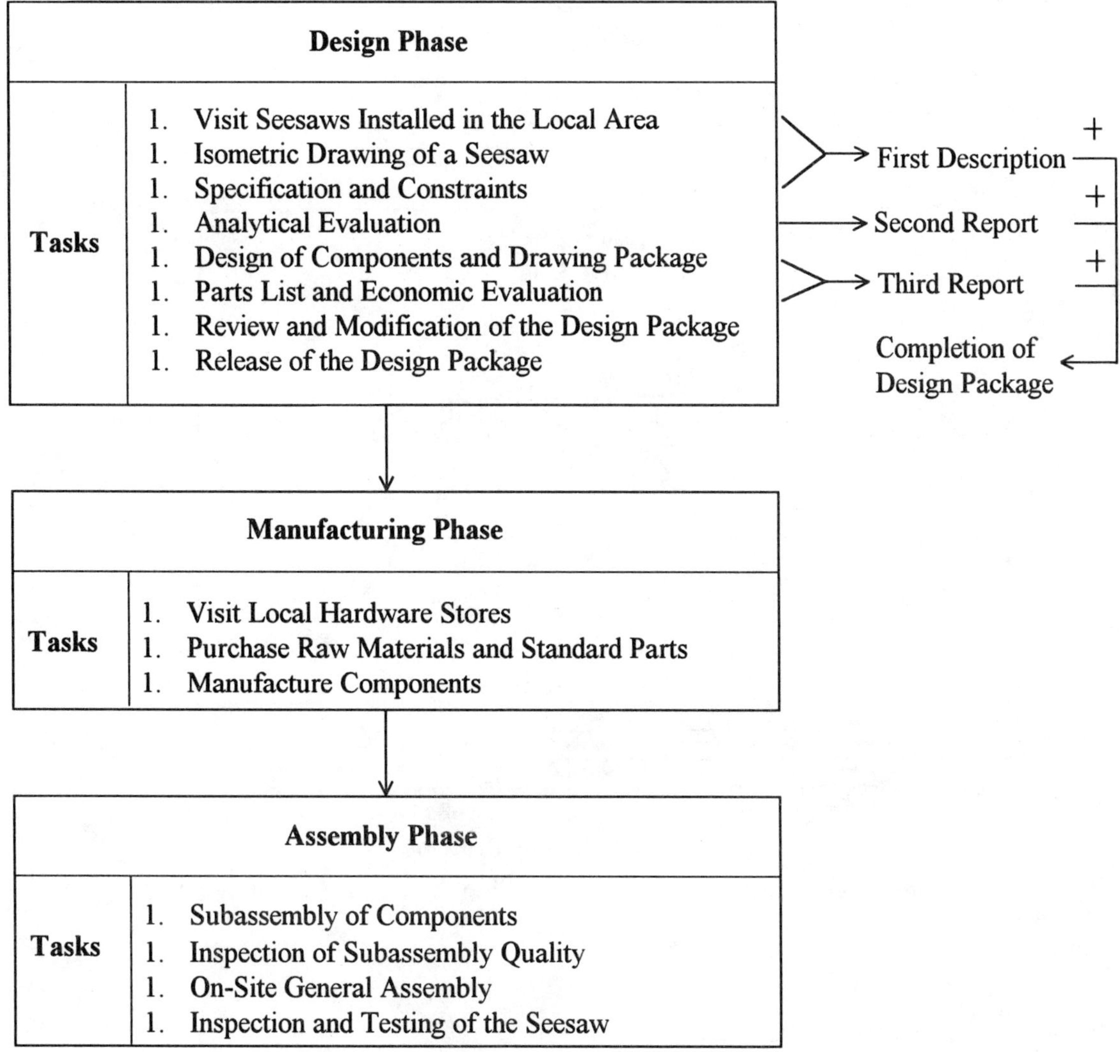

Figure 2-2 Flow Chart Illustrating the Three Phases of the Product Design Process.

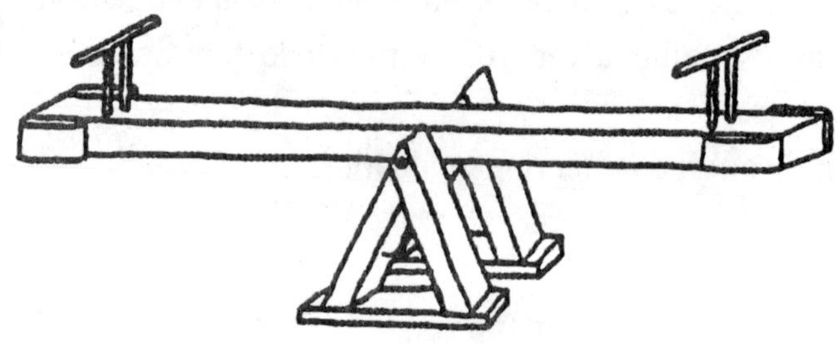

a. Sketching

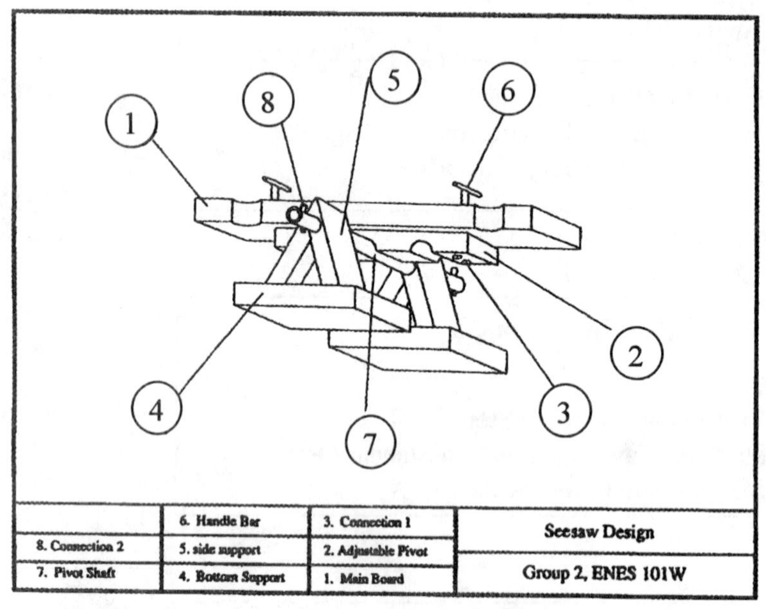

b. Assembly drawing

Figure 2-3 Comparison between a Sketch and an Assembly Drawing

2.4 Orthographic Drawings

An engineer deals with the design of machines and structures --- in three dimensions. However, to represent a structure on paper or a computer screen, he or she must display their drawings in two dimensions. For a three dimensional object to be defined and understood using a two dimensional medium is a problem. To solve the problem, we usually employ the method of orthographic projection to represent the exact shapes and sizes of objects. This method is based on the assumption that the object remains fixed as the observer changes position to obtain different views of the object. Using orthographic projection, drawings are prepared to document the shape and size of each component in a complete assembly. Dimensions and tolerances are added to these drawings to complete the definition of a component.

The shape of an object is represented in orthographic projection by drawing perpendiculars for two or more sides of the object on a projection plane. To demonstrate this projection procedure, let's consider the block like object shown in Fig. 2-4. We illustrate the concept of orthographic projection of the object in Fig. 2-4 by showing the projection planes in Figs. 2-5 and 2-6. The generation of a single orthographic projection to give the front view of the object is presented in Fig. 2-4. Parallel lines are projected forward in this illustration to locate all of the points necessary to draw the front view of the stepped block. Note that the front view does not indicate the complete shape of the object nor the dimensions from the front to the rear. For this reason, additional projections are required to completely describe the stepped block.

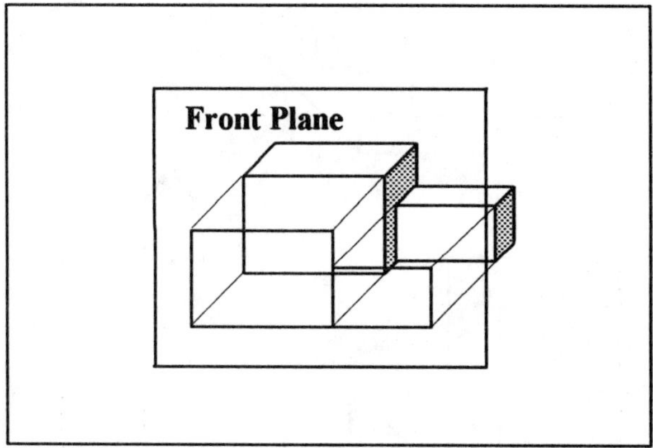

Figure 2-4 Stepped Blocks and the Frontal Plane Which Shows the Front View.

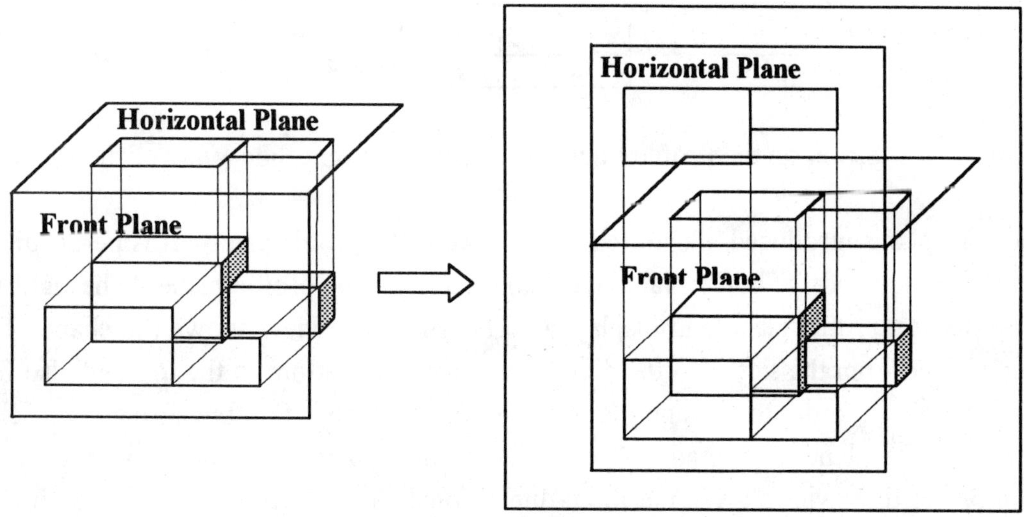

Figure 2-5 Projection onto the Horizontal Plane Generates the Top View of the Blocks.

The generation of the top view of the block is presented in Fig. 2-6. We introduce a horizontal plane in this figure to provide a surface on which the top view is displayed. Note that the frontal and horizontal planes are perpendicular to each other. The top view gives the appearance of the object as viewed from above, and shows the dimensions from front to rear. In order to arrange the two views of the object on a two dimensional surface, the horizontal plane is rotated through 90° to become coplanar with the frontal plane.

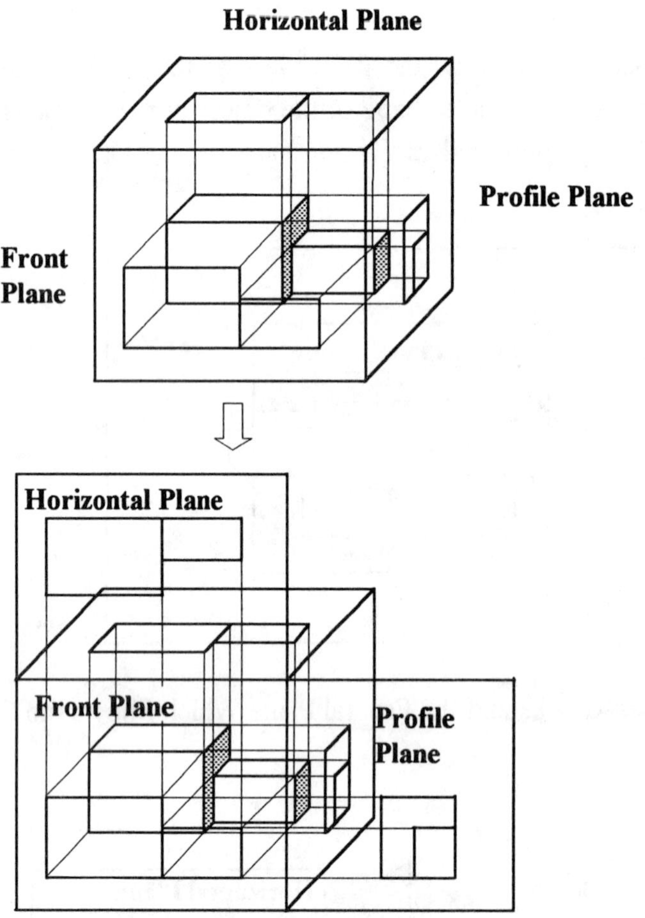

Figure 2-6 Projection on the Profile Plane Generates the Right Side View of the Blocks.

The third plane introduced, illustrated in Fig. 2-6, is the profile plane. It is perpendicular to the both the frontal and the horizontal planes, and is named the profile plane. The right side view is projected on profile plane and displayed. This right side view shows the shape of the object when viewed from the side, and the dimensions from the bottom to the top and also from the front to the rear. In order to arrange the three views of the object on the surface of a screen or paper, the profile plane is rotated 90° onto the frontal plane. As shown in Fig. 2-6, a combination of the three views gives the three-dimensional shape of the object and each of its dimensions.

In an orthographic projection, the picture planes are called "planes of projection", and the perpendiculars, "projecting lines" or "projectors." In examining these projections, or views, we should not think of the views as flat surfaces on the transparent planes, but instead imagine that you are examining the object through transparent planes.

You should be aware that the projection procedure demonstrated in the above illustrations represents the convention widely adopted by design engineers in the United States and other countries with close economic ties with the U. S. The design engineers in Europe and much of Asia employ a different procedure. There are some significant differences between the two conventions. Examining Fig. 2-7a, shows four projection quadrants each available for projecting a view of the object to a picture plane. As illustrated in Fig. 2-7a, a cubic object located in both the third quadrant and the first quadrant has been projected. The first difference in projecting these cubes is two squares formed by solid lines compared to two squares formed by dashed lines. The second difference in using quadrant I for projection, is that the front view of the object occurs above the horizontal plane, not below it as is the case when the cube is in quadrant III. Although this class uses third-angle projection (quadrant III placement of the object) exclusively, the concept of first-angle projection (quadrant I placement of the object) is important. In the international business world, we will from time to time be called upon to interpret engineering drawings drawn in the first-angle system. Being aware of both system conventions will assist you in reading the drawings correctly.

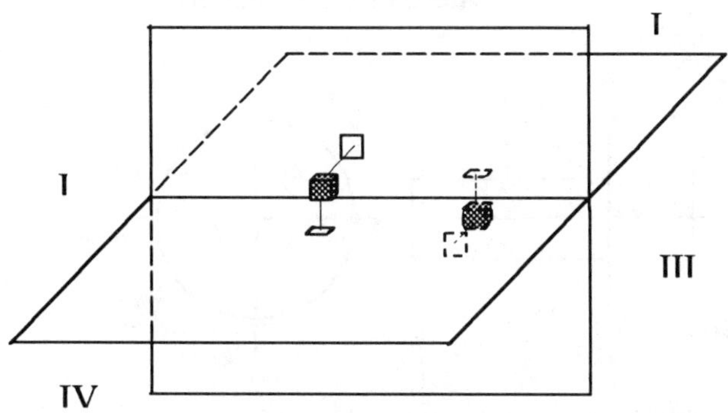

Figure 2-7a. Placement of Cubes in Quadrants I and III to Illustrate Different Reference Systems.

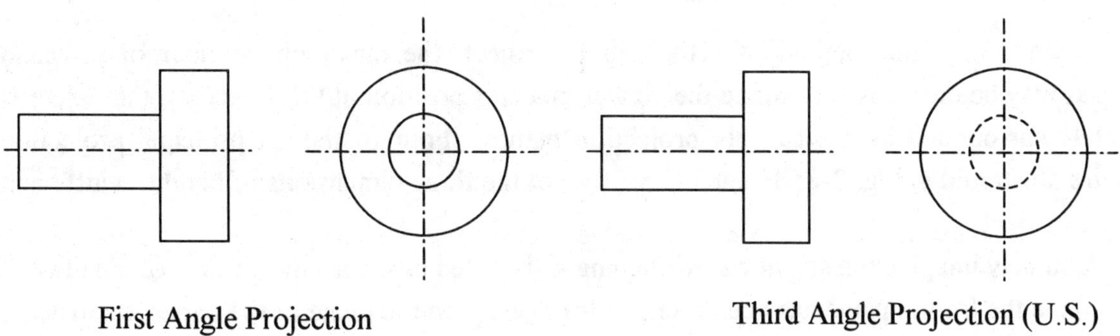

First Angle Projection

Third Angle Projection (U.S.)

Figure 2-7b. Examples of First and Third Angle Projections.

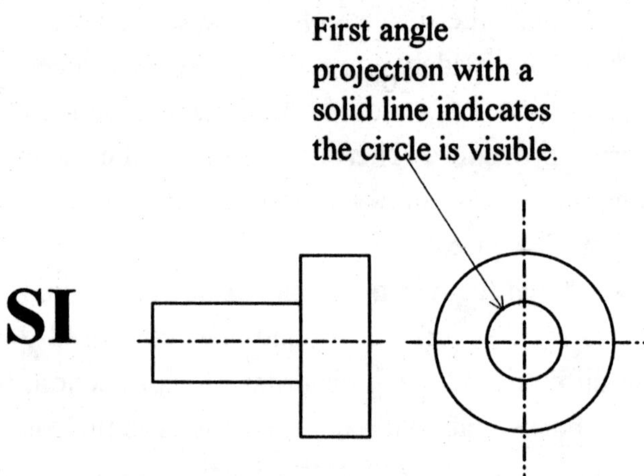

a. Metric Units and First-Angle Projection.

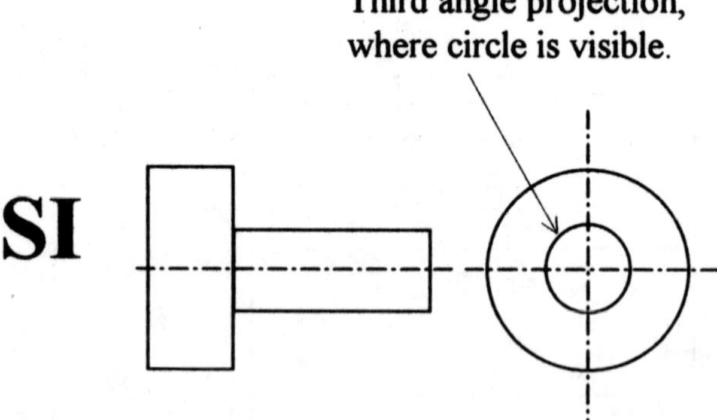

b. Metric Units and Third-Angle Projection.

Figure 2-7c The Reference System Used to Construct Engineering Drawings Using Orthographic Projection

If we extend the method of orthographic project, the maximum number of principal views that may be drawn is six. Since the viewer changes position at 90^0 intervals, the object is completely surrounded by a set of six projection planes. Four of the six principal projection planes are illustrated in Fig. 2-8. In each view, two of the three dimensions of height, width, and depth are evident.

You may imagine the six views by opening a six-sided box onto the surface of a pad of paper. Since the frontal plane has been selected for display on the plane of the paper, the front view automatically occurs on this plane. The other sides are hinged at the corners and rotated onto the surface of the paper. The projection onto the horizontal plane provides the top view. The projection onto the right profile plane gives the right side view. The projection onto the

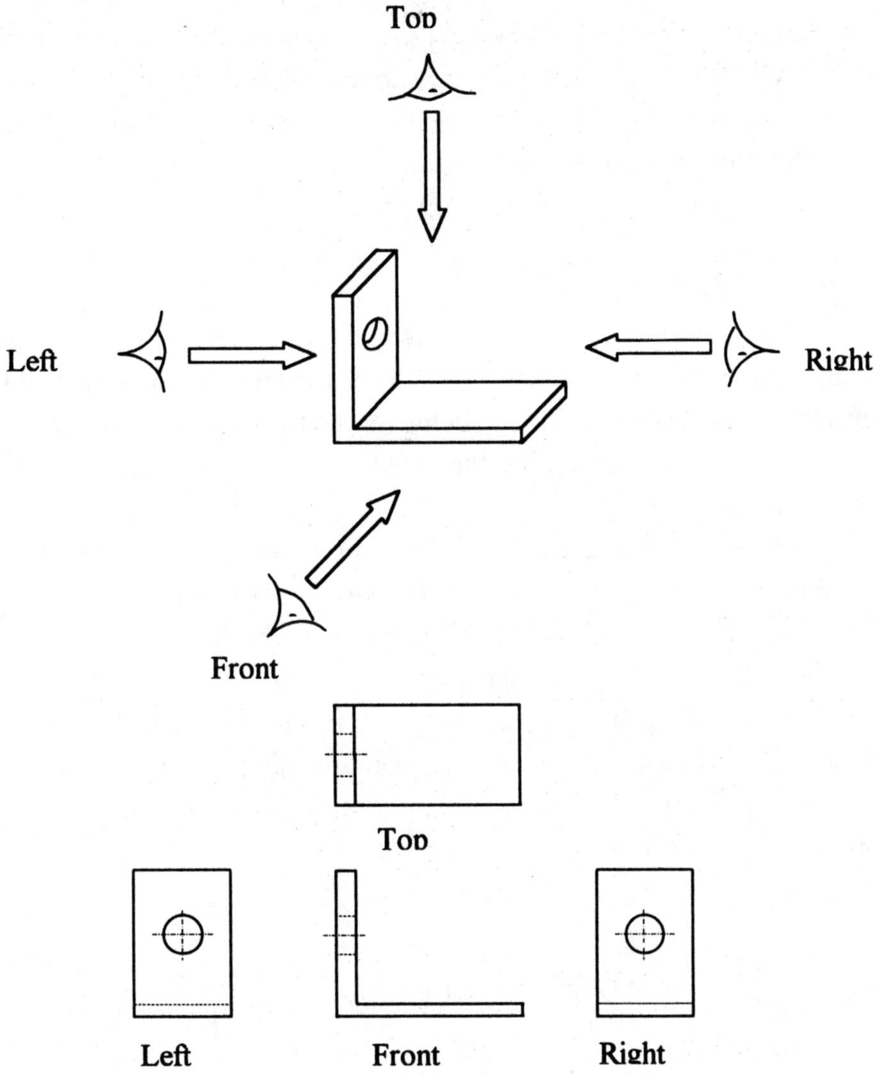

Figure 2-8 Four of the Six Principal Views of an Angle Bracket.

left profile plane is the left side view. By reversing the direction of observation, a bottom view is obtained instead of a top view, or a rear view instead of a front view. In some cases, a bottom view or rear view or both may be required to show some detail of the shape of a very complex component. In practice, three views usually are sufficient to represent the shape and dimensions of typical components. For objects with relatively simple geometric shapes, one or two views may be sufficient to completely define the shape and dimensions of the component. are also used for objects in engineering designs.

2.5 Descriptive Geometry

The illustrations of drawings prepared with orthographic projection indicated that, points, lines, and planes are the basic geometric elements used in describing three-dimensional spatial geometry. To assure the precision of a given component, relationships between its geometric features and the basic geometric elements must be established in an explicit and unique manner. Descriptive geometry is the study of these relationships, and it serves as the basis for algorithm development in all of the CAD programs.

2.5.1 Projection of Points

A point defines a position in space, and it is dimensionless. However, a two or more points can be employed to establish the length of a lines, the area of a surface, and the volume of a three dimensional shape. To uniquely define a point in space, at least two adjacent orthographic views are required that give the three dimensions needed to position the point relative to some reference point.

In descriptive geometry, quantitative measurements are used to define the location of a point in space. We illustrate these measurements in Fig. 2-9 where point (a), confined in a six-sided box, is shown. Three projections of the point onto the frontal, horizontal, and profile planes are indicated. We also illustrate the process of opening the box to arrange the three projection plans onto a single plane, i.e., the frontal plane. The corners of the box are labeled from A to F to aid you in visualizing the hinge lines on the single plane representation. Finally, we show that point (a) is f units to the left of the profile plane, k units below the horizontal plane, and h units behind the frontal plane.

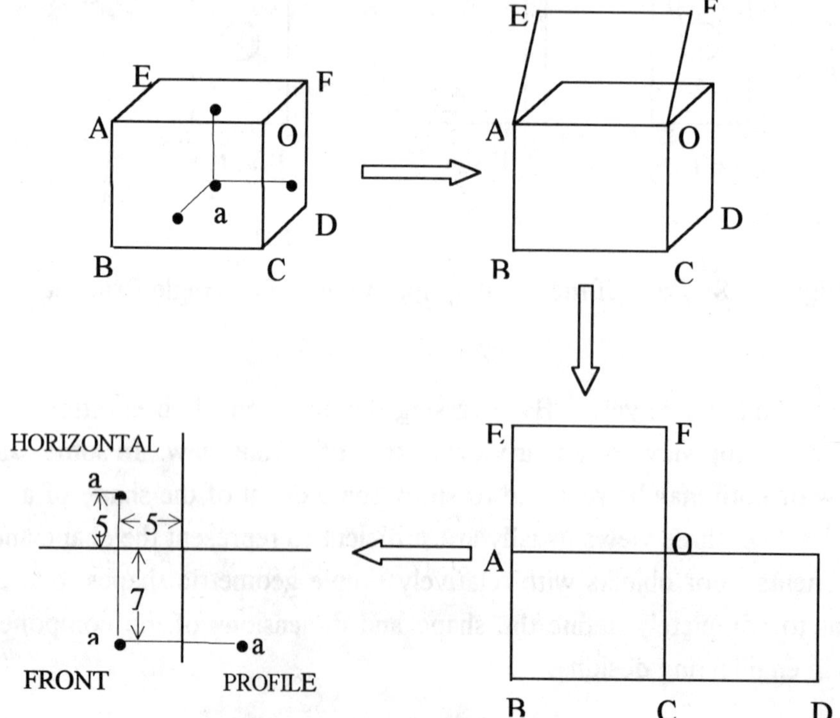

Figure 2-9 Point Projection onto the Horizontal, Frontal, and Profile Planes.

2.5.2 Projection of Lines

A line is the straight path between two points in three-dimensional space. In drawings constructed using orthographic projection, a line may be viewed in its true length, foreshortened, or as a point. These three representations of a line are illustrated in Fig. 2-10.

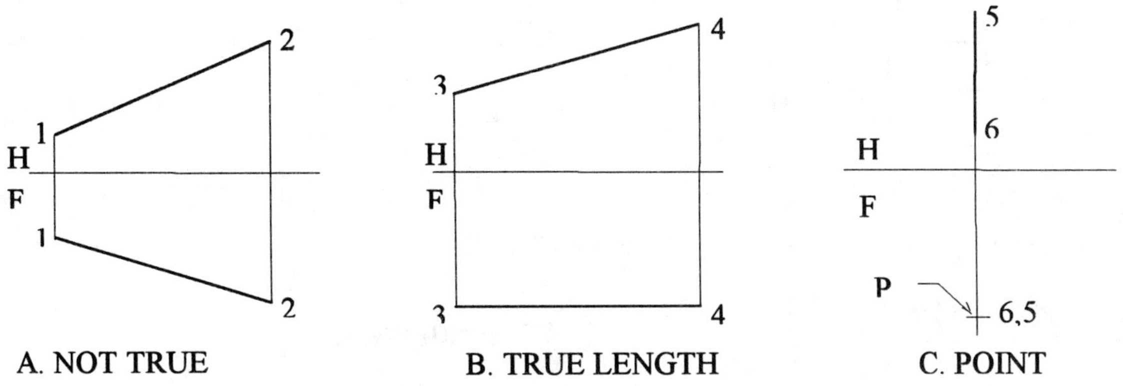

Figure 2-10 Three Views of a Line in Space Projected onto Various Principal Planes.

(1) A foreshortened line is viewed if it is not parallel or perpendicular to a principal projection plane.

(2) A true-length line is viewed when the line is parallel to at least one of the principal projection planes. The three principal planes (horizontal, frontal, and profile) are shown in Fig. 2-11. In this illustration, we show a horizontal line, which is parallel to the horizontal projection plane. It appears in true length in the horizontal projection plane that displays the top view.

(3) A line that projects as a point occurs when the line of sight is parallel to that line. In orthographic projection, a line that projects as a point is parallel to two principal planes. For example, a line projecting as a point in the front view is parallel to both the horizontal and profile planes.

The standard three orthographic views of an oblique line in three-dimensional space are presented in Fig. 2-12. The true length of the line is not reflected in any of the views. We must perform a calculate to determine the true length of an oblique line..

2.5.3 Projection of Planes

A plane is a flat surface. Any two points on this surface may be connected by a straight line that lies entirely on the surface. The positions of planes in space may be designated by the positions of two intersecting lines, two parallel lines, a point and a line, or three points not in a straight line, as illustrated in Fig. 2-13.

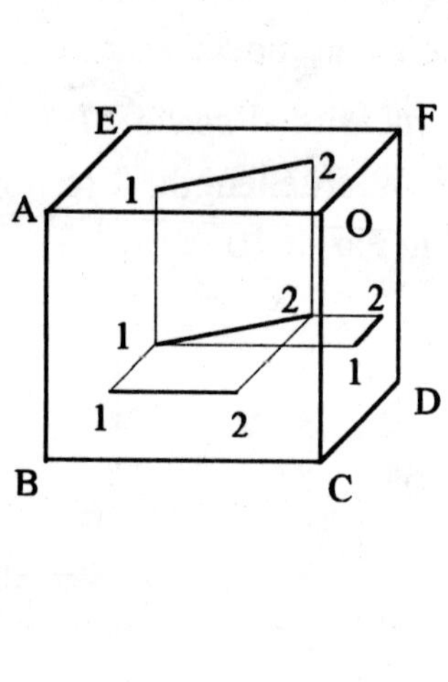

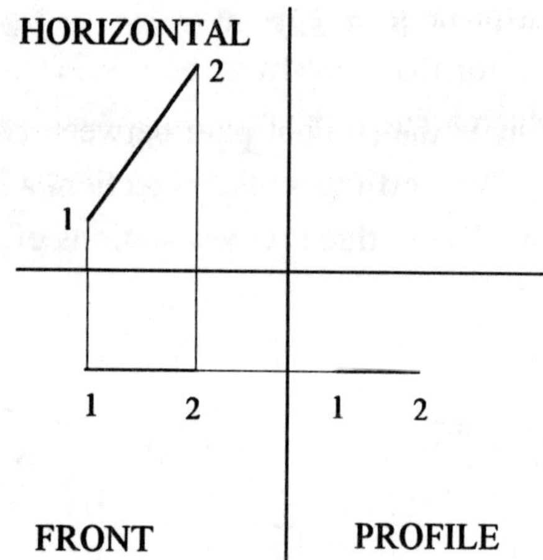

A. Horizontal Line

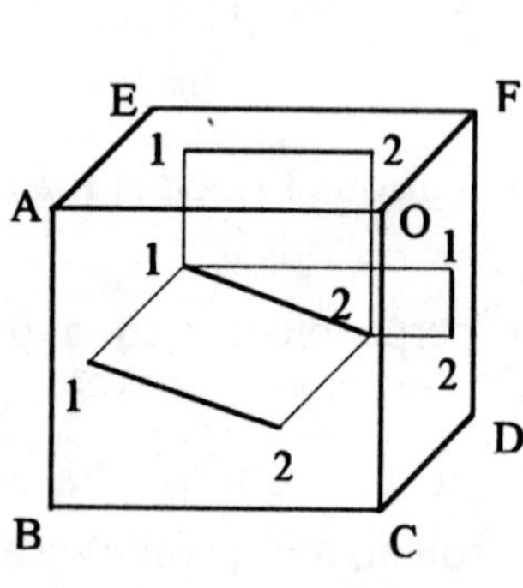

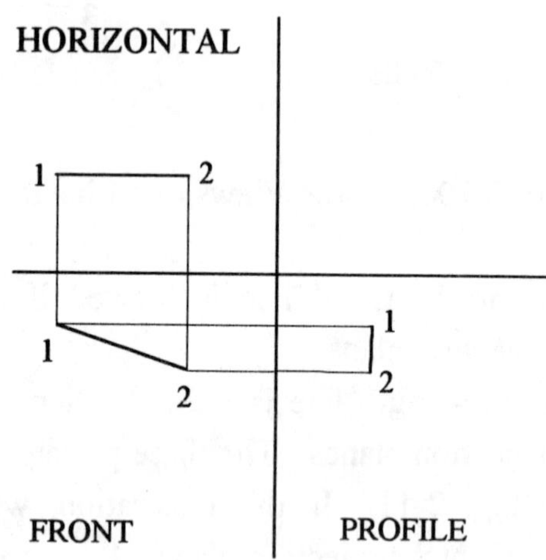

B. Frontal Line

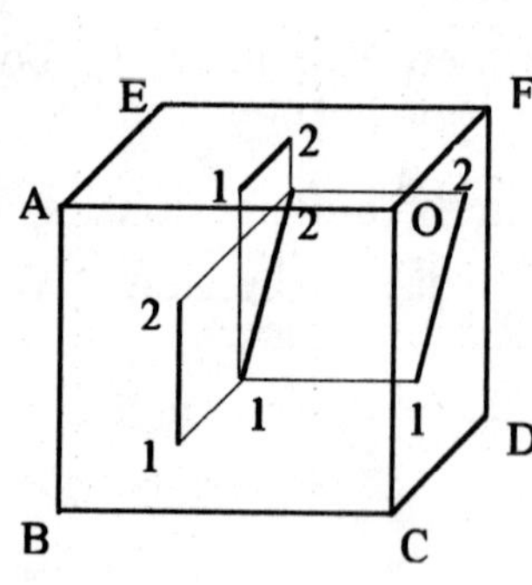

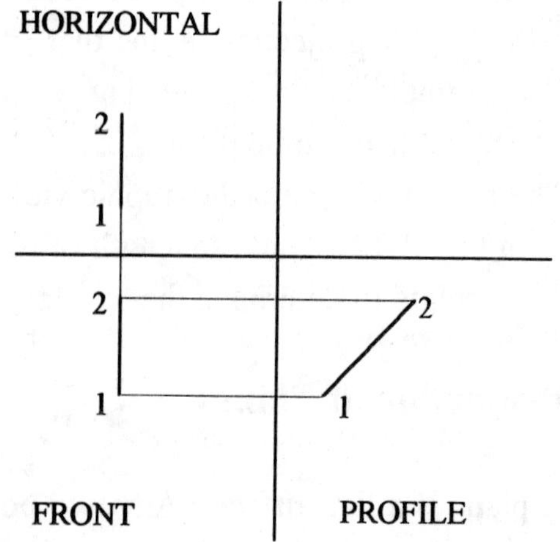

C. Profile Line

Figure 2-11 Illustrations of three types of principal lines

If a plane is to appear as an edge, or a line, in a principal projection view, the line of observation for that view must be parallel to the plane. A plane must appear as an edge in any view in which a line on that plane appears as a point. A plane will be seen in its true shape and size in any view for which the line of observation is perpendicular to the plane.

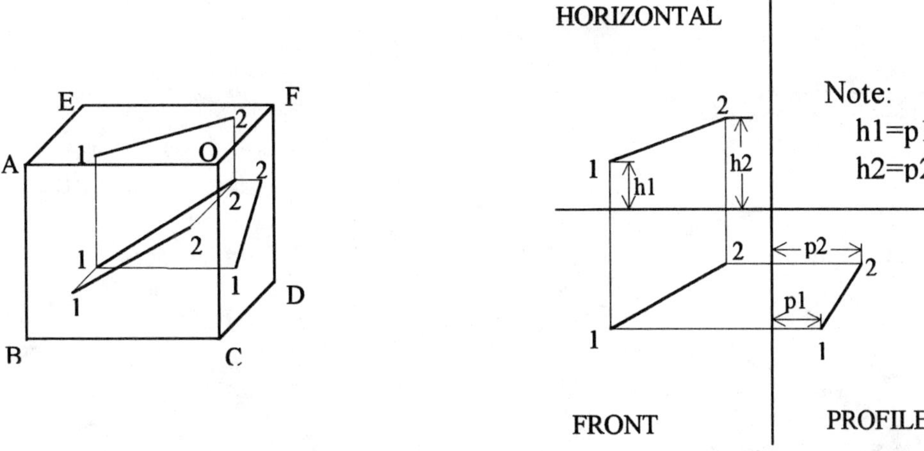

Figure 2-12 The Three Orthographic Views of an Oblique Line in Space.

2.5.4 Visibility

To draw a correct representation of a feature in an orthographic projection, it is necessary to indicate whether a line is visible or hidden. Any view of an object is bounded by visible lines, so visibility needs to be determined only for those lines that fall within the outline of the view. When nonintersecting lines cross in certain views, determining which line is above or in front of the other is referred to as establishing a line's visibility. Establishing line visibility is a requirement in preparing orthographic drawings representing many complex three dimensional components. Fortunately, many of the CAD programs are capable of determining the visibility of the lines within the object's boundaries. In a computer aided environment, the command "show hidden" is usually available on the design screen. Switching on "show hidden" results in the display of all hidden lines as dashed lines. Turning off "show hidden" eliminates all the dashed lines on the design screen.

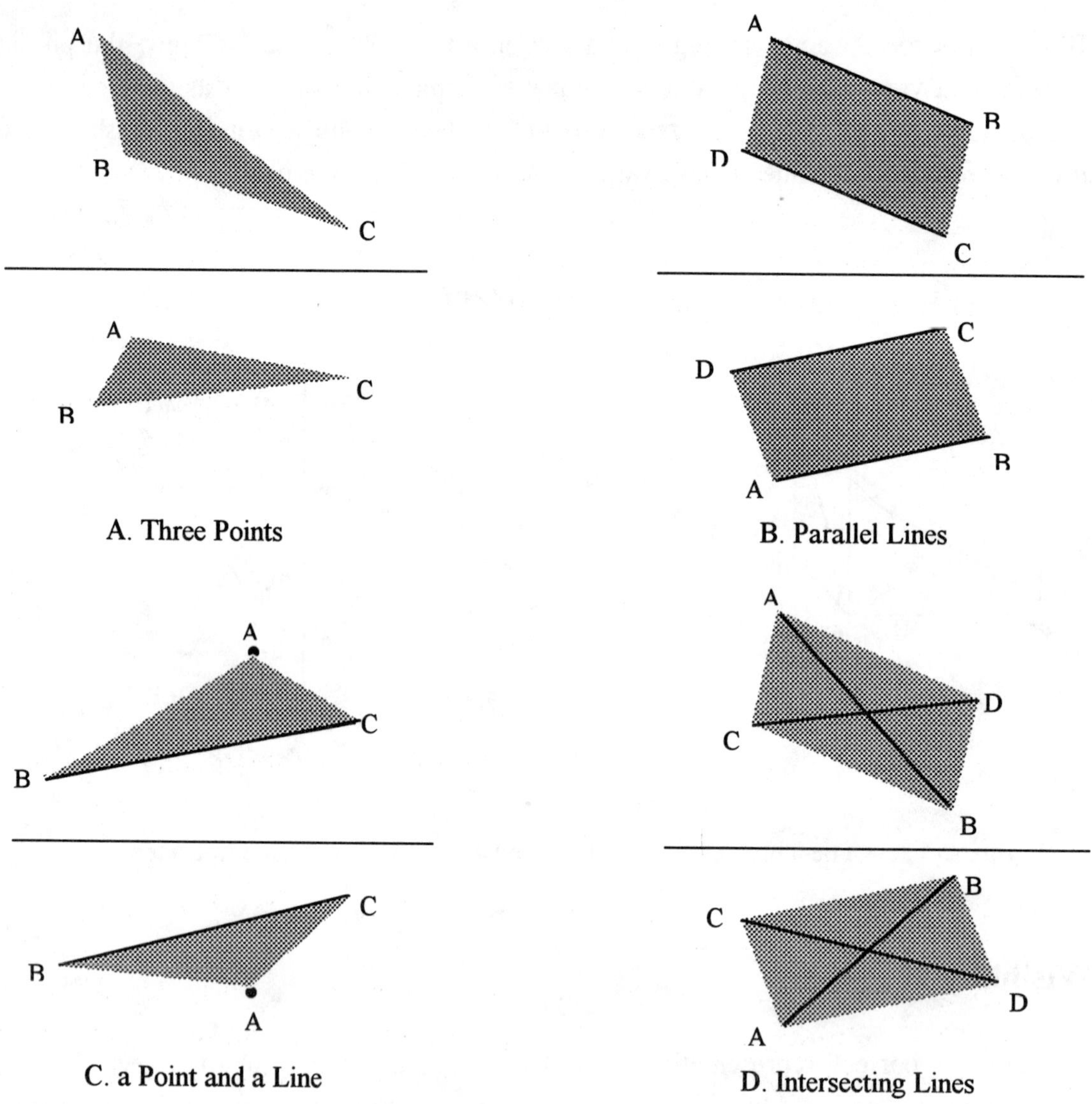

Figure 2-13 Defining Planes with, Points and Lines and Combinations Thereof.

2.5.5 Types of Lines Used in Engineering Drawings

Engineering drawings are made with lines and symbols. In general there are 9 types of line used in engineering drawings. They are shown in Fig. 2-14. Each line is used for a definite purpose.

Visible lines or object lines: They are solid and thick lines to represent outlines of parts or visible edges of parts.

Hidden lines: They are dashed lines with medium width to represent edges that are not directly visible in a view.

Center lines: They are thin lines and are composed of alternate long and short dashes to represent the axes of symmetry.

Section lines: They are solid and thin lines, and are used for sectional views at an angle with the outlines.

Dimension lines: They are the same line type with section lines, but used for dimensioning sizes of features.

Cutting plane lines: They are solid and thick lines to represent the location of cross section.

Phantom lines: They are similar to center lines except with double dots, and are used to represent adjacent parts.

Long break lines: They are similar to center lines except with break signs to indicate the end of a partially illustrated feature.

Short break lines: They are drawn by hand to indicate the boundary of a small detail.

Engineering drawings make use of standard lines to aid in showing the details associated with a part or parts. When a CAD system is used, the software system may take care of properly selecting line types. However, it is important for a design engineer to know using proper line types when designing components.

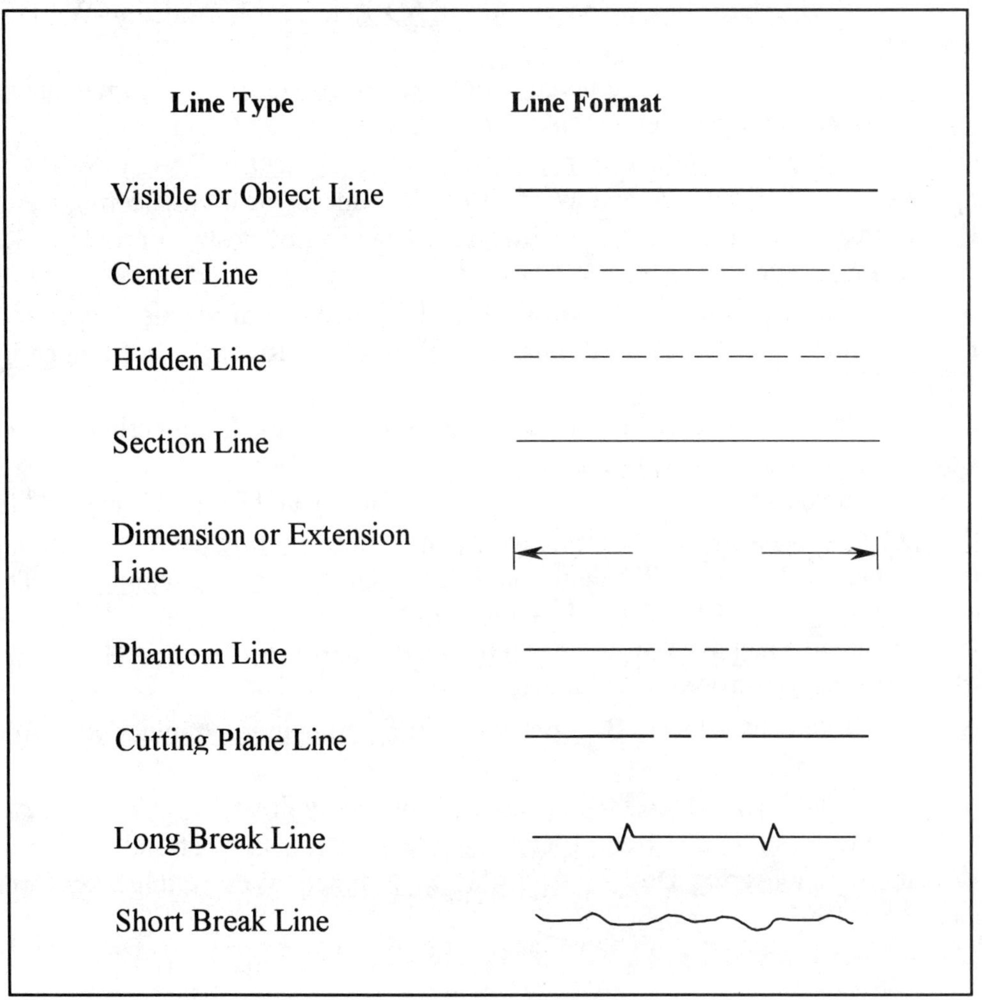

Figure 2-14 Types of Lines Used in Engineering Drawings

REFERENCES

1. H. R. Buhl, Creative Engineering Design, Iowa State University Press, Ames, Iowa, 1960.
2. A. Burstall, A History of Mechanical Engineering, M.I.T. Press, Cambridge, MA, 1965.
3. J. W. Dally and T. Regan, Introduction to Engineering Design, Book1: Solar Desalination, College House Enterprise, Knoxville, TN, 1996.
4. J. W. Dally and T. Regan, Introduction to Engineering Design, Book2: Weighing Machines, College House Enterprise, Knoxville, TN, 1997.
5. J. W. Dally and G. M. Zhang, "A Freshman Engineering Design Course," Journal of Engineering Education, 83-91, April, 1993.
6. B. L. Davids, A. J. Robotham and Yardwood A., Computer-aided Drawing and Design, London, 1991.
7. J. Dieter, Engineering Design, McGraw-Hill, New York, 1983.
8. C. E. Douglass, and A. L. Hoag, Descriptive Geometry, Holt, Rinehart and Winston, Inc., 1962.
9. J. H. Earle, Graphics for Engineers, AutoCAD Release 13, Addision-Wesley, Reading, Massachusetts, 1996.
10. J. Encarnacao, E. G. Schlechtendahl, Computer-aided Design: Fundamentals and System Architectures, Springer-Verlag, New York, 1983.
11. G. Farin, Curves and Surfaces for Computer-aided Geometric Design, New York, 1988.
12. S. Fingers, J. R. Dixon, A review of research in mechanical engineering design, Part I: Descriptive, prescriptive and computer-based models of design processes, Research in Engineering Design, 1(1), 51-68, 1989.
13. S. Fingers, J. R. Dixon, A review of research in mechanical engineering design, Part II: Representations, analysis, and design for the life cycle, Research in Engineering Design, 1(2), 121-38, 1989.
14. R. M. Glorioso, and S. H. Francis, Introduction to Engineering, Prentice Hall, Englewood Cliffs, New Jersey, 1975.
15. C.S. Krishnamoorthy, Finite Element Analysis, Theory and Programming, 2nd Ed., 1995.
16. P. Ingham P. CAD System in Mechanical and Production Engineering, London, 1989.
17. E. B. Magrab, Integrated Product and Process Design and Development: The Product Realization Process, CRC Press, Boca Raton, NY, 1997.
18. R. L. Norton, Machine Design: An Integrated Approach, Prentice Hall, Upper Saddle River, New Jersey, 1996.
19. A. A. G. Requicha and H. B. Voelcker, IEEE Computer Graphics and Applications, 2(2), 9-24, 1982.
20. N. P. Suh, The Principles of Design, Oxford University Press, New York, NY, 1990.
21. C. L. Svensen and W. E. Street, Engineering Graphics, Van Nostrand Company, 1962.
22. J. Walton, Engineering Design: from Art to Practice, West Publishing Company, St. Paul, MN, 1991.

EXERCISES

1. What are the three engineering projections that are most commonly used? Prepare the three projections for an object you would like to present. You may sketch the projections by hand.
2. Under what situations do you need more than three projections? Under what situations do you need 2 projections or one projection? Illustrate the principle that a minimized number of projections are preferred in engineering documentation.
3. Prepare the engineering drawing of an object where one projection is needed.
4. Prepare the engineering drawing of an object where two projections are needed.
5. Prepare the engineering drawing of an object where three projections are needed.
6. Prepare the engineering drawing of an object where more than three projections are needed.
7. Go to an engineering library and conduct a search for textbooks under the title "Engineering Graphics". (1) Print a list of 20 books. (2) Identify three or four chapters, the contents of which most of the 20 books share. (3) Write a one page summary report on the fundamentals of engineering graphics.
8. Two tubes AB and BC are to be connected by a third tube AC, thus completing the circulation of air among the three locations, A, B and C. Note that location B is at the top of a hill. The top of the hill is 8 ft east of A and 5 ft north of C. The slope of AB and BC are 30 degrees and 45 degrees , respectively.

   (1) Draw the front, top and right side views of the design configuration.
   (2) Find the true length of the third tube AC.
   (3) Find the true size of the angle for the special fitting.

9. A vertical channel from a pump is to be constructed. It is to be connected to a horizontal channel 3 meters above and 2 meters west of the pump outlet. The connecting channel is to be fastened to the vertical channel with a 45 degrees elbow and to the horizontal channel with a 60 degrees elbow. How far north of the pump outlet should the 60 degrees elbow be placed?

CHAPTER 3

DIMENSIONING ENGINEERING DRAWINGS

3.1 Geometric Dimensioning

An engineering drawing without dimensions only characterizes the geometric shape of an object. The drawing is not completed and the design process cannot continue without including the following information on the drawing:

- The size of each feature.
- The relative positions between features.
- The surface condition required.
- The accuracy of sizing and positioning.

Geometric dimensioning is a process to specify dimensions describing the size and position of all of the features comprising a component, to include notes relevant to the manufacturing process, and to provide information related to the product realization process, such as purchase, process planning, etc. With the addition of dimensions and notes, engineering drawings serve as manufacturing documents and are included as part of legal contracts. The process of dimensioning is so critical that the American National Standards Institute (ANSI) has established the standards, namely, Y14.5M, Dimensioning and Tolerancing for Engineering Drawings.

3.1.1 Size or Position Dimensions

Dimensions are classified as those of size or position. An example presented in Fig. 3-1, shows that each geometric feature of a component is accounted for in terms of its size and position. There are eight dimensions required to describe size and four dimensions for position. Without the four dimensions to define the position, the part design is not uniquely defined. On the other hand, the use of additional dimensions is not allowed. Additional dimensions over constrain the geometry and introduce uncertainty in the interpretation of the size and position of some features. It is not uncommon to see that alternative techniques for dimensioning a certain component exist. The criterion of selecting an appropriate technique for dimensioning is to facilitate the manufacturing process. Learning the skills needed for dimensioning is enhanced with hands-on experience on the shop floor. In fact, the ANSI standards embody the accumulation of design experience, and provide excellent guidelines for appropriate dimensioning.

There are two units of measurement that are currently employed in dimensioning, namely the decimal inch in the U. S. Customary system (English), and the millimeter in the SI system (metric). Numerical values shown in engineering drawings are assumed to be in the unit of millimeter or inch unless a different unit is specified. The conversion factor between inch (in.) and millimeter (mm) is 25.4 mm = 1 in.

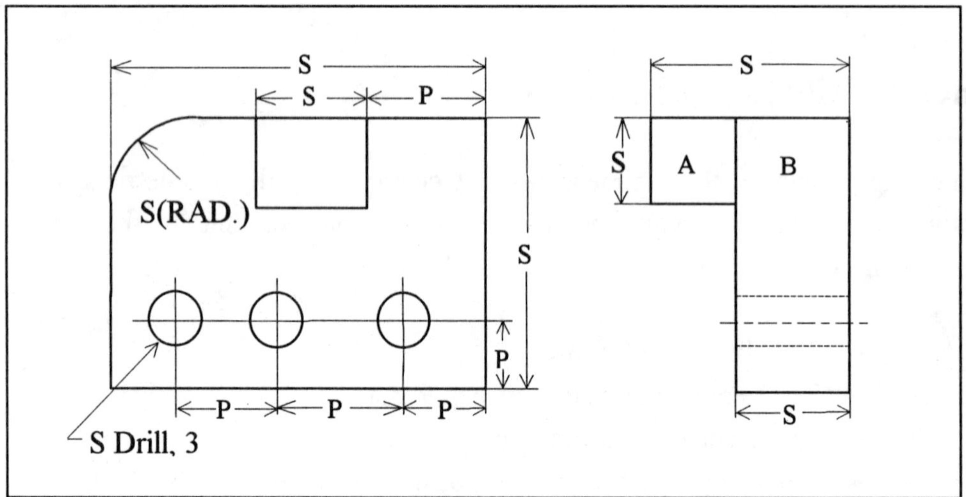

Figure 3-1 Examples of Dimensions of Size and Position. S indicates size and P position.

A comparison of units used to dimension a single view of a block is presented in Fig. 3-2. Note for the SI system, dimensions in millimeters are rounded to the nearest whole number. For the U. S. Customary system, dimensions in inches are indicated with two decimal places, even for whole number such as 3.00. In general, dimensions of fractions of an inch are expressed in terms of decimals rather than common fractions to reduce the effort required for the arithmetic.

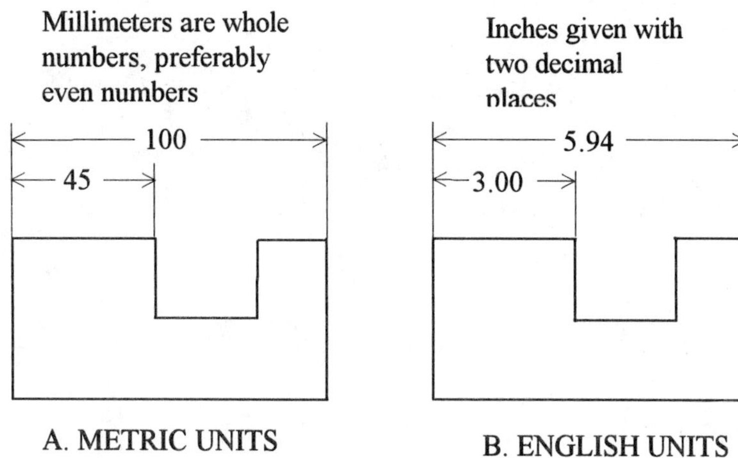

Figure 3-2 Examples of dimensioning in SI and U. S. Customary units.

3.1.2 Basic Examples of Dimensions of Size and Position

Since every solid has three dimensions, each of the geometric features in the component must have its height, width, and depth defined by a dimension. The prism is the most common shape. As illustrated in Fig. 3-3A, three dimensions are required to define square, rectangular, or triangular prisms. For regular polygonal shapes, usually only two dimensions are given --- the length and either the distance "across corners" or the distance "across flats".

The cylinder is the second most common shape. A cylinder requires only two dimensions --- diameter and length, as illustrated in Fig. 3-3B. Partial cylinders, such as fillets and rounds, are dimensioned by defining the radius instead of the diameter. A good general rule is to dimension complete circles with the diameter and circular arcs (partial circles) with the radius.

Right cones can be dimensioned by specifying the altitude and the diameter of the base. Cones usually occur as frustums, and require the diameters of the both ends be specified in addition to the length, as illustrated in Fig. 3-3C.

Right pyramids are dimensioned by giving the dimensions of the base and the altitude. Right pyramids often occur as frustums, requiring dimensions for both bases, as illustrated in Fig. 3-3D.

Oblique cones and pyramids are dimensioned in the same manner as right cones and pyramids, but a dimension parallel to the base is added to define the offset of the vertex.

Spheres are dimensioned by specifying the diameter. Other surfaces of revolution are dimensioned by specifying the dimensions of the generating curve.

After the basic geometric shapes have been dimensioned for size, the position of each relative to the others must be specified. Position must be established in height, width, and depth directions. Rectangular shapes are positioned with reference to their faces. Cylindrical and conic shapes are positioned with reference to their center lines and their ends.

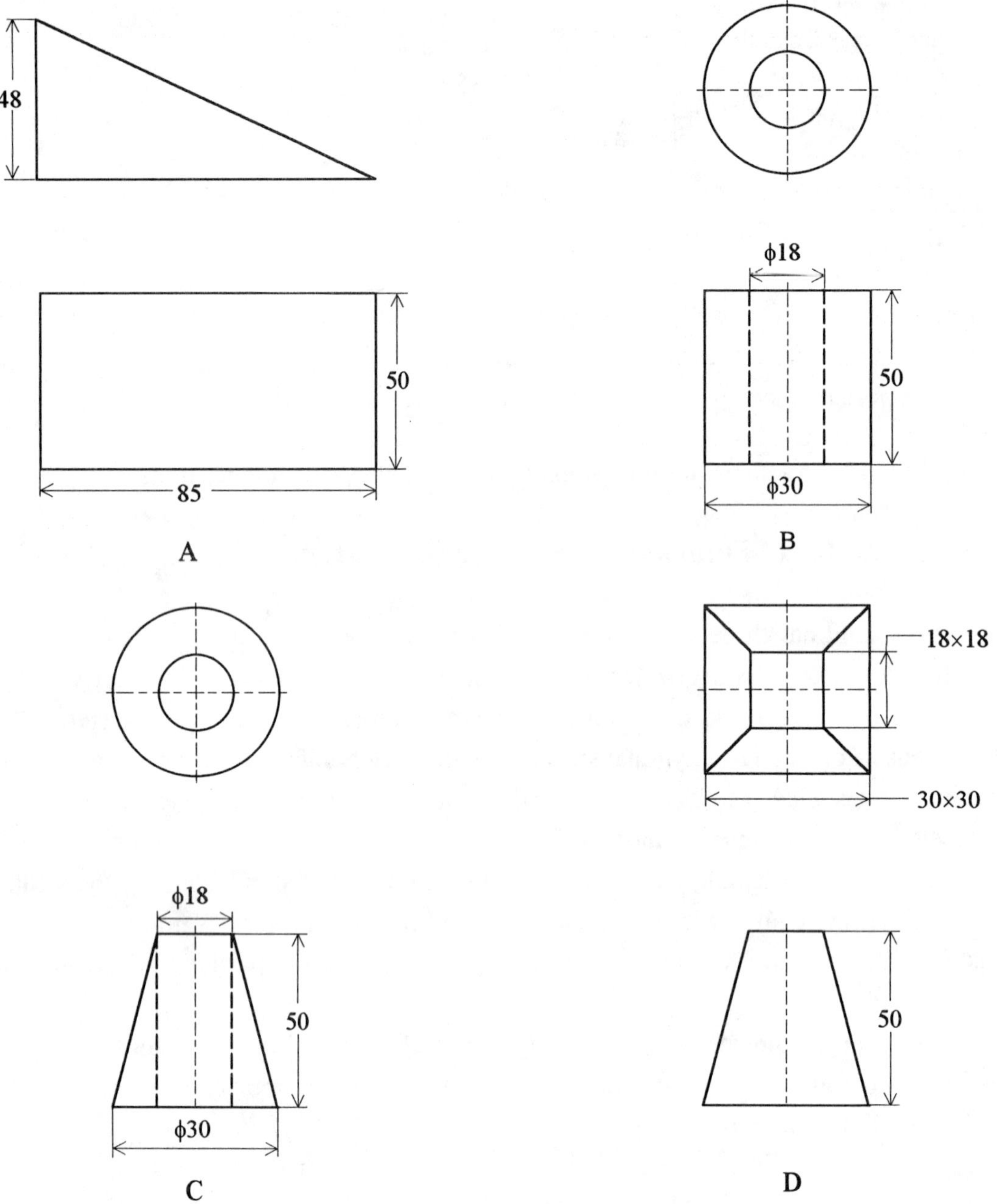

Figure 3-3 Dimensioning to define the size of a prism (A), cylinder (B), cone (C), and pyramid (D)

One basic shape will often coincide or align with another on one or more faces of the component. In such cases, this alignment serves partially to locate the shapes. For example in Fig. 3-1, the prism A requires only one dimension for complete positioning with respect to prism B because two surfaces are in alignment and two in contact.

Coincident center lines often eliminate the need for position dimensions. In the cylinder in Fig. 3-3B, the center lines of the hole and of the cylinder coincide, and no position dimensions are needed. The three holes in Fig. 3-1 are on the same centerline, and the dimension perpendicular to this common centerline positions all three holes in the vertical direction.

3.1.3 Selecting Dimensions

Specifying dimensions for the size of certain shapes is dependent upon the requirements of a particular process or operation (e.g., machine shop, pattern making operation, forging, sheet-metal, or steel fabrication, die casting, welding, etc.). These basic shapes are produced by using one or more of these manufacturing methods. However, a manufacturing drawing is often prepared showing the dimensions of size as a note rather than as regular dimensions. This practice of using notes is common when shop processes such as drilling, reaming, counter-boring, and punching are involved.

Selecting dimensions for position usually requires more consideration than selecting dimensions of size because there are usually several options for specifying a position. In general, positional dimensions are specified between finished surfaces, center lines, or a combination thereof, as illustrated in Fig. 3-4.

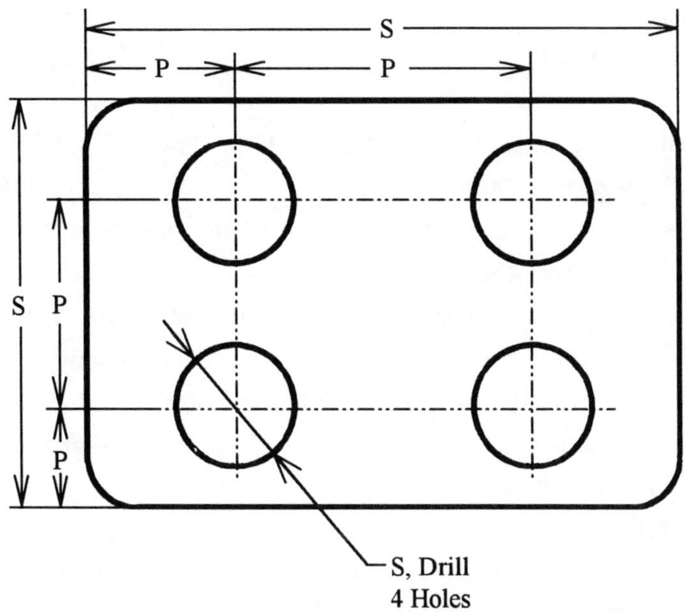

Figure 3-4 Example of Dimensioning Position and Size (S = size and P = position).

The position of a point or center is established by offset dimensions from two center lines or surfaces, as shown in Fig. 3-5A. This approach is preferable rather than the use of angular dimensions, as shown in Fig. 3-5B, unless angular dimensions are more practical to employ in the fabrication process.

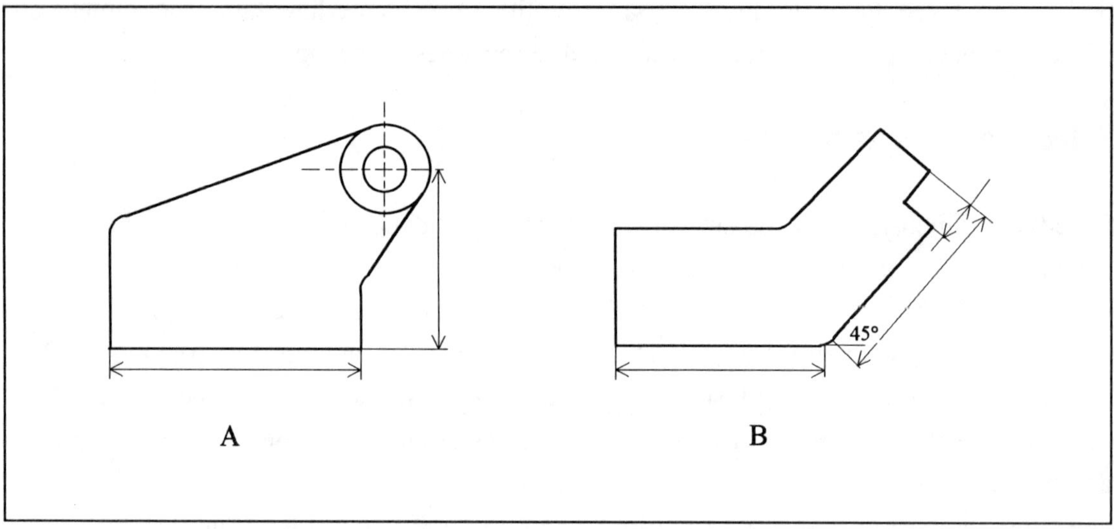

Figure 3-5 Examples of Dimensioning Position by Offsets (A) and by Angle (B).

3.1.4 The Contour Principle

In all cases, dimensioning methods stress clarity. One view of a part will usually describe the shape of some feature better than other views. A feature should be dimensioned in the view that best represents the shape of the feature. Selection of the best view for either position or shape dimensions is known as the contour principle.

It is natural to search for a dimension of a certain feature wherever that feature appears with the most clarity. Reading the drawing is facilitated when the dimension is placed on these views. The rounded corner, the drilled hole, and the lower notched corner are all features that are most clearly depicted on the front view in Fig. 3-6. It is evident that the positions of these features be dimensioned in this front view. On the other hand, the shape of the object is more evident on the top view and dimensions defining the shape are provided in this view.

The use of the contour principle implies that duplicate and/or unnecessary dimensions are to be avoided. When a drawing is changed or revised, a duplicate dimension may go unnoticed. As a result, a particular dimension could have two different values, one of which is incorrect! In computer-aided design environment, software tools are capable of detecting such errors and remind the designer to make the necessary corrections.

In practice, redundant and unnecessary dimensions always occur when all the individual dimensions are specified together with the overall dimension, as illustrated in Fig. 3-7A. One dimension in the series positioning dimensions for the holes must be omitted if the overall dimension is used. It is imperative that only one dimension be used to locate the horizontal position of each hole, as illustrated in Fig. 3-7B.

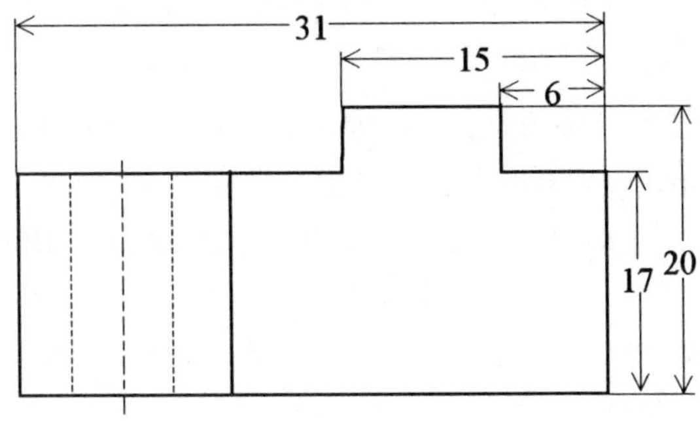

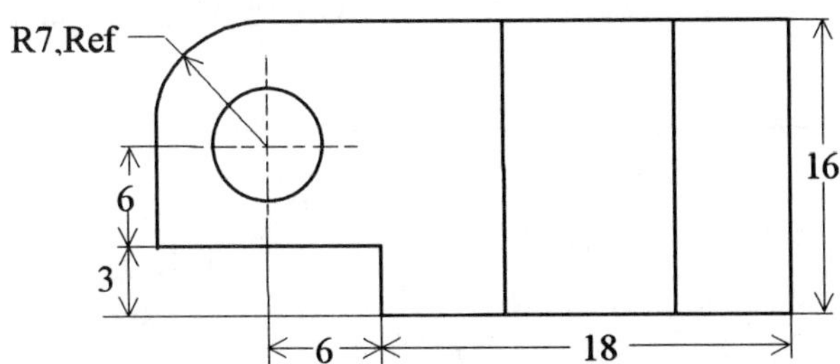

Figure 3-6 An Example of Dimensioning Using the Contour Principle.

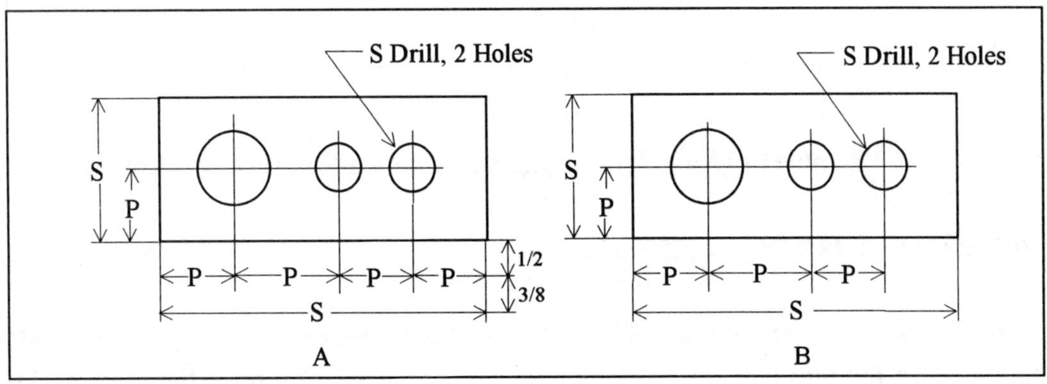

Figure 3-7 Example of Unnecessary Dimensions and the Correction to Remove the Redundancy.

There are several rules for the placement of dimensions that make the contour principle easier to apply. These rules are listed below:

(1) Dimensions placed outside the view are preferred, unless added clearness, simplicity, and ease of reading will result from placing a few dimensions inside the view. For good appearance, dimensions should be kept off the surfaces revealed by sections cuts.
(2) Dimensions between the views are preferred unless there is some compelling reason for placing them elsewhere.
(3) Dimension lines should be spaced, in general, 1/2 inch from the outlines of the view. This applies to a single dimension or to the first dimension of several in a series, as

illustrated in Fig. 3-7A. Parallel dimension lines should be spaced uniformly with at least 3/8 inch between lines.

(4) Always place a longer dimension line outside a shorter one to avoid crossing dimension lines with the extension lines from other dimensions. An overall dimension (maximum size of the component in a given direction) will be outside all other dimensions.

Occasionally, when a drawing is to be used for reference or checking, all dimensions in a series are given together with the overall dimension. In such cases, one dimension is defined as the reference dimension, and is enclosed within parentheses as shown in Fig. 3-8. Alternatively, the designer may use the abbreviation REF, as in 3.50 REF instead of using parentheses.

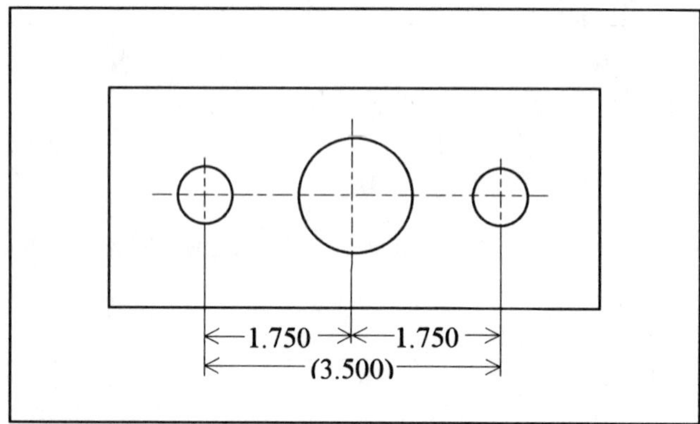

Figure 3-8 An Example of Using Parentheses to Identify a Reference Dimensions.

3.1.5 Dimensional Clarity Using a Datum

Datum points, lines, and edges of surfaces of a component are all features that are assumed to be exact for purposes of computation or reference. As such they are preferred features from which the positions of other features are established. For example, in Fig. 3-9A the left side and bottom edge of the component are used as the references in dimensioning. In Fig. 3-9B, the centerlines of the central hole provide the datum lines. When the positions of one feature are specified by dimensions from a datum, other features are also positioned from the same datum, and not with respect to one another.

A feature selected to serve as a datum must be clearly identified and readily recognizable. Note in Fig. 3-9 that the edges are obvious datum lines. The datum surfaces must be accessible during manufacturing so that difficulties will not be encountered in measuring the dimensions.

A datum surface (on a physical part) must be more accurate than the allowable variation on any dimension that is referred to the datum. To achieve the accuracy required of the datum surfaces, it may be necessary to specify their flatness, straightness, roundness, etc.

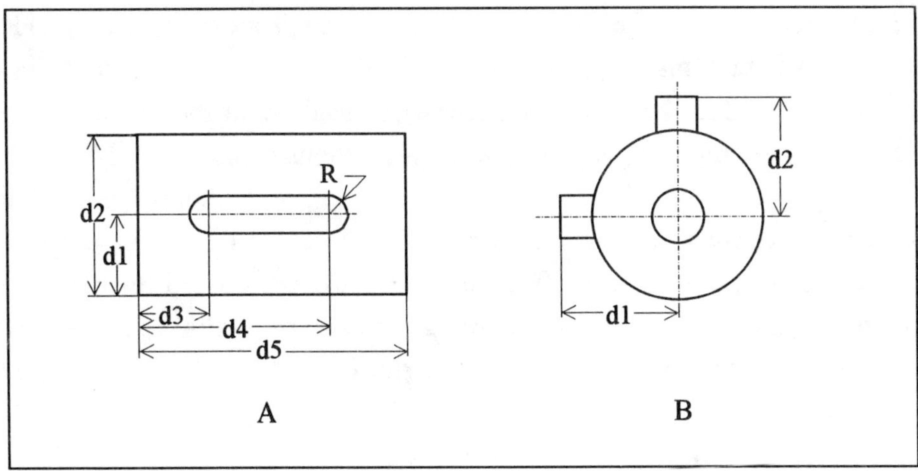

Figure 3-9 Dimensioning from a Datum Surface (A) and a Datum Center (B).

3.1.6 Practical Considerations

Use of Scale

Special attention must be given to the scale of the component that is to be dimensioned. A part may be drawn to any scale convenient to the draftsperson. However, the scale must be indicated on the working drawing, often in the title block. It is important to realize that the dimensions placed on the drawing are always dimensions of the actual size of the component. The value of the dimension for a given length never changes, regardless of the scale used to prepare the drawing. For example, if a length is 500 mm on the component, the value 500 will appear on the drawing regardless of the scale used in representing the component.

Notations adopted by Y14.5M

The two basic methods used to specify a distance on a drawing are dimensions and notes, as illustrated in Figs. 3-10A and 10B. Notice the use of R10 and ϕ12 DRILL in these two figures. Using the notation of R10 and ϕ12 DRILL conforms to SI system of units as commonly employed in Europe and Asia. Such notation was adopted as an acceptable standard in the U. S. by the American National Standards Institute in the ANSI Y14.5M standard entitled, Dimensioning and Tolerancing. However, one may discover the radius and diameter, R10 and ϕ12 DRILL, specified as 10R and 12 DRILL on many drawings. Referring to Fig. 3-10A, we observe that a dimension is used to define the distance between two points, lines, planes, or some combination thereof. Extension lines lead to the particular feature defined by the dimension. On the other hand, notes are word statements giving information that cannot be conveyed easily by orthogonal views or dimensions. Notes almost always refer to some standard shape, operation, or material. Notes are classified as general or specific, where a general note applies to the entire part, and a specific note applies only to an individual feature. Occasionally a note will save making an additional view, for example, by indicating right- and left-hand parts. General notes do not require the use of a leader and should be grouped together above the title block. Examples of general notes include: "Finish all over", "Fillets 1/4R, rounds 1/8R, unless otherwise specified," "All draft angles 7^0 ," "Remove burrs," etc. Specific notes almost always require a leader and should be placed fairly close to the feature to which they apply. Notes are often employed to specify sizes of tools to be used and the number of features to be formed in a manufacturing operation (e. g. "ϕ12 thru, 4 holes"). Notes often use abbreviations such as thru instead of through to save space.

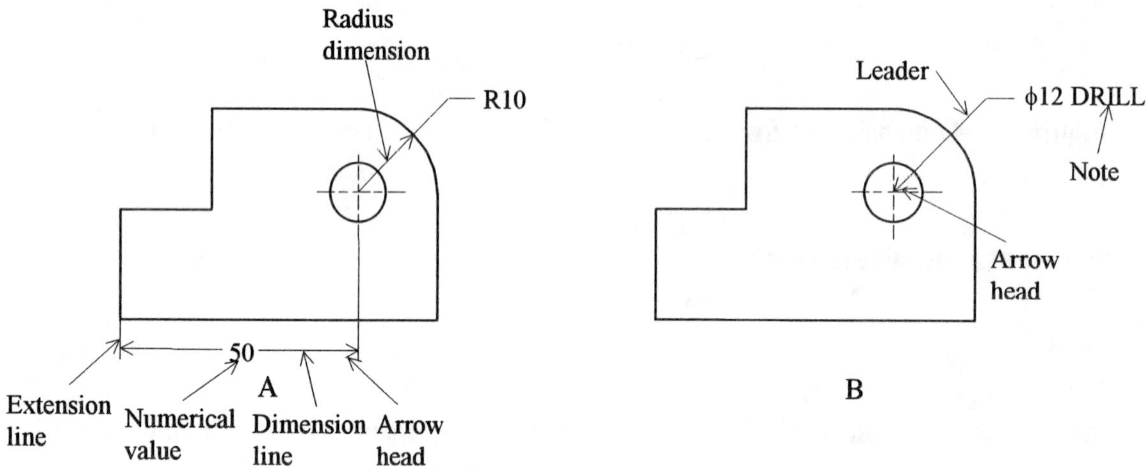

Figure 3-10 Dimensioning with the SI System (A), and a Note (B)

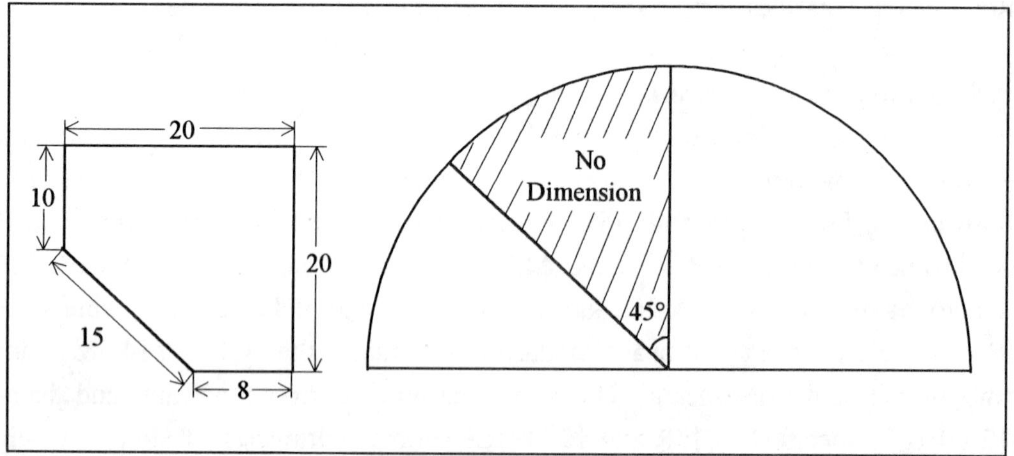

Figure 3-11 Orientation of Text and Numbers in the Aligned System.

Orientation of Text and Numbers

Two different approaches are employed in placing text and numbers on engineering drawings, namely the aligned or the unidirectional. The aligned system where the numbers are parallel to the dimension line, and the fraction bar is in line with the dimension line. The figures should be arranged so that the drawing is read from the bottom or the right side. Avoid placing dimensions in the note blocks included in the shaded regions like the one illustrated in Fig. 3-11.

3.2 Tolerancing

Products we use on a daily basis are designed to operate at particular performance levels. To operate at these levels, products are made with various degrees of precision. Take as an example two very different products, each with a feature that performs a similar function --- a wheel on an axle. Suppose the wheel and axle assembly is installed on a child's wagon and another on a Boeing 737 airplane. In both cases the purpose of the wheel and axle assembly is to allow the vehicle to roll. Obviously the performance levels are quite different for these two applications, and the degree of precision needed for each assembly differs greatly. The need for precision and careful machining in manufacturing the components is far greater for the airplane than for the child's wagon.

Technology and business today require that parts be specified with increasingly exact dimensions. Many parts made by different companies at widely separated locations must be interchangeable. This requirement demands precise size specifications and achieving these exact dimensions in production. The technique of dimensioning parts within a required range of variation to ensure interchangeability requires the designers to be knowledgeable in tolerancing. Tolerances are the allowable deviations from a specified dimension. For instance, a component with a dimension specified as 120 mm with a tolerance of ± 1 mm is acceptable provided that the measurement of this dimension is within the range from 119 mm to 121 mm. Certainly, tolerancing is directly related to the manufacturing process involved in forming the component. In general, manufacturing costs increase as tolerances become smaller, and it is prudent to specify tolerances as large as possible to minimize production costs without interfering with the function of the component.

Tolerancing of components allows the designer to control the dimensions, namely the maximum and minimum sizes of each feature and their position. The component shown in Fig. 3-12 has a hole with tolerances specified for its diameter. The hole may vary from 0.500 to 0.508 in diameter. The range of allowed variation (0.008 inch) is the tolerance. The tolerance of a feature can affect the tooling selected to produce the component. Numerically higher tolerances can use less precise tooling than lower ones. The hole shown in Fig. 3-12 could be made with a sharp drill, but if the tolerance is reduced to 0.005 inch, higher cost processes such as reaming or grinding might be required.

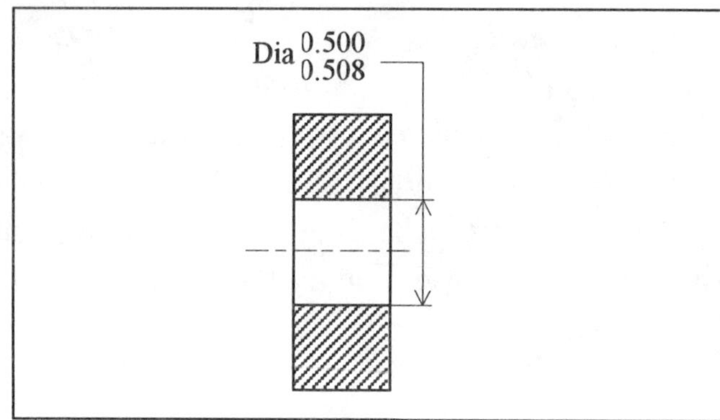

Figure 3-12 Example of a Method Used to Specify Tolerances for a Hole.

3.2.1 The Fit of Contacting Parts

The interaction between a hole and a shaft is called a fit. The fit of a shaft in a hole is very common in a wide range of products. Wheels rolling on shafts and pistons sliding in cylinders are just two examples. The performance of these contacting systems is vitally dependent on the fit of the contacting components. There are three classifications used to specify different types of fit:

- Clearance fit
- Interference fit
- Transition fit

Clearance Fit

In a clearance fit, the diameter of the shaft is always smaller than the diameter of the hole. The shaft "clears" the hole, leaving an annular gap. The amount of clearance varies with the tolerance value placed on the hole and shaft. An example presented in Fig. 3-13 indicates that shaft diameters may vary from 0.493 to 0.497 and the hole diameters may vary from 0.500 to 0.508 in. The tolerance is 0.004 in. on the shaft and is 0.008 in on the hole. The maximum clearance is determined by subtracting the smallest shaft diameter from the largest hole diameter (0.508-0.493 = 0.015 in.) The minimum clearance is determined by subtracting the largest shaft diameter from the smallest hole diameter (0.500-0.497 = 0.003 in.)

Interference Fit

An interference fit occurs when the shaft diameter is larger than the hole diameter prior to assembly. Interference fits are also known as force fits since the shaft is forced into the hole during an assembly operation. A simple tap of a shaft into a hole may be sufficient to assemble a light force fit, but a heavy force fit often requires the shaft to be cooled and/or the hole to be heated to facilitate assembly. Once assembled, an interference fit provides a permanent coupling of the two components. Disassembly occurs only if the shaft is forced out of the hole. In severe cases, the housing containing the hole may have to be cut to permit the shaft to be withdrawn. Interference fits are an excellent way to assemble parts which are to remain fixed in position relative to each other. A bearing at a fixed location on an axle is an example. A drawing of a shaft and a disk, presented in Fig. 3-14, illustrates an interference fit. The shaft diameter is always larger than the hole diameter. In this instance, the tolerance on the shaft diameter is 0.003 in. and 0.008 in. on the hole diameter. The maximum clearance for an interference fit is determined using the same approach employed for the clearance fit. The smallest shaft diameter is subtracted from the largest hole diameter to give 0.508-0.509=-0.001 in. Note the minus sign, implies a negative clearance which is in fact the amount of interference. For the maximum clearance, the largest shaft diameter is subtracted from the smallest hole diameter, giving 0.500-0.512=-0.12 in. Again the clearance is a negative value indicating the amount of interference. For interference fits, the terms minimum interference and maximum interference are often used. Maximum interference occurs with minimum clearance and minimum interference occurs with maximum clearance.

In the assembly specified in Fig. 3-14, the maximum clearance was -0.001 inch, which is the least (minimum) interference. The minimum clearance of -0.012 inch was the most (maximum) interference.

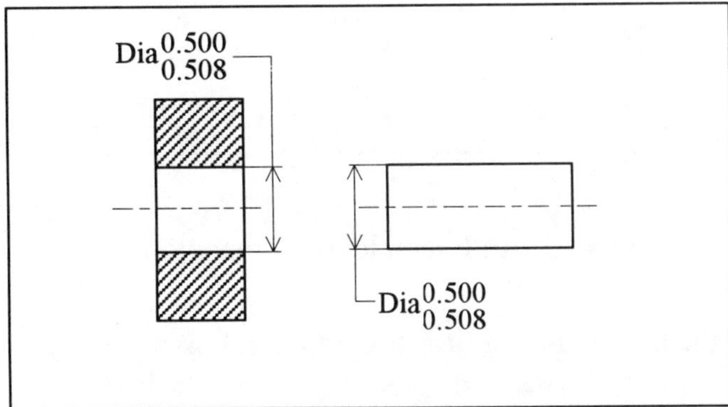

Figure 3-13 An example of a Clearance Fit

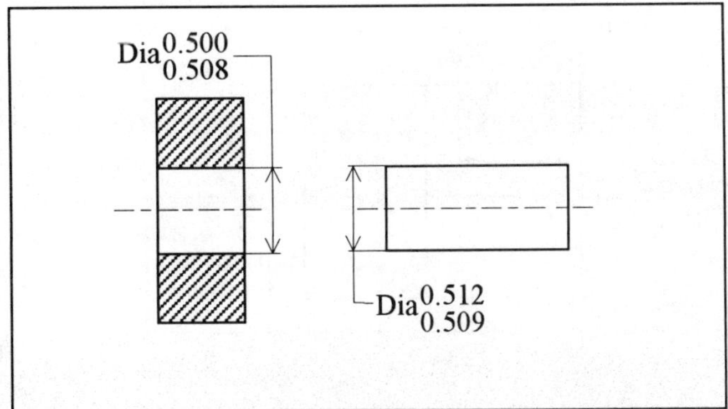

Figure 3-14 An Example of an Interference Fit.

Transition Fit

A transition fit exists when the maximum clearance is positive and the minimum clearance is negative. Therefore, the shaft may clear the hole with an annular gap or it may have to be forced into the hole. Transition fits are used only for locating a shaft relative to a hole, where accuracy is important, but either a clearance or an interference is permitted. An example of a transition fit is presented in Fig. 3-15. The tolerance on the shaft diameter is 0.011 in. and the tolerance on the hole diameter is 0.008 in. The maximum clearance is 0.508 = 0.498 = 0.010 in, a positive value indicating clearance. The minimum clearance is 0.500 - 0.509 = -0.009 in., a negative value indicating interference.

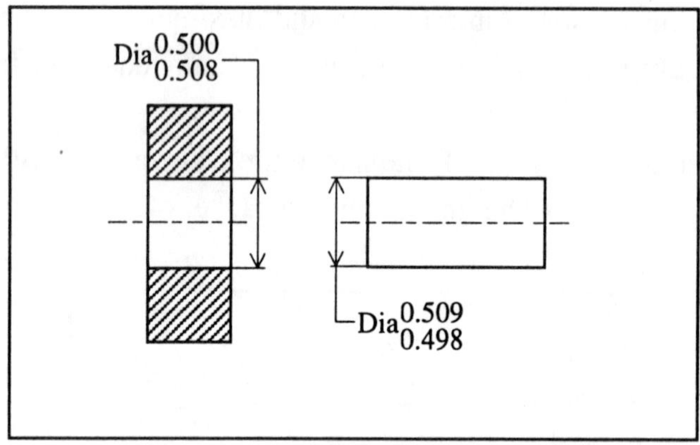

Figure 3-15 An Example of a Transition Fit

A comparison of the three types of fit is presented in Fig. 3-16.

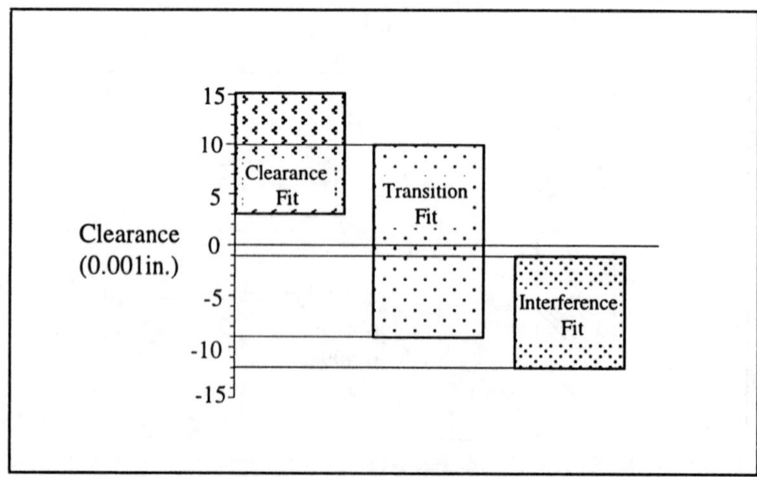

Figure 3-16 Comparison of the Range of Clearance for the Three Types of Fit

3.2.2 Definitions Used in Tolerancing

(1) Nominal size: The size used for a general description. Example: 3/4 inch diameter.

(2) Basic size: The size used when the nominal size is converted to a decimal from which deviations are made to produce limit dimensions. Example: a 0.7500 inch shaft is the basic size for a 3/4 inch nominal size shaft. When tolerances are shown in limit form, the basic size is unknown.

(3) Tolerance: The total variation permitted in the size of a feature. It can be specified in three ways:

- Unilateral tolerances where dimensions vary in only one direction from the basic size (either larger or smaller).
- Bilateral tolerances where dimensions vary in both directions from the basic size (larger and smaller)
- Limit tolerances with dimensions representing the largest and smallest sizes permitted for a feature.

Figure 3-17 presents these three methods of specifying tolerances on dimensions for both general spaces and tight spaces.

	GENERAL SPACE	TIGHT SPACE	DIAMETER FORM
UNILATERAL TOLERANCE	1.750 $^{+.000}_{-.004}$	.450 +.004 −.000	ϕ.600 $^{+.000}_{-.004}$
BILATERAL TOLERANCE	1.750±.003	.450 +.003 −.001	ϕ.600±0.04
LIMIT FORM	1.640 1.635	+.003 −.001	ϕ.600–0.604

Figure 3-17 Unilateral, Bilateral, and Limit tolerances.

(4) Allowance: An alternative term for tightest possible fit with a minimum clearance or maximum interference.

(5) Hole basis system. A system in which the basic size appears as one of the limit dimensions (usually the minimum dimension) of the hole, but not of the shaft. As an example, for a basic size of 2.000, the limit dimensions of a hole might be 2.000 and 2.007 in. For the shaft, which is to fit into the hole, the limit dimensions might be 1.994 and 1.989 in. providing a clearance fit. This example gives the same tolerance, minimum clearance, and maximum clearance as for the shaft basis system. A graphic representation of the hole basis system is presented in Fig. 3-18.

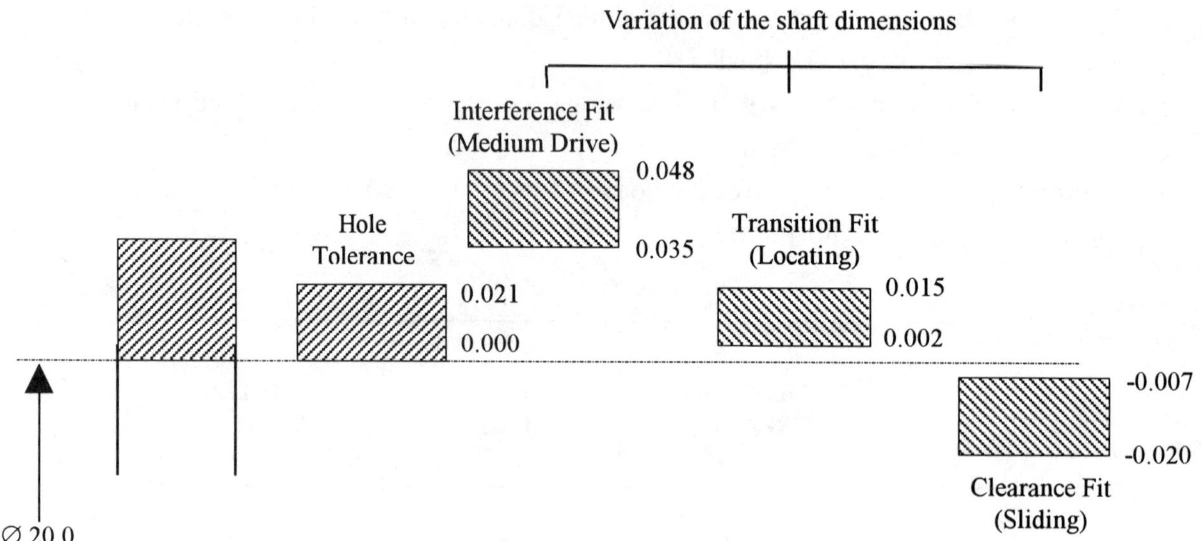

Figure 3-18 Graphic Representation of the Hole Basis System.

(6) Shaft basis system. A system in which the basic diameter appears as one of the limit dimensions (usually the maximum diameter) of the shaft, but not on the hole. As an example, for a basic size of 2.000, the limit dimensions of a shaft for a shaft basis system might be 2.000 and 1.995 inch. For the corresponding hole, the limit dimensions might be 2.006 and 2.013 inch. This example gives the same tolerance, minimum clearance, and maximum clearance as for the basic-hole system. A graphic representation of the hole basis system is presented in Fig. 3-19.

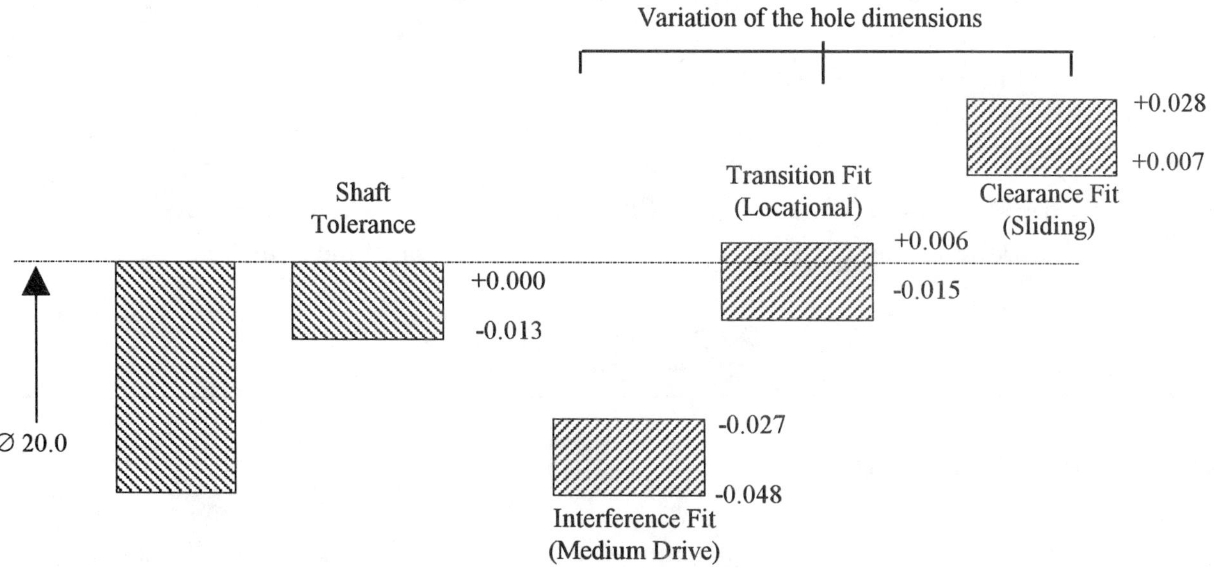

Figure 3-19 Graphic Representation of the Shaft Basis System

3.2.3 Examples of Determining Tolerances for Various Fits

How do we determine the tolerances to specify when we design a complex machine with clearance, transition and interference fits? Standards for tolerances have been established that guides our selections. Appendices 1 through 10 at the end of this textbook list many of the standards used in tolerancing.

Example 3-1 Suppose you are given instructions to specify tolerances on two holes with diameters of 20 and 50 mm. The tolerances are to be selected from the IT table (International Tolerance Grades --- ANSI B4.2, given in Appendix 10). If the tolerance grades are IT7, IT9, and IT11, determine the tolerances to be specified for each hole and for each grade.

(1) For a hole 20 mm in diameter with an IT7 tolerance, we find IT7 = 0.021 mm.
(2) For a hole 20 mm in diameter with an IT9 tolerance, we find IT9 = 0.052 mm.
(3) For a hole 20 mm in diameter with an IT11 tolerance, we find IT11 = 0.130 mm.
(4) For a hole 50 mm in diameter with an IT7 tolerance, we find IT7 = 0.025 mm.
(5) For a hole 50 mm in diameter with an IT9 tolerance, we find IT9 = 0.062 mm.
(6) For a hole 50 mm in diameter with an IT11 tolerance, we find IT11 = 0.160 mm.

Example 3-2 Let's suppose that we are dealing with a hole basis system as illustrated in Fig. 3-18. The basic diameter of the hole is 20 mm, its tolerance grade is H7, and it is to be used in an assembly involving a medium drive interference fit. Determine the maximum and minimum diameter of the hole.

From Appendix 7, find $D_{MAX} = 20.021$ mm and $D_{MIN} = 20.00$ mm.

Next, consider a shaft basis system as illustrated in Fig. 3-19. The basic diameter of the shaft is 20 mm, and for a medium drive fit a tolerance grade of s6 is specified. Find the maximum and minimum diameter of the shaft.

From Appendix 7, find $D_{MAX} = 20.048$ mm and $D_{MIN} = 20.035$ mm.

3.3 Assembly Tolerances

In this section, we discuss the maximum-minimum method for determining the tolerance associated with the assembly of two or more parts. Assembly tolerances are described with a series of four examples, which have been selected to demonstrate several techniques that are employed in the analysis of tolerances.

Example 3-3: As illustrated in the following figure, the dimensions and tolerances of Part A and Part B are $50^{+0.12}_{-0.06}$ mm and $75^{+0.20}_{-0.04}$ mm, respectively. If Parts A and B are assembled together, determine the maximum and minimum dimensions of the assembly.

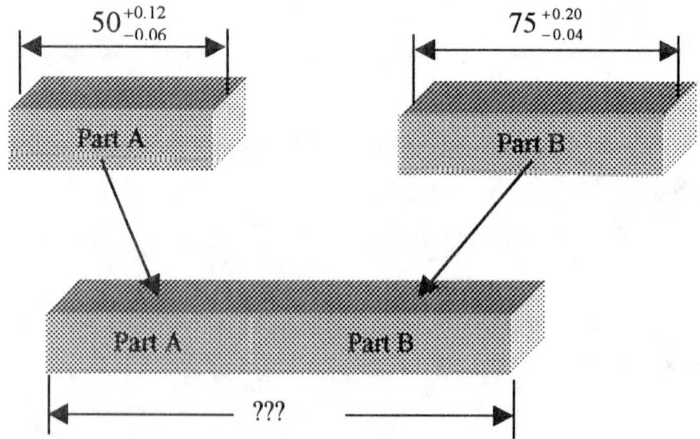

Solution:

$$?_{max} = A_{max} + B_{max} = 50.12 + 75.20 = 125.32 = 125 + 0.32 \quad mm$$

$$?_{min} = A_{min} + B_{min} = 49.94 + 74.96 = 124.90 = 125 - 0.10 \quad mm$$

$$Tolerance = ?_{max} - ?_{min} = 125.32 - 124.90 = 0.42 \quad mm$$

Example 3-4 A company is manufacturing boxes to be employed to carry and store computer diskettes. The width of the diskette is 90 ± 0.10 mm. A designer has measured the actual dimension of disks, as illustrated in the following figure. To insure that disks can be inserted into the box easily, the minimum clearance should be no less than 0.1 mm. To avoid excessive motion of the disks within the box, the maximum clearance between the inner wall of the container and a disk is to be no more than 0.4 mm. Determine the maximum and minimum dimensions of the interior of the box to meet these specifications?

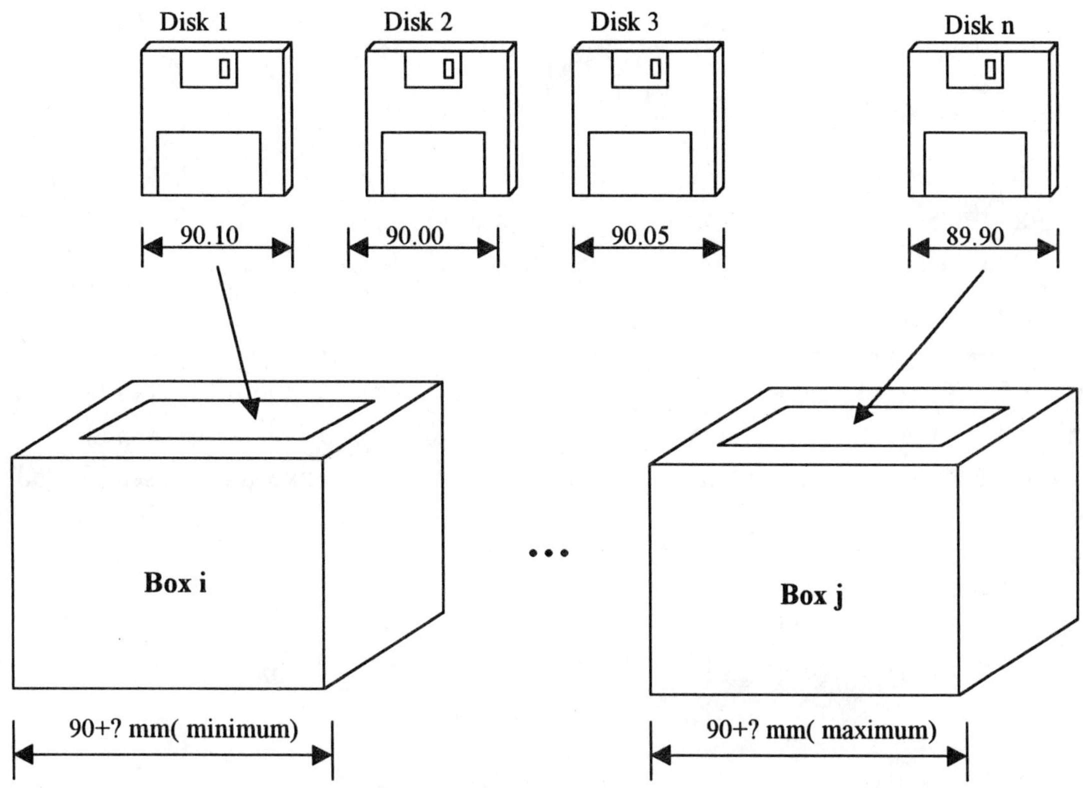

Box i with the minimum dimension should have the minimum clearance(0.1 mm) when a disk with the maximum dimension is inserted.

Box j with the maximum dimension should have the maximum clearance(0.4 mm) when a disk with the minimum dimension is inserted.

Solution:

?min - 90.10 = 0.10

⇒ ?min = 90.10+0.10 = 90.20 mm

?max – 89.9 = 0.40

⇒?max=89.9 +0.4 = 90.30 mm

Tolerance = ?max- ?min = 90.30 –90.20 = 0.10 mm

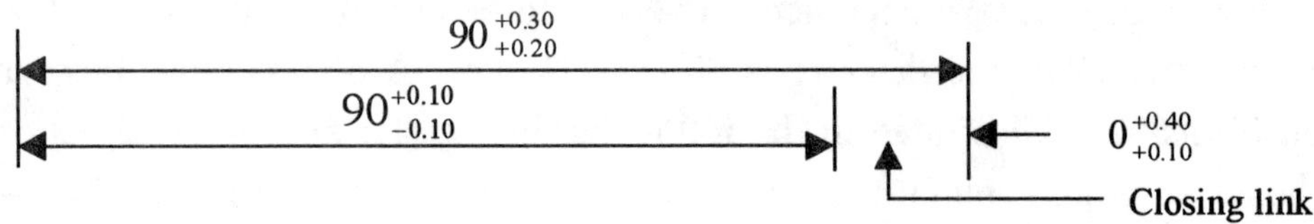

The tolerance of the closing link C_L is given by:

$$C_L = \sum_{i=1}^{2} \text{Tolerance of individual dimension composing link}_i$$

$$C_L = (0.30\text{-}0.20) + [+0.10\text{-}(-0.10)]$$
$$C_L = 0.10 + 0.20 = 0.30 \text{ mm}$$

In this example, we have introduced a new concept called the closing link. The closing link C_L is a variable dimension formed during the assembly process. Since C_L is a variable dimension it has a tolerance that is equal to the sum of the tolerances of all individual parts that make up the assembly.

Example 3-5: As illustrated in the following figure, the dimensions and tolerances for parts A and B are $50^{+0.12}_{-0.06}$ mm and $75^{+0.20}_{-0.04}$ mm, respectively. Parts A and B will be assembled into part C, which is a container. The maximum and minimum clearances between the assembly and the container are 0.50 and 0.02 mm, respectively. Determine the maximum and minimum dimensions of the container's width.

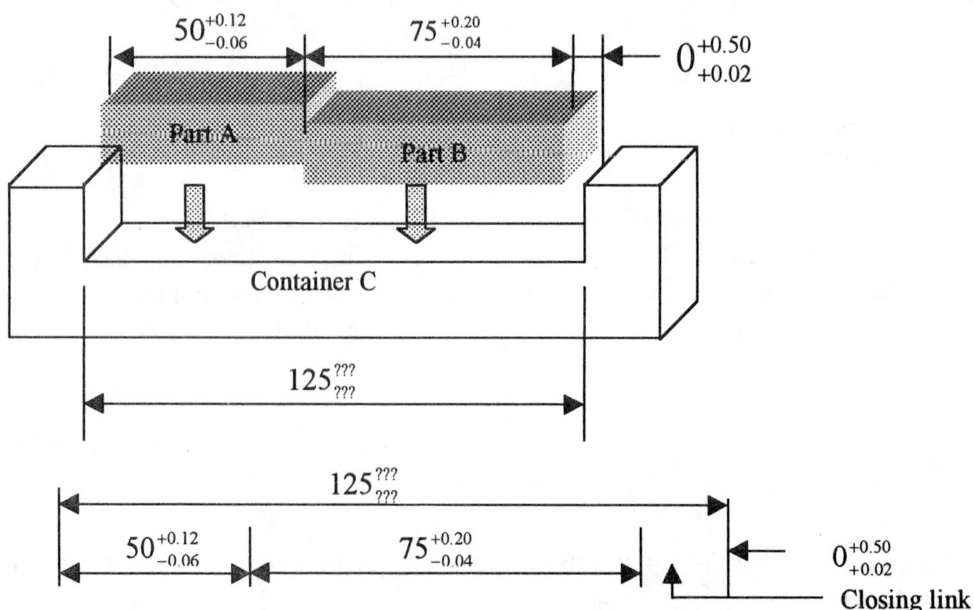

Solution: It is important to identify the closing link in this assembly. The three dimensions of Part A, Part B and Container C exist before the assembly. The dimension formed during assembly is the clearance with its maximum value of 0.50 mm and minimum 0.02 mm. Therefore, the closing link is the clearance, as indicated in the dimension chain shown above.

$0.50 = ?\text{max} - (A_{min} + B_{min})$ $\qquad$ $?\text{max} = 0.50 + 49.94 + 74.96 = 125.40\text{mm}$

$0.02 = ?\text{min} - (A_{max} + B_{max})$ $\qquad$ $?\text{min} = 0.02 + 50.12 + 75.20 = 125.34\text{mm}$

Example 3-6: An assembly of 8 components including a shaft, two bearings, a gear and a sleeve are shown in the following illustration. The dimensions and their associated tolerances are also provided.

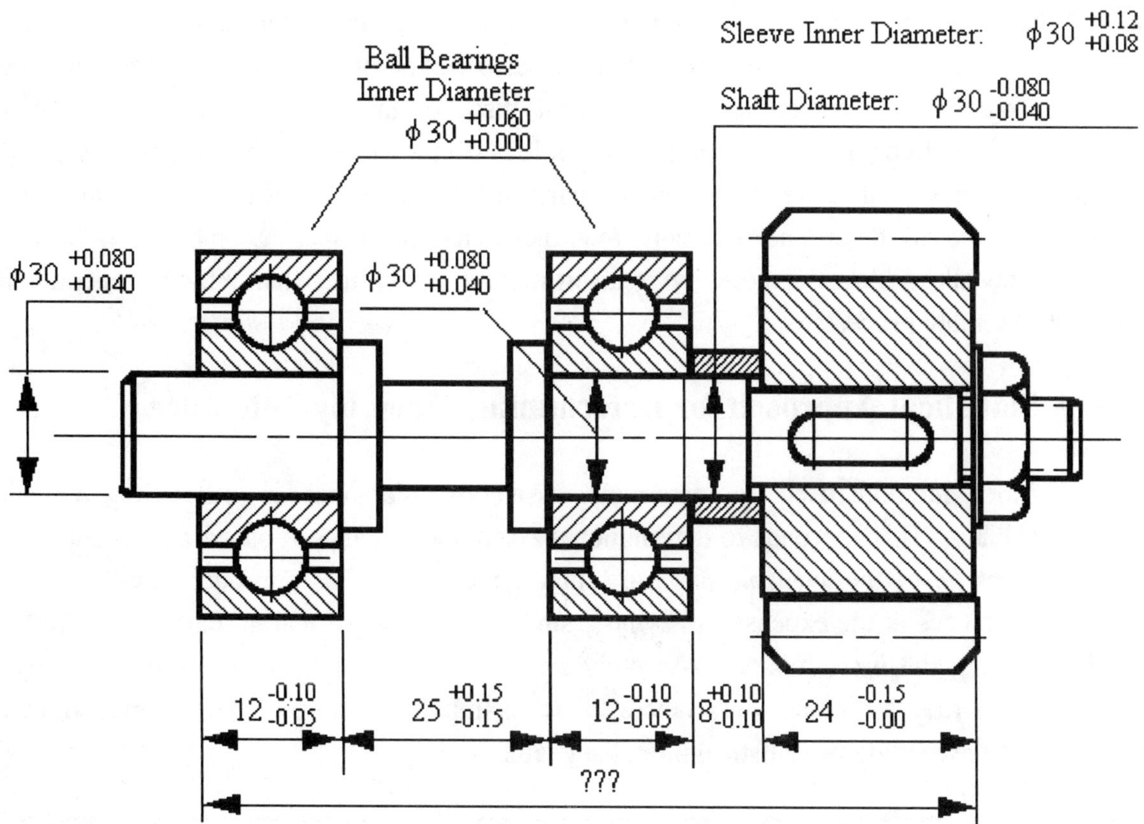

(1) Determine the maximum and minimum dimensions of the length between the outside edges of the bearing races after assembly.

Max = (12 - 0.05) + (25 + 0.15) + (12 - 0.05) + (8 + 0.1) + (24 - 0) = 81.15 mm
Min = (12 - 0.10) + (25 - 0.15) + (12 - 0.10) + (8 - 0.1) + (24 - 0.15) = 80.40 mm

(2) Determine the maximum and minimum clearances between the sleeve and the shaft.

Max = (30 + 0.12) - (30 - 0.08) = 0.20 mm
Min = (30 + 0.08) - (30 - 0.04) = 0.12 mm

(3) Determine the maximum and minimum clearances and interference between the bearings and the shaft.

Max = (30 + 0.06) - (30 + 0.04) = 0.02 mm (clearance)
Min = (30 + 0.00) - (30 + 0.08) = -0.08 mm (interference)

The maximum-minimum method for determining dimensions and clearances described in this example is known as tolerancing for 100% interchangeability. The tolerance of the closing link C_L, or an assembly tolerance, is based on the worst case scenario. The worst case occurs when all the minimum dimensions or the maximum dimensions, are encountered in the assembly of a particular component. The worst case scenario might occur if two components were assembled together. It is not likely to occur if three or more components are assembled together. In the unlikely event the worst scenario does occur, the assembly problem is solved by selecting a different part from the stock of parts available. It is clear that the maximum-minimum method to determine assembly tolerances insures proper assembly, but it overly constrains the allowable tolerances by over-estimating difficulties that may be encountered during the assembly process.

3.3.1 A Statistical Approach for Determining Assembly Tolerances

Another method used to determine assembly tolerances is based on statistical interchangeability. This approach to determine an assembly tolerance is based on a scenario that a large percentage of the available parts are interchangeable. This approach results in larger allowable tolerances at the expense of having a small percentage of mating parts that cannot be assembled during the first attempt. Assembly problems that arise with this approach may be minimized until they are of no importance. Two examples are presented to demonstrate the application of the method of statistical interchangeability.

Example 3-7 As illustrated in the following figure, the dimensions and tolerances of a hole and a shaft are 20±0.15 and 19.85±0.09 mm, respectively. The hole and the shaft are to be assembled together to provide a "clearance fit". Determine the range of the clearance between the hole and the shaft after assembly.

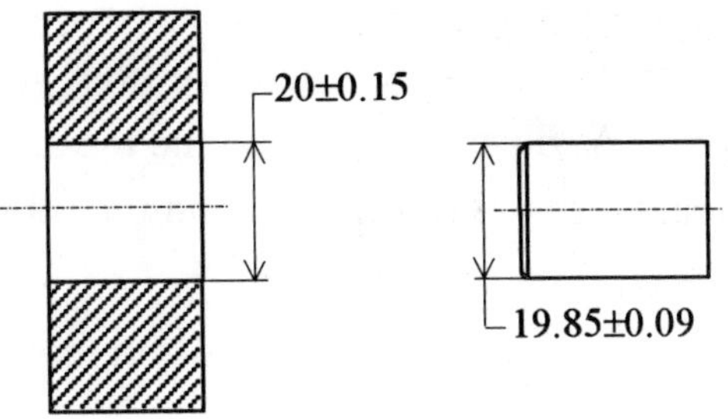

The tolerance on the hole hole is illustrated in below. The maximum dimension and minimum dimension of the hole are 20.15 and 19.85 mm, respectively.

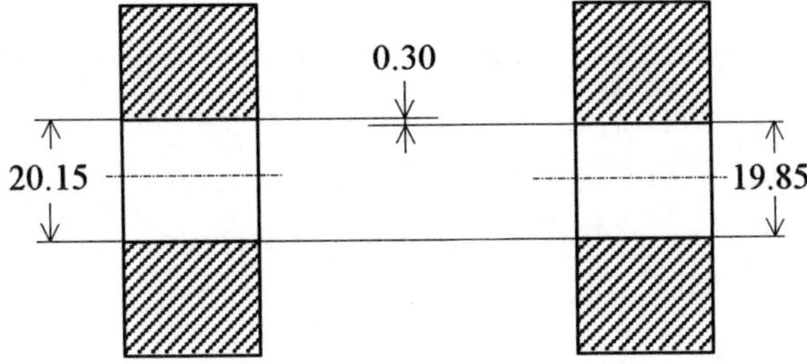

If we characterize the diameter of the hole using a statistical approach, we assume that it follows a normal distribution with its central position μ_{hole} set at 20.00 mm and a standard deviation σ_{hole} which is calculated from:

$$\sigma_{hole} = \frac{2 * 0.15}{6} = 0.05 \quad \text{(mm)}$$

In the following, we depict the normal distribution characterizing the variation of the dimension of the hole diameters:

Normal Distribution for the Hole Diameters
(assume: 6σ = the tolerance)

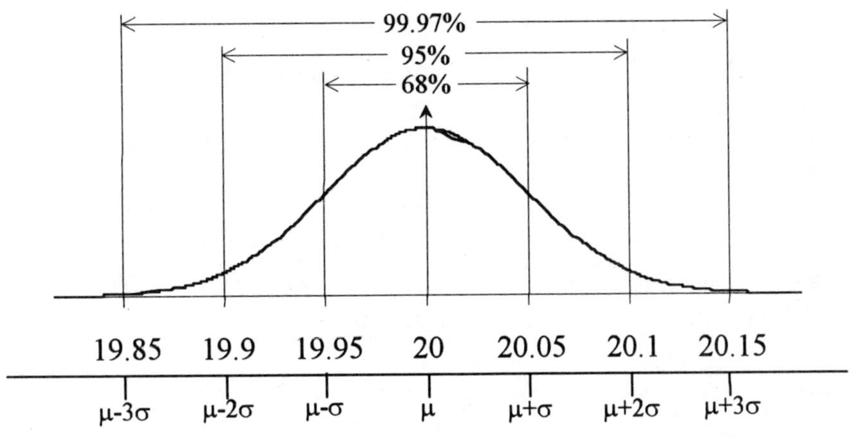

Following a similar procedure, we characterize the statistical distribution for the dimensions of the shaft diameters. As illustrated below, the maximum and minimum dimensions of the shaft diameter are 19.94 and 19.76, respectively, and the tolerance of the shaft diameter is 0.18 mm.

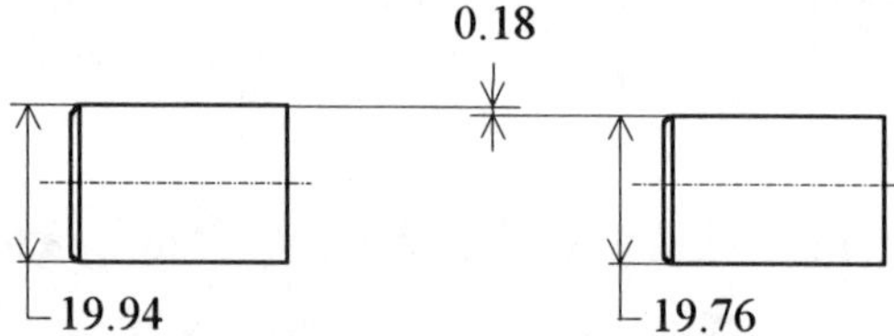

We determine the standard deviation σ_{shaft} of the normal distribution for the shaft from the relation below:

$$\sigma_{shaft} = \frac{2 * 0.09}{6} = 0.03 \quad \text{(mm)}$$

The normal distribution of the dimensional variation of the shaft diameters is shown in the following figure:

Normal Distribution for the Shaft Diameters

(assume: 6σ = the tolerance)

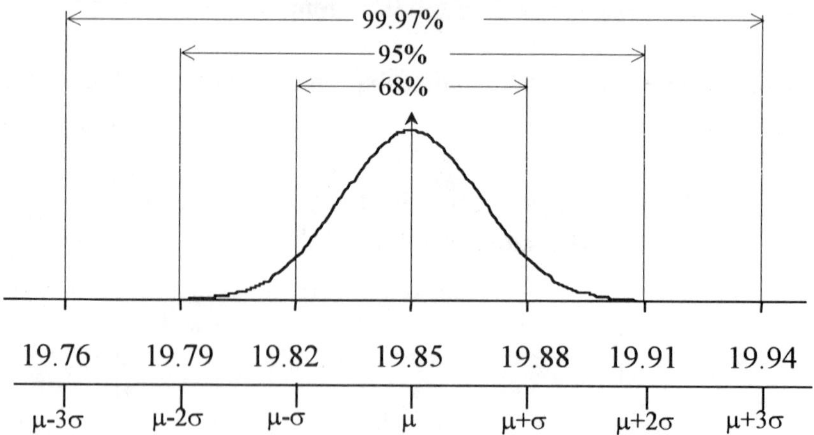

In this case study, involving the assembly of a hole and a shaft a clearance will develop so that the shaft can slide into the hole. The numerical value of the clearance is a random variable that can be characterized by a normal distribution. The central position of this normal distribution is:

$$\mu_C = 20\text{-}19.85 = 0.15 \quad \text{(mm)}$$

The standard deviation of the clearance distribution is given by:

$$\sigma = \sqrt{\sigma_{hole}^{2} + \sigma_{shaft}^{2}} = \sqrt{0.05^2 + 0.03^2} = 0.058 \quad \text{(mm)}$$

The normal distribution characterizing the clearance following the assembly of a randomly selected shaft into a randomly selected hole is shown below:

Normal Distribution for the Clearance after Assembly
(assume: 6 σ = the tolerance)

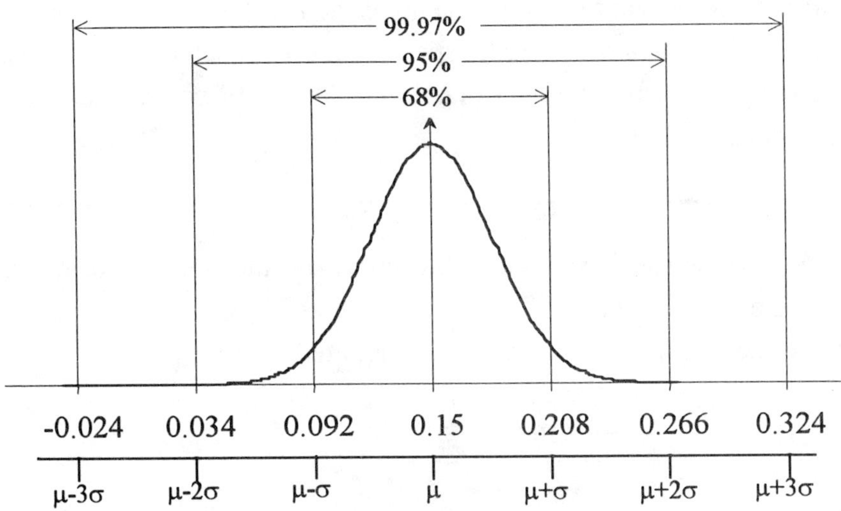

The data listed in Table 3.1 gives an interpretation of the clearances after the assembly. 67% of the assemblies will have a clearance between 0.092 and 0.208 mm (± 1σ). 95% of the assemblies will have a clearance between 0.034 and 0.266 mm (± 2σ). 99.75% of the assemblies will have a clearance between –0.024 and 0.266 mm (± 3σ).

Table 3.1
Probability of the Occurance of Various Clearances after Assembly

The Variation of the Clearances after Assembly	Probability (%)
0.092 – 0.208	68%
0.034 – 0.266	95%
-0.024 – 0.324	99.75%

It is critical that the meaning of a negative value for the clearance be clearly understood. A negative clearance means that the shaft diameter is greater than the hole diameter. This situation indicates that interference (not clearance) will occur in the assembly of a few of the components. When this circumstance occurs, a selection of either a different shaft or a different hole may be sufficient to overcome the assembly problem. Recall that the diameter of the hole and the shaft are random variables, and for this reason another selection is likely to assure a clearance fit in assembly. “Is it possible that interference will occur again after another selection of components?” It is very unlikely to select a second combination of shaft and hole that will produce an interference fit because the probability of this interference occuring on the first selection was only 5%. The probability of two consecutive selections producing interference is only (5%)(5%) = 0.25%. It is clear that a clearance fit can be assured without significant difficulty if the tolerances of the shaft and the hole are established at $\pm 3\sigma$. In fact, this example demonstrates the importance of using the statistical method to determine the distribution of clearances, and predicting the consequences of random selection of components when assembly occurs on the production line.

Example 3-8: Let’s next revisit Example 3-3 but approach the problem using the statistical method for the analysis of the tolerances.

The length and the tolerances for part A are illustrated below:

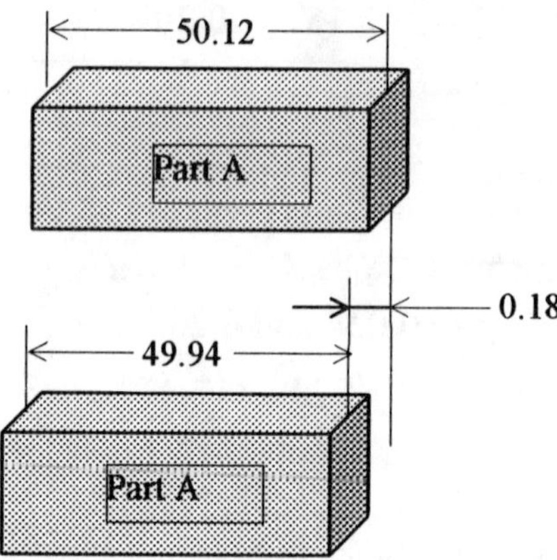

With the statistical method, the standard deviation for the length of part A is determined from the following relation:

$$\sigma_A = 0.18/6 = 0.03 \text{ mm}$$

The normal distribution representing Part A is centered at:

$$\mu_A = (50.12 + 49.94)/2 = 50.03 \text{ mm}$$

We can depict the normal distribution of the length of part A as shown in the frequency distribution curve presented below. It is very important to note that the central position of the dimension distribution is not at 50 mm, but at 50.03 mm. The tolerance specified is not symmetric about 50 mm. Since the distribution for part A is not properly centered, we can anticipate more assembly difficulties than would occur with a perfectly centered distribution for the lengths.

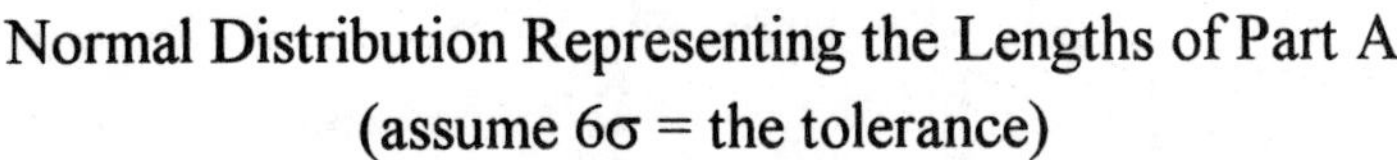
Normal Distribution Representing the Lengths of Part A
(assume 6σ = the tolerance)

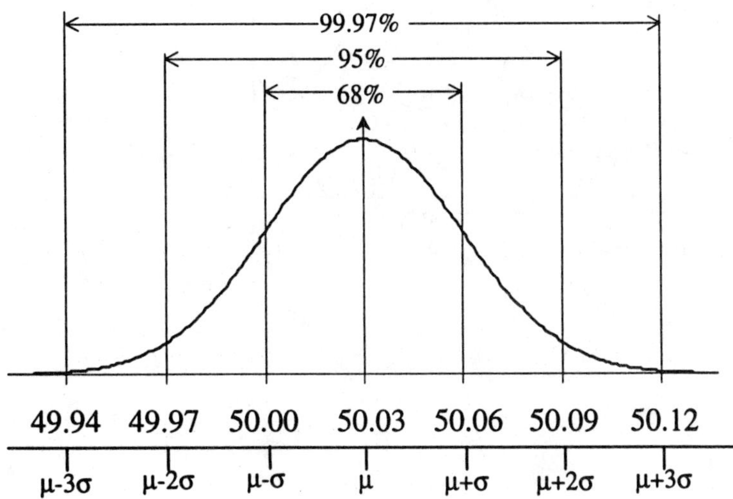

Following exactly the same procedure, we determine that part B is statistically characterized by:

$$\mu_B = 75.08 \text{ mm}$$

and

$$\sigma_B = 0.04 \text{ mm}$$

We assume that the length of the assembly of parts A and B can be represented with a normal distribution. Then the normal distribution is centered at:

$$\mu_{A+B} = 50.03 + 75.08 = 125.11 \text{ mm}$$

The standard deviation of the normal distribution of the assembly is given by:

$$\sigma_{A+B} = [\,\sigma_A^2 + \sigma_B^2\,]^{1/2} = [\,(0.03)^2 + (\,0.04)^2\,]^{1/2} = 0.05 \text{ mm}$$

The normal distribution characterizing the length of the assembly of parts A and B is shown below. Note that the bell shaped curve is centered at 125.11 mm. In examining the extremes of this curve, we can conclude that 99.97% of the assemblies will be between 124.96 and 125.26 mm in length.

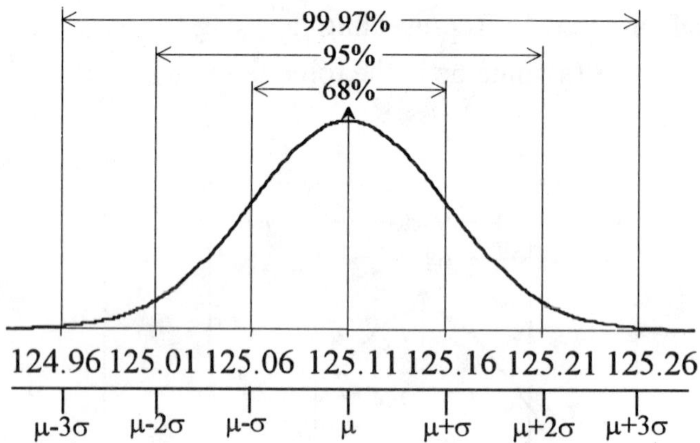

REFERENCES

1. American National Standards Institute, Y14 series of standards: e.g. Y14.1 Drawing Sheet Size and Format; Y14.2 line Converting and Lettering; Y14.3 Multi and Sectional View Drawing; Y14.5 Dimensioning and Tolerancing.
2. P. E. Allaire, Basics of the Finite Element Method, Solid Mechanics, Heat Transfer, and Fluid Mechanics, University of Virginia , 1985.
3. R. E. Barnhill, IEEE Computer-Graphics and Applications, 3(7), 9-16, 1983.
4. C. A. Born, W, J, Rasdorf, and R. E. Fulton, Engineering Data Management: The Technology for Integration, Boston MA, 1990.
5. J. W. Dally and T. Regan, Introduction to Engineering Design, Book1: Solar Desalination, College House Enterprise, Knoxville, TN, 1996.
6. J. W. Dally and T. Regan, Introduction to Engineering Design, Book2: Weighing Machines, College House Enterprise, Knoxville, TN, 1997.
7. J. W. Dally and G. M. Zhang, "A Freshman Engineering Design Course," Journal of Engineering Education, 83-91, April, 1993.
8. B. L. Davids, A. J. Robotham and Yardwood A., Computer-aided Drawing and Design, London, 1991.
9. J. Dieter, Engineering Design, McGraw-Hill, New York, 1983.
10. J. Encarnacao, E. G. Schlechtendahl, Computer-aided Design: Fundamentals and System Architectures, Springer-Verlag, New York, 1983.
11. G. Farin, Curves and Surfaces for Computer-aided Geometric Design, New York, 1988.
12. S. Fingers, J. R. Dixon, A review of research in mechanical engineering design, Part I: Descriptive, prescriptive and computer-based models of design processes, Research in Engineering Design, 1(1), 51-68, 1989.
13. S. Fingers, J. R. Dixon, A review of research in mechanical engineering design, Part II: Representations, analysis, and design for the life cycle, Research in Engineering Design, 1(2), 121-38, 1989.
14. M. P. Groover and E. W. Zimmers, Computer-aided Design and Manufacturing, Englewood Cliffs, NJ, 1984.
15. D. Hearn, M. P. Baker, Computer Graphics, Englewood Cliffs, NJ, 1986
16. P. Ingham P. CAD System in Mechanical and Production Engineering, London, 1989.
17. E. B. Magrab, Integrated Product and Process Design and Development: The Product Realization Process, CRC Press, Boca Raton, NY, 1997.
18. R. L. Norton, Machine Design: An Integrated Approach, Prentice Hall, Upper Saddle River, New Jersey, 1996.
19. N. P. Suh, The Principles of Design, Oxford University Press, New York, NY, 1990.

EXERCISES

1. The space shuttle is equipped with a battery system. The reliability of a selected battery manufactured by XYZ company is 99%, indicating that the possibility of malfunctioning during the service is one out of one hundred. To secure the space shuttle mission, three batteries are connected in parallel to supply the power to the space shuttle. You are asked to design a battery container to hold the three batteries. The required design specifications are listed below:

   (1) A minimum clearance of 0.1 mm is required to ensure that three batteries can be easily inserted into the container.

   (2) A maximum clearance of 0.4 mm is required to ensure that the inserted batteries remain in a stable without shaking during launch, flight and landing.

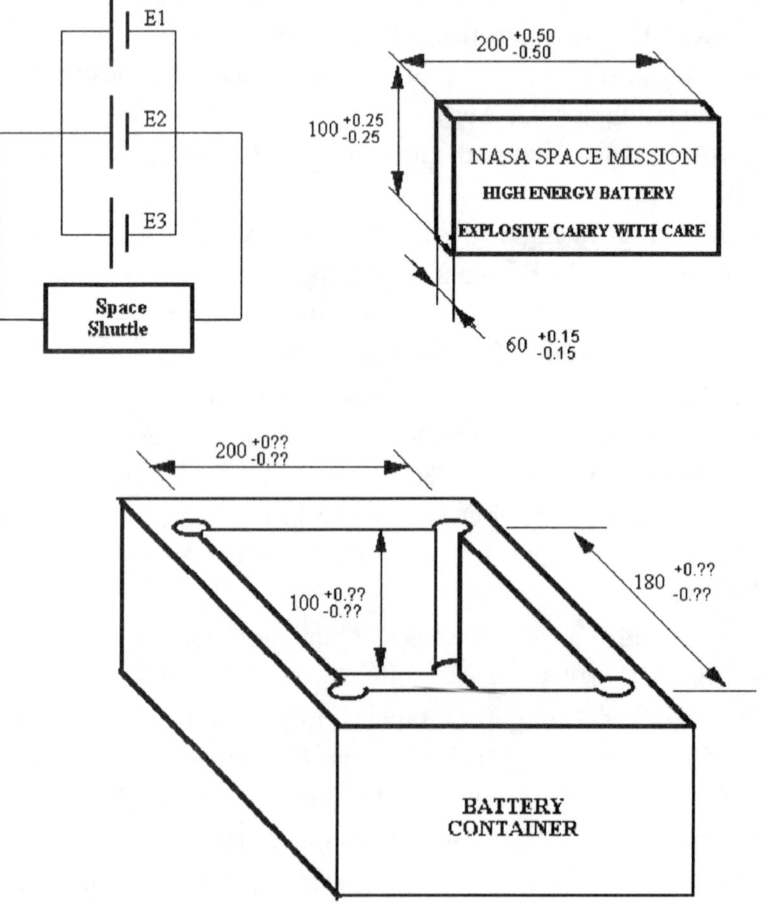

Your responsibility is to determine the tolerance range for each of the three dimensions. This is an open end problem. Therefore, try your best. You may apply both the method of 100% interchangeability and the method of statistical interchangeability.

2. Tolerances can be specified in three formats depending on their nature. They are unilateral format, bilateral format, and limit format. For each of the tolerances listed below, write the other two formats if these formats exist.

3. Use the tables provided in Appendixes to identify:
   (1) The maximum and minimum dimensions of the hole and/or shaft;
   (2) The tolerances for the hole and the shaft, respectively; and
   (3) The maximum clearance and the minimum clearance.
   (1) basic diameter: 5 inches, Basic-Hole System, Clearance Fit = LC2.
   (2) basic diameter: 5 inches, Basic-Hole System, Interference Fit = LN3.
   (3) basic diameter: 5 inches, Basic-Hole System, Transition Fit = LT3.
   (3) Draw a comparison diagram which illustrates these three fits.

4. Use the tables provided in Appendix to identify:
   (1) The maximum and minimum dimensions of the hole and/or shaft;
   (2) The tolerances for the hole and the shaft, respectively; and
   (3) The maximum clearance and the minimum clearance.
   (a) basic diameter: 5 inches, Basic-Shaft System, Clearance Fit = SLIDING.
   (b) basic diameter: 5 inches, Basic-Shaft System, Interference Fit = FORCE.
   (c) basic diameter: 5 inches, Basic-Shaft System, Transition Fit = LOCAT. TRANS.
   (d) Draw a comparison diagram which includes the above three cases.

5. The maximum and minimum dimensions of a shaft are 20.18 and 19.82, respectively. The maximum and minimum dimensions of a hole are 20.24 and 20.00, respectively. Use the statistical method to construct a normal distribution figure characterizing the variation of the clearance after the assembly.

6. A senior engineering is reviewing the tolerance information specified by a young engineer who just graduated from college. The tolerance value is 0.24 mm for the diameter of a hole on a component. The senior engineer knows that the smallest standard deviation among the drills available in the company is 0.06 mm. There are 10,000 pieces of the component with the hole to be manufactured. He is trying to tell the young engineer how many of these components will be defective if the tolerance is kept at 0.24 mm. You are asked to make the determination. How accurate is your calculation?

CHAPTER 4

Pro/ENGINEER DESIGN SYSTEM

4.1 Fundamentals of Pro/ENGINEER Design system

Pro/ENGINEER, as a CAD system, came to the market in the late 1980's. It was strictly a UNIX-based system running on a mainframe computer. However, during the past five years, significant improvements have been made in the system design of Pro/ENGINEER. Recent releases now run on both WINDOWS and UNIX based systems. The Pro/ENGINEER system is now capable of capturing the design intent while producing a quality database that can be used for several purposes including documentation, engineering analysis, and the integration of design and manufacturing.

Just as babies must learn to crawl before they can walk and walk before they run, to use Pro/ENGINEER efficiently, one must first understand the design requirements and conduct the design through a systematic sequential process in which features are created step by step. With the material presented in this textbook, the concept of feature based design is demonstrated with many examples and illustrations. Now let's start the crawling process and begin to understand the capabilities of Pro/ENGINEER. Take it easy!

4.1.1 Feature-Based Part Design

In a computer aided design environment, a part is viewed as a combination of features, such as blocks, cylinders, holes, slots, etc. A model is developed to represent the component being designed using these features, which comprise the geometrical shape and size of the part. The Pro/ENGINEER design environment offers a design engineer a unique and powerful tool to visualize his/her idea(s) and facilitates creativeness in the design process using feature-based component modeling. The modeling process is characterized by:

(1) The flow of modeling a part in Pro/ENGINEER follows the process used in producing the part in a manufacturing environment. Figure 4-1a illustrates a procedure followed in manufacturing a block with a centrally located hole. Essentially, there are four steps in the operation:

(a) An operator goes to a storeroom to select the bar stock that best fits the requirements.
(b) A cutoff saw is employed to produce the length required for manufacturing the part.
(c) The surfaces are machined to meet the dimensional requirements (tolerances) and finish requirements (roughness).
(d) A drill bit is selected and a drilling machine is employed to drill the hole at the central position in the machined block.

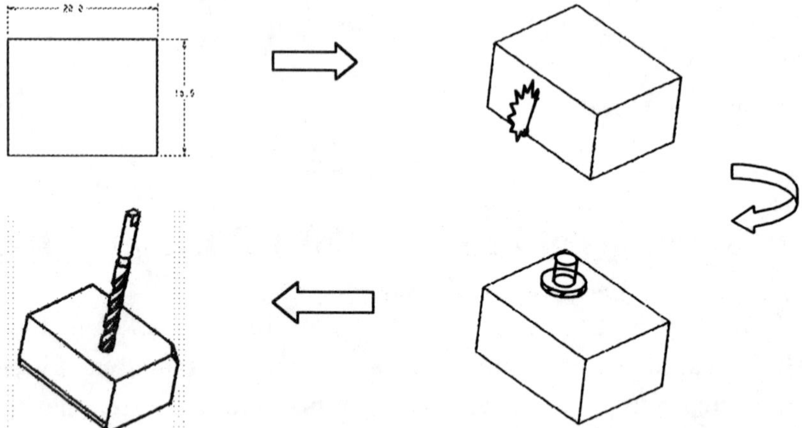

a. A Process Used in Manufacturing a Block with a Centrally Located Hole.

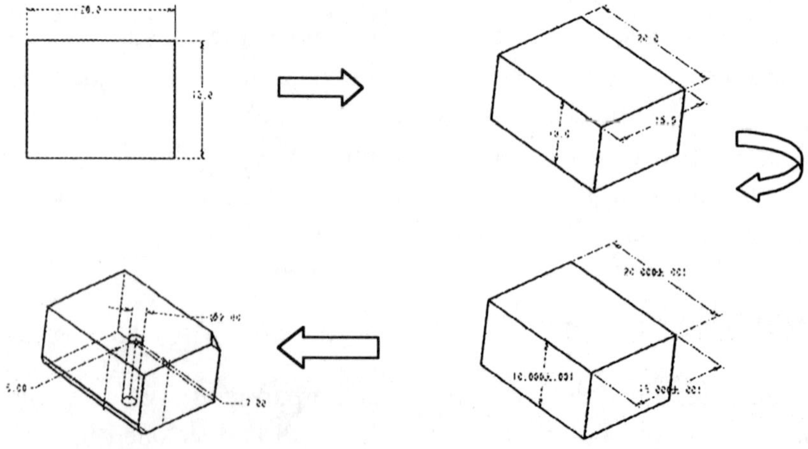

b. Part Modeling in Pro/ENGINEER

Figure 4-1 Comparison between the Part Modeling and Part Manufacturing.

The modeling procedure under the Pro/ENGINEER design environment closely follows the process of manufacturing the part being designed. The four basic steps in the modeling process under the Pro/ENGINEER design environment are listed below and illustrated in Fig. 4-1 b:

a. Making a two-dimensional (2D) sketch of the cross section area of the bar stock.
b. Extending the 2D sketch in the third direction by a dimension equal to the length of the bar stock to be cut by the saw.
c. Giving the proper tolerances to the dimensions to ensure that the part meets specifications.
d. Cutting a centrally located hole at the top of the block, and specifying its diameter.

One advantage of using Pro/ENGINEER is that changes can be made to one or more components at any time in the design process with little difficulty. A process plan develops accordingly to manufacture the component.

(2) Defining geometrical features follows the procedure of making engineering sketches on the shop floor, design office or the lunch table. An engineer designer usually uses a pencil and a piece of paper, even scratch paper, to initiate his/her design process. When we are on the shop floor, a ruler and/or a compass is not used in preparing the drawings. Instead sketches are made initially, and modifications are made when we find time to get back to the office. Under the Pro/ENGINEER design environment, a built-in function, called REGENERATE, is available to the designer for sketching and making design changes to the engineering drawings thereafter.

(3) The advanced built-in computer graphics under the Pro/ENGINEER design environment significantly facilitates the design process. Visualization in three dimensional space can be made at any time during the design to verify the concept or to confirm that the design meets the specifications.

4.1.2 Pro/ENGINEER Working Environment – Three Windows

This section introduces the basic steps employed in designing a component. The command to start Pro/ENGINEER depends on the computer system you are using. (i. e. a PC, Sun workstation, or SGI machine.) However, the following procedure is applicable to initiate the design process regardless of the computer system used to run the program. To start Pro/ENGINEER, simply select the icon displayed on your computer screen, or type **proeng**. (The exact command may be different depending on the system installation). After you start Pro/ENGINEER, you use your **keyboard** and/or **mouse** to input your data, select commands, and save your files, as is the usual practice when operating a computer.

Pro/ENGINEER runs in a multiple window environment. After you start up Pro/ENGINEER, your screen will display the three windows illustrated in Figure 4-2.

(1) **Main drawing window.** It is the largest window shown on the computer screen. This window is used to create the geometry of the part. The asterisks in

series along the top bar of the working window indicate that the main drawing window is the active one at the time the screen was imaged.

(2) **Message window.** It is the window displayed below the main drawing window. The user will read the messages displayed to guide the modeling process, and answer the prompts by typing the required information in this window. The user is encouraged to constantly review the information displayed in this window, especially any time help is sought.

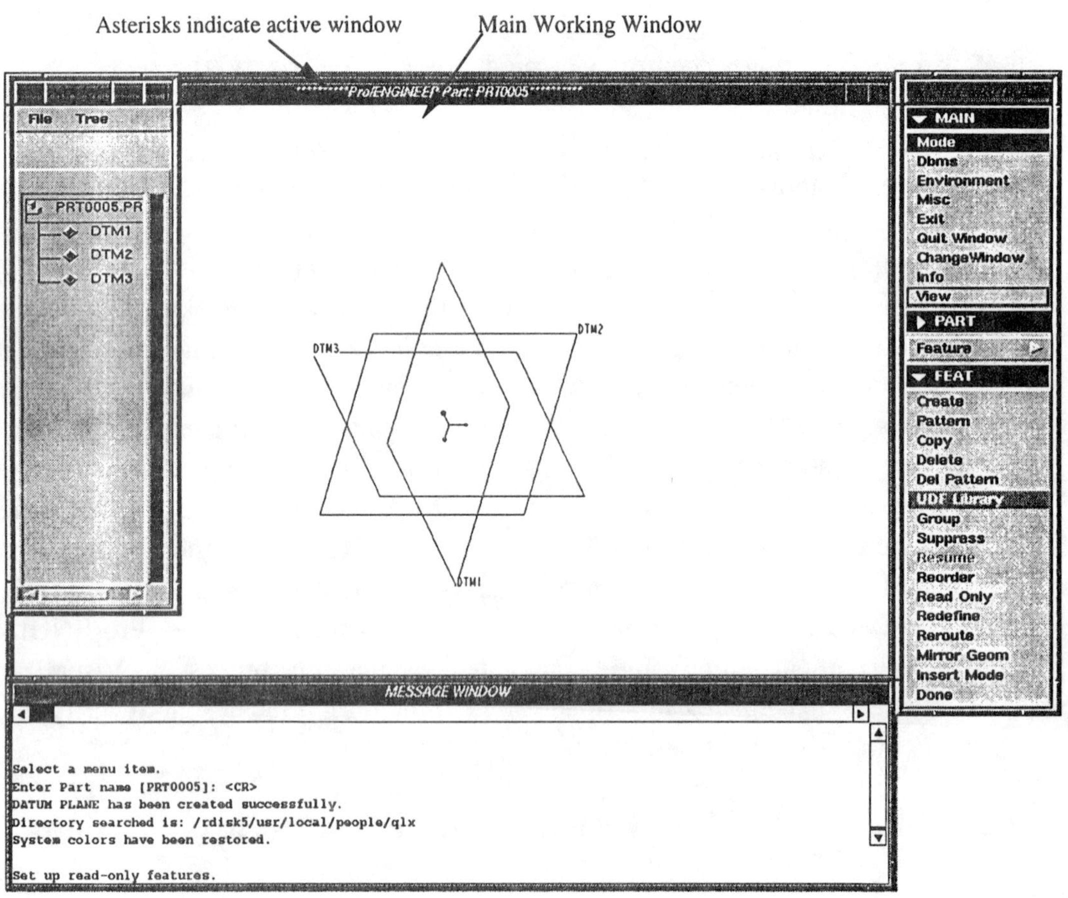

Figure 4-2 Pro/ENGINEER Windows.

(3) **Pop-up windows for commands.** These are small windows displaying menu lists. The Pro/ENGINEER design system is operated by a menu-driven system. Windows for selecting commands arrives just on time. These pop-up windows function as Menu Manager, as shown in Figure 4-2. The user enters most of the commands through the Menu Manager by navigating the menu listing with a mouse (see Example 4-2 for details).

To exit from Pro/ENGINEER, choose **Exit** under the Main Menu and click "YES" when the confirmation window appears.

Establishing a Reference System and a Coordinate System

When a designer begins to create a part, he or she selects **Part** from the **MODE** menu in the pop up window. Then we choose **Create** from the **PART** menu. This action will prompt the designer to enter the name of the part to be designed in the message window:

Enter Part Name []:

Enter a file name to associate with the part and depress the **Enter** key. Now a part file exists with an appropriate name that was entered.

Create a Reference System and a Coordinate System before Designing any Part!

Orthographic drawings have been widely used as the engineering language for communication by the design and manufacturing communities. To prepare multi-view orthographic drawings, we need to define a reference system; specifically, a system with three planes perpendicular to one another. To create drawings in this reference system, we also need a coordinate system to quantify the dimensions and positions of the various features. By convention, a coordinate system is established with its origin at the intersection point of the three orthogonal planes. Under the Pro/ENGINEER working environment, it is a general design practice to establish a reference system and a coordinate system before beginning to designing a part. The command sequence to create a reference system and a coordinate system is listed below:

Feature, Create, Datum, Plane, Default

Note that this sequence represents the following steps:

(1) Choose **Feature** from the **PART** menu.
(2) Choose **Create** from the **FEAT** menu.
(3) Choose **Datum** from the **FEAT CLASS** menu.
(4) Choose **Plane** from the **DATUM** menu.
(5) Choose **Default** from the **MENUDTM OPT** menu.

On the main drawing window, DTM1, DTM2, and DTM3 are displayed, as illustrated in Fig. 4-3. To create the coordinate system, the following command sequence is used:

Create, Datum, Coord Sys, Default, Done

Note that this sequence represents the following steps:

(1) Choose **Create** from the **FEAT** menu.
(2) Choose **Datum** from the **FEAT CLASS** menu.
(3) Choose **Coord Sys** from the **DATUM** menu.
(4) Choose **Default** and **Done** from the **OPTIONS** menu.

On the main drawing window, in addition to DTM1, DTM2, and DTM3, three axes, namely, x, y, z and CS0 (representing the coordinate system origin) are shown (see Figure 4-3).

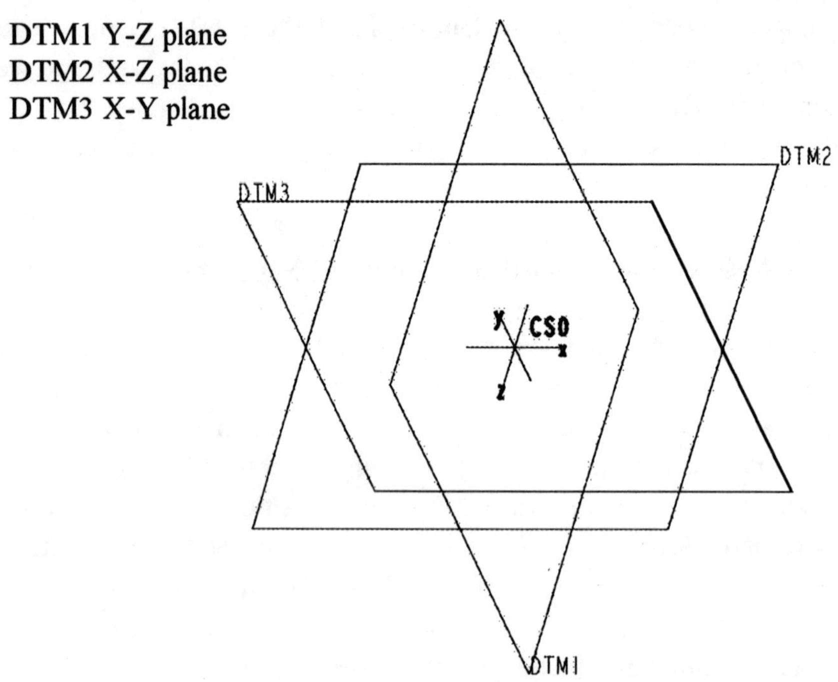

Figure 4-3 A Reference System and its Associated Coordinate System

4.2 Solid Modeling of an Object in the 3D Space

Under the Pro/ENGINEER design environment, we model a part being designed by creating individual features in three dimensional space. These features are arranged to define the geometrical size and shape of the part. This modeling approach provides a logical easy to follow procedure for the designer when creating a part.

4.2.1 Example 4-1: Construction of a Sliding Basket

To illustrate the procedure for designing a component, let's consider as an example the sliding basket shown in Figure 4-4. The three maximum dimensions along the x, y, and z directions are 55 mm, 25 mm, and 40 mm, respectively.

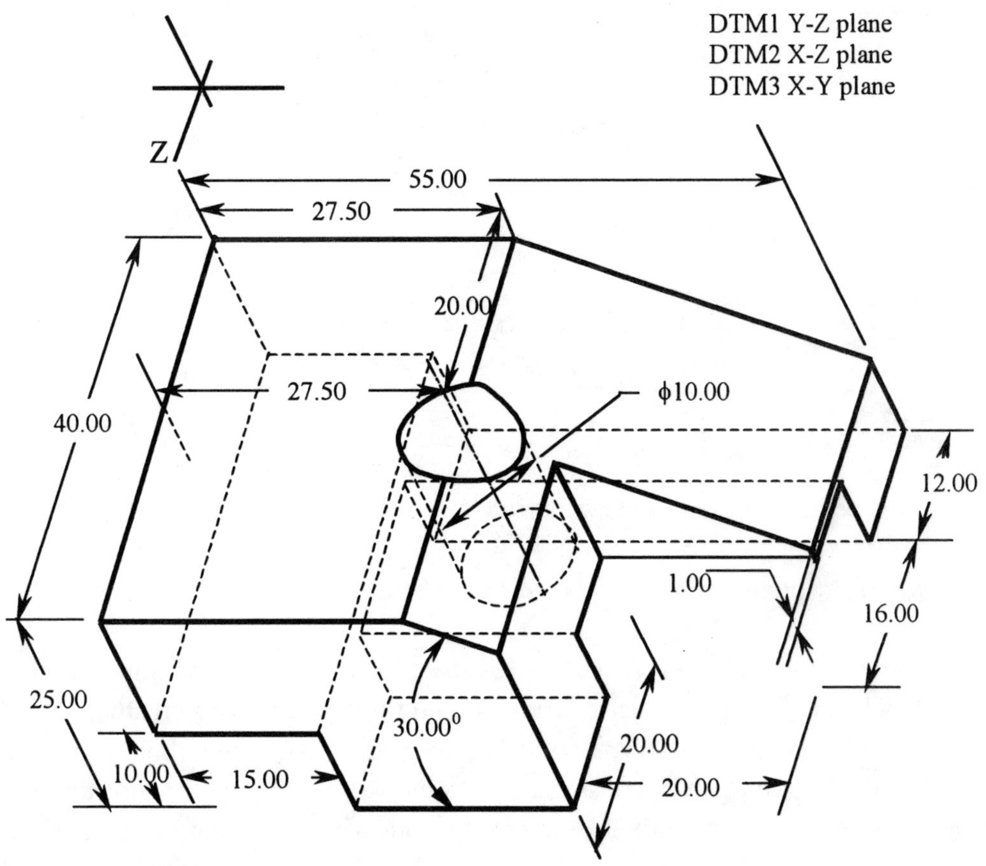

Figure 4-4 Geometry and Dimensions of a Sliding Basket

Feature-based Construction Process

As we have emphasized, the solid modeling process of a component follows the manufacturing that will be employed in fabrication. A 7-step manufacturing procedure to produce the sliding basket is listed below and illustrated in Figure 4-5:

1. Create a block using protrusion as the basic command for features in this solid model.
2. Cut a hole in the center of the block.
3. Form the corner slot feature by cutting the block at the left bottom corner and extend the cut from the front surface completely through to the back face.
4. Form the slope feature by cutting off a triangle shaped block at the upper right corner and extend the cut from the front surface completely through.
5. Form the sliding slot centered on the right side of the block by cutting off a rectangular shaped cubic and cutting it from the right side completely through.
6. Cut off the rectangular shaped slot from the bottom of the block to form the feature at the front-right corner of the block.

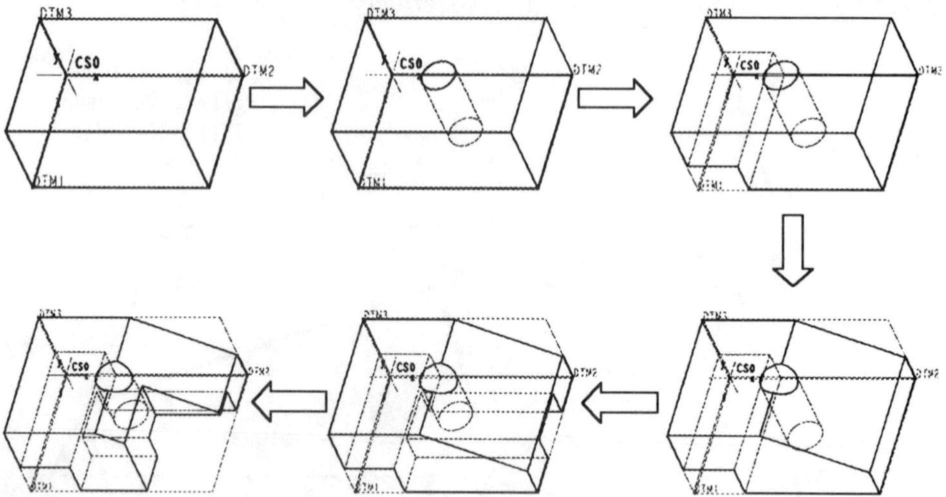

Figure 4-5 Feature-based 3D Geometry Generation Process for a Sliding Basket

Create features in 2D Space and extend them to 3D Space

As indicated previously, Pro/ENGINEER is a feature-based design system. Each feature is first sketched in 2D space then extended in the third direction to create a 3D solid.

Let us following the 7-step manufacturing procedure to create the sliding basket in Pro/ENGINEER. The first feature to construction is a block with the three major dimensions of 55 mm, 25 mm and 40 mm, respectively. Select the following command sequence, from the **FEAT** menu:

Create, Solid, Protrusion, Extrude, Solid, Done.

Now choose the **One Side** option from the **SIDES** menu, and then select **Done**, indicating that the feature will be extruded to only one side of the sketching plane.

Choose **Plane** from the **SETUP PLANE** menu and pick DTM3 as your **Sketching Plane**. In other words, a 2D sketch will be created in the x-y plane (reference to Figure 4-3). Note that an arrow is displayed on the screen. If the arrow direction is towards you, or if it is in the direction along the positive z direction, this indicates that the 2D sketch created in the x-y plane will be extend along the positive z direction. Choose **Okay** if you want a solid to be created this way. For a solid to be created in the negative z direction, you need to choose **Flip**, then **Okay**. In this example, let's choose the positive z direction as the extrusion direction for the third dimension.

To permit the screen to show a sketch plane in 2D space, you must choose a command from the **SKET VIEW** menu. There are four choices, **Top**, **Bottom**, **Right**, and **Left**. Choose **Top** at this moment, and then select DTM2 on the main drawing window. The 2D layout shown in Figure 4-6 is displayed on the monitor screen. This display indicates that a design engineer is holding his/her sketch paper so that the positive x direction is to the right and the positive y direction upward. Up to this point, we have created a piece of paper (a 2D plane) with a specified orientation. You are now ready to sketch the part geometry.

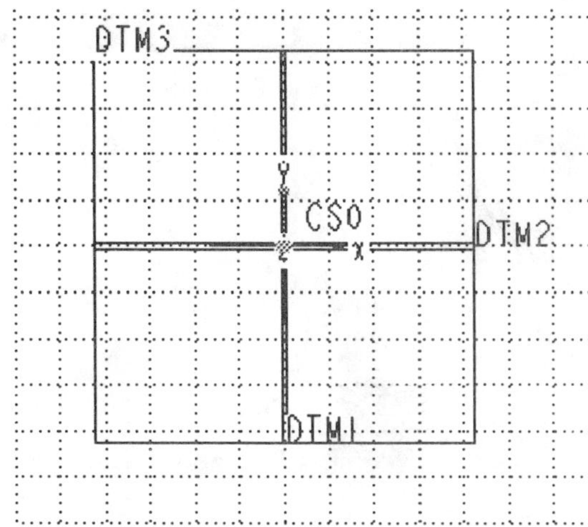

Figure 4-6 Define the Orientation of a Sketch Plane

To begin sketching, choose **Sketch** from the **SKETCHER** menu and **Rectangle** from the **Geometry** menu. Select two points on the 2D screen to indicate the diagonal of the rectangle, thus defining the width and height dimensions of the rectangle. As illustrated in Figure 4-7, one point is selected on the positive y-axis and the other point is selected on the positive x-axis.

Choose **Alignment** from the **SKETCHER** menu, and double click the left side of the rectangle and double click the bottom side of the rectangle. The message "—ALIGNED —" appears in the message window. The rectangle has been positioned along the x-axis and y-axis, or sd0 = 0 and sd1 = 0 where sd stands for size dimension.

Choose **Dimension** from **SKETCHER**. Use the left mouse button to click the left and right sides. Move the cursor to a place above the upper side of the rectangle and press the middle mouse button to place the width dimension, sd2. Repeat the process by clicking the upper and lower sides, moving the cursor to a place of your preference to place the height dimension and pressing the middle mouse button, sd3.

The 2D screen shows there are no specific numerical numbers in terms of dimension, except for the signs, sd2 and sd3. Now choose **Regenerate** from the **SKETCHER** menu and select **Modify** from **SKETCHER.** On the main drawing window, first select the width dimension, type 55 in the message window, and hit the **Enter** key. Note that sd2 has been changed to 55. Select the height dimension, type 25 in the message window, and hit the **Enter** key. Choose **Regenerate** from **SKETCHER** for the second time. The 2D sketch after regeneration should be similar to that presented in Figure 4-8.

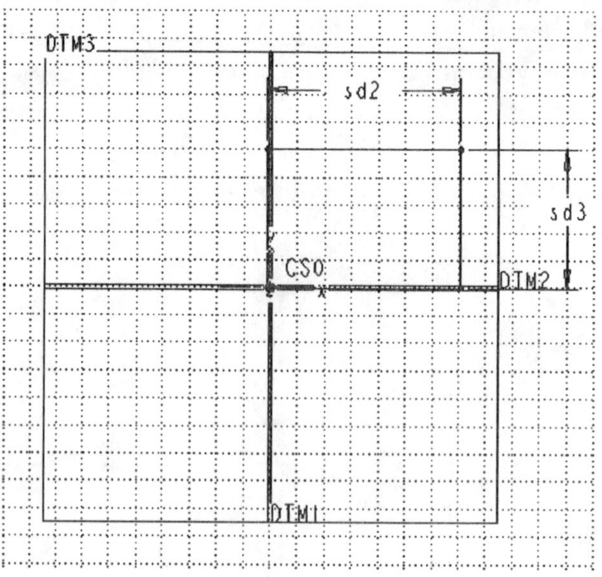

Figure 4-7 Sketch a Rectangle in the 2D Screen

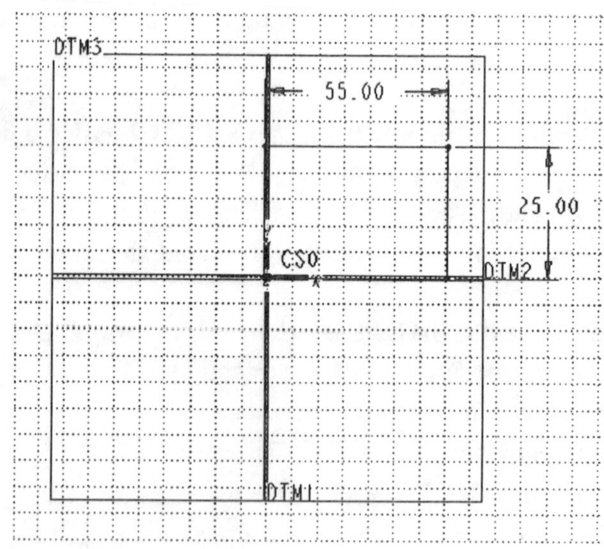

Figure 4-8 Completion of a 2D Sketch of the Block Feature

We have now completed the sketch process in 2D space. Choose **Done** from **Sketcher**. To extend the 2D sketch to a solid model requires that we define the third dimension --- the length of the block feature. It is equal to 40 mm.

Choose **Blind** option from the **SPEC TO** menu and then **Done**. In the message window, enter the third dimension by typing 40 and hitting the **Enter** key. The main purpose of the **SPEC TO** menu is to allow you to choose where to end the extrusion. With the **Blind** option, the depth of the extrusion must be specified. In other words, a length of 40 for the block feature has to be input by the user.

Note that there is another window shown in the computer screen. It is the Element Information Window. The Element Information Window monitors the process of defining the feature step by step. Choose **OK** from the Element Information Window, thus completing the construction of the block feature with the three dimensions equal to 55, 25 and 40 mm, respectively. To view the 3D feature, type dv and hit the Enter key. Note that dv stands for default view, which changes the geometry from 2D (X-Y plane) to 3D view as illustrated in Figure 4-9.

It is important to note that the dv command must be set up during installation before it can be used as the shortcut in the previous procedure. Under the **Pro/ENGINEER** design environment, the routine procedure to visualize a designed feature on the computer screen is listed below:

(1) From the **MAIN** menu, select **View**;
(2) From the **MAIN VIEW** menu, select **Orientation**;
(3) From the **ORIENTATION** menu, select **Default**; an isometric view of the designed feature or features will be displayed on the drawing window; and
(4) From the **MAIN VIEW** menu, select **Done-Return** to complete the 3D visualization.

It is suggested that the reader should examine the block feature carefully. Identify the sketch plane (DTM3) and the location of the origin of the coordinate system with respect to the block. The origin is at the lower left corner of the block, and is assigned to the corner on the sliding basket, specially marked as CSO.

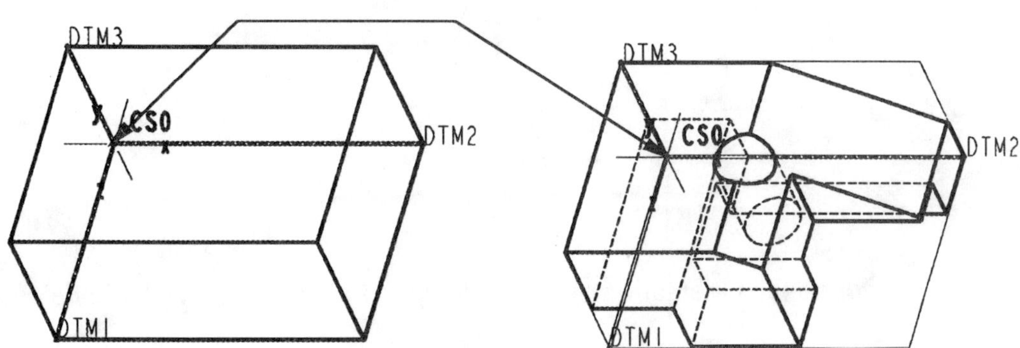

Figure 4-9 Completion of a Solid Model for the Block Feature

To continue with the modeling process, we want to drill a hole with a diameter of 10 mm. Follow the procedure previously used for creating the block feature, except change one command, i.e., replace **Protrusion** by **Cut**. Therefore, the command sequence to create a hole, starting from the **FEAT** menu, is listed below:

Create, Solid, Cut, Extrude, Solid, Done

Note the replacement in the above sequence, i.e., using the command **Cut**, instead of **Protrusion**. The reason is evident due to some material of the block is being taken away when the hole is made. Now choose the **One Side** option from the **SIDES** menu. The feature will be extruded to only one side of the sketching plane.

Choose **Plane** from **SETUP PLANE** menu and pick the top surface of the block feature as your **Sketching Plane**. This means that you want to drill the hole starting from the top surface or a plane parallel to DTM2 (the x-z plane). Select the plane as the 2D-sketch plane. When selecting the plane, an arrow downward is shown. This indicates that the drilling direction is towards the block material. So, choose **Okay**.

To permit the screen to show the sketch plane, select **Top** from **SKET VIEW** first. Now spin the constructed block so that the x-y plane can be viewed in locating the hole. To spin the block, hold down the CTRL key and the middle mouse button, and then move the mouse. When mouse moves, the block starts spinning. Stop at the position displaying the x-y plane. The 2D layout shown in Figure 4-10 is displayed on the monitor screen. Note that the origin of the coordinate system is located at the upper and left corner. This layout gives a very clear idea of the orientation of the block and is comparable to holding a sketch pad to make a drawing of the hole.

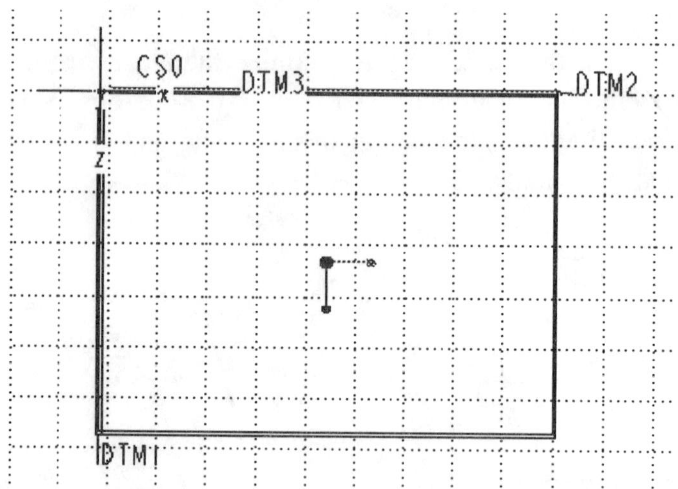

Figure 4-10 Define the Orientation of a Sketch Plane for Modeling the Hole Feature

Choose **Sketch** from the **SKETCHER** and **Circle** from **Geometry** menu. Under the **Circle** command, select **Ctr/Point**. Move the cursor inside of the rectangle and press the left mouse button to place the center position. When you are moving the cursor away from the center position, a circle is shown. Push the left mouse button to complete the 2D sketch of the hole, as illustrated in Figure 4-11. To specify the size of the circle, place a diameter dimension on the circle. Select **Dimension** from the **SKETCHER** menu and click the circle twice with the left mouse button. Move the cursor away from the circle and press the middle mouse button to place the dimension. To define the center position, we need to specify its x and z values. Click the center of the circle and the upper side of the rectangle to define the z position. Click the center of the circle and the left side of the rectangle to define the x position. When the three dimensions are shown, select **Regenerate**.

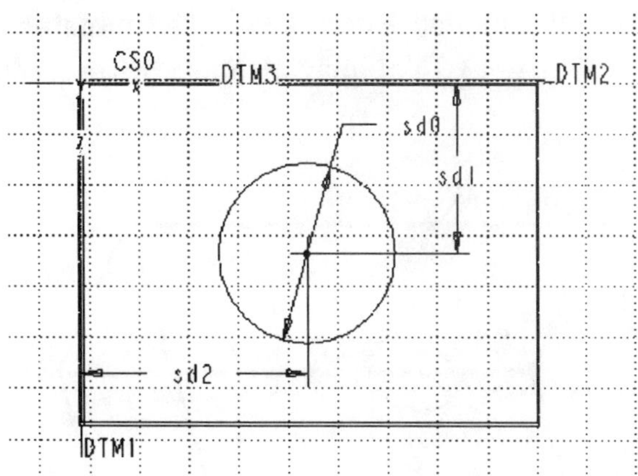

Figure 4-11 Sketch a Circle in the 2D Screen

After regeneration, pick **Modify** from **SKETCHER** and select the diameter dimension text. In the message window, Pro/ENGINEER prompts for a new dimension value. Enter 10 in the message window, and hit the **Enter** key. Select the dimension for the x position and enter 55/2. Select the dimension for the z position and enter 40/2. Choose **Regenerate** from the **SKETCHER** menu. The screen shows the dimensioned hole as indicated in Figure 4-12.

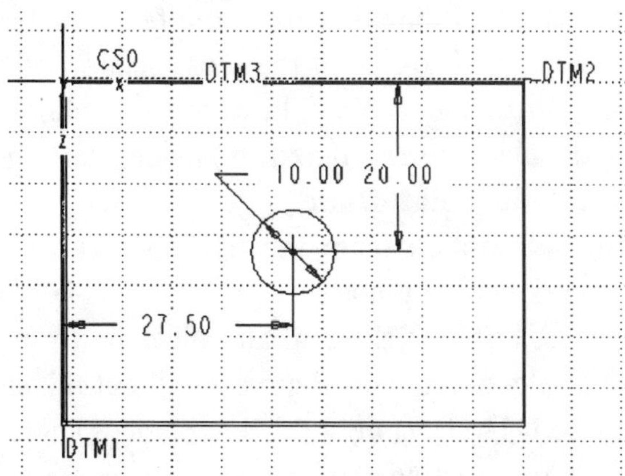

Figure 4-12 Completion of a 2D Sketch of the Block Feature

Choose **Done** from **Sketcher**. An arrow appears on the screen to indicate the part of the material that is being removed. If the direction shown is correct, select **Okay**. The message window will indicate that a 2D sketch of the hole feature has been successfully completed.

Now extend the 2D sketch to a solid model in the third dimension (i.e., the depth of the hole feature).

Choose **Thru All** option from the **SPEC TO** menu and then **Done**. This means that the hole is a passes completely through the block. Choose **OK** from the Element Information Window, thus completing the construction of the hole feature with a diameter of 10 mm. To view the 3D feature, type dv and hit the **Enter** key. The monitor displays a block with a though hole as indicated in Figure 4-13. You may also use the procedure described on page 4-11 to view the 3D feature.

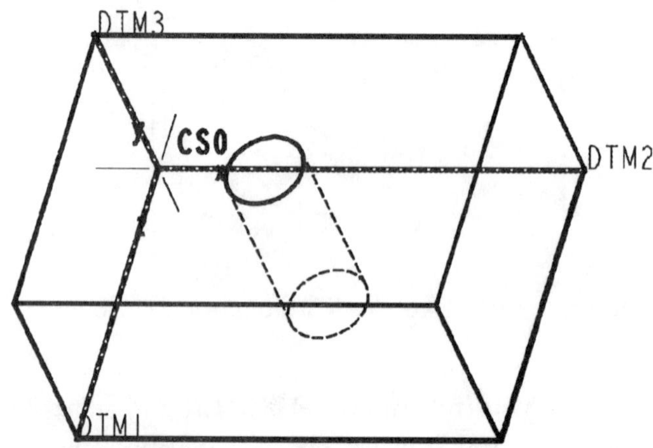

Figure 4-13 Completion of a Solid Model for the Hole Feature

After completing the creation of the block and hole features, let's continue the modeling process and cut the slot at the corner. The size of the cross section of the slot is 15 mm in the x direction and 10 mm in the y direction. We follow the procedure previously employed in creating the hole feature. First, select the command sequence, starting from the **FEAT** menu:

Create, Solid, Cut, Extrude, Solid, Done

Now choose the **One Side** option from the **SIDES** menu so that the feature will be extruded to only one side of the sketching plane.

Choose **Plane** from **SETUP PLANE** menu and pick the front side of the block feature as your **Sketching Plane**. This selection means that you will cut the slot starting from the front side, which is a plane parallel to DTM3, namely the x-y plane. Therefore, we select this plane to perform the 2D-sketch. When selecting the plane, an arrow towards the inside is shown, indicating that the direction of the milling cutter is going into the block material. This is the correct direction so we choose **Okay**.

To permit the screen to show a sketch plane, first select **Top** from **SKET VIEW**. Just choose the top surface of the block feature, as it is ready for selection. The 2D layout presented in Figure 4-14 is shown on the monitor screen. Note that the origin of the coordinate system is located at the lower and left corner. This layout gives a very clear picture of the orientation of the model, and is similar to positioning a sketch pad before making a drawing of the slot.

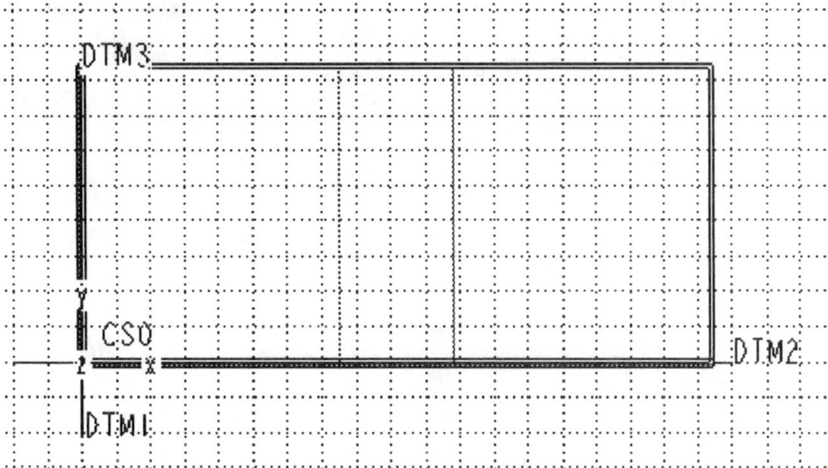

Figure 4-14 Orientation of a Sketch Plane for Modeling the Corner Slot Feature

Choose **Sketch** from the **SKETCHER** and **Rectangle** from **Geometry** menu. Select two points on the 2D screen to indicate the diagonal of the rectangular slot, thus defining the width and height dimensions of the slot as illustrated in Figure 4-15.

Choose **Alignment** from **SKETCHER**, and double click the left side of the slot and double click the bottom side of the slot. The message "—ALIGNED —" appears in the message window. The slot now is positioned along the left and bottom sides of the block feature.

Choose **Dimension** from **SKETCHER**. Use the left mouse button to click the left and right sides of the slot. Move the cursor to a place to position the dimension. Repeat the process by clicking the upper and lower sides of the slot, moving the cursor to position the dimension.

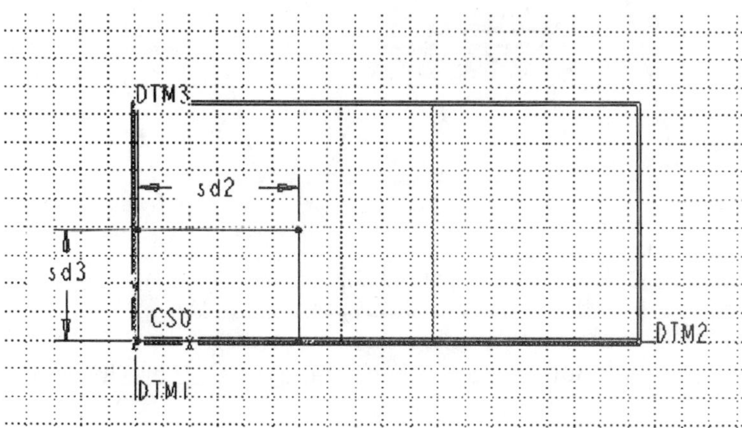

Figure 4-15 Sketch a Rectangle in the 2D Screen

Now choose **Regenerate** from the **SKETCHER** menu and select **Modify** from **SKETCHER.** On the main drawing window, select the width dimension, type 15 in the message window, and hit the **Enter** key. Select the height dimension, type 10 in the message window, and hit the **Enter** key. Choose **Regenerate** from the **SKETCHER** menu for the second time. The 2D sketch that appears on the monitor screen is illustrated in Figure 4-16.

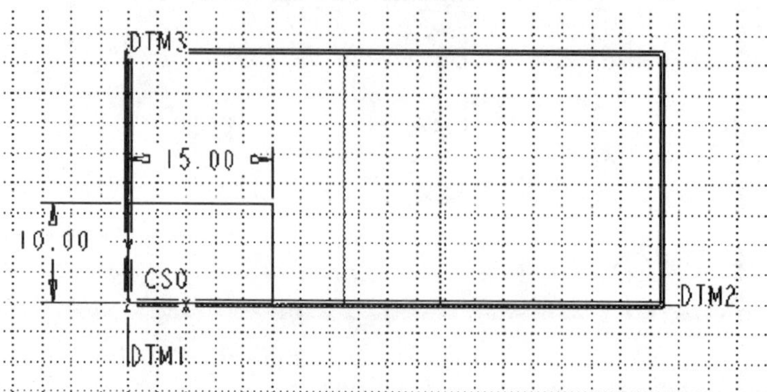

Figure 4-16 Completion of a 2D Sketch of the Corner Slot Feature

Choose **Done** from **Sketcher**, indicating that a 2D sketch of the slot feature has been successfully completed. Now extend the 2D sketch to a solid model in the third dimension by employing the through all slot feature.

Choose **Thru All** option from the **SPEC TO** menu and then **Done**. This means that the slot is cut completely through the block. Choose **OK** from the Element Information Window, to complete the construction of a slot with a cross section size of 15 x 10 mm. To view appearance of the slot in 3D, type dv and hit the **Enter** key (see Figure 4-17).

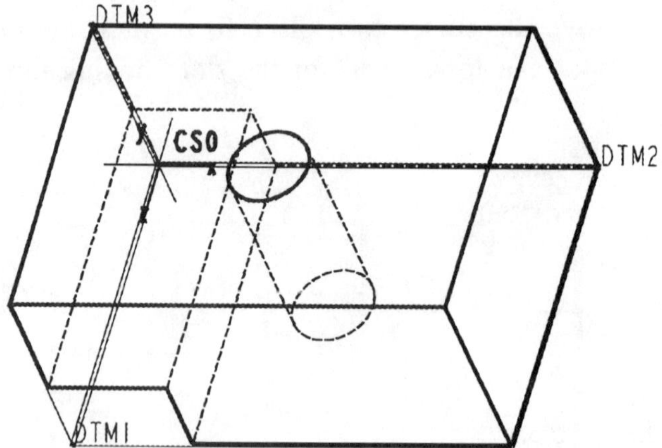

Figure 4-17 Completion of a Solid Model for the Corner Slot Feature

Let's continue our modeling process. This time we will cut a surface with a slope of 30° relative to the x axis. The procedure is the same as previously used for cutting the slot feature. First, select the command sequence, starting from the **FEAT** menu:

Create, Solid, Cut, Extrude, Solid, Done.

Now choose the **One Side** option from the **SIDES** menu. The feature will be extruded to only one side of the sketching plane.

Choose **Plane** from **SETUP PLANE** menu and pick the front side of the block feature as your **Sketching Plane**. This means the slope of the cut starts from the front side, in the same manner the cut for the slot began from the front surface. Select the front plane for the 2D-sketch. When selecting the plane, an arrow towards the inside is shown, indicating that the direction of the milling cutter is into the block material. This is correct so we choose **Okay**.

Select **Top** from **SKET VIEW** to choose the top surface of the block feature. The 2D layout presented in figure 4-18 is shown on the monitor screen. Note that the origin of the coordinate system is located at the lower left corner. This layout gives you a very clear visualization of the orientation and is similar to holding a sketch pad prior to making a drawing of the slope.

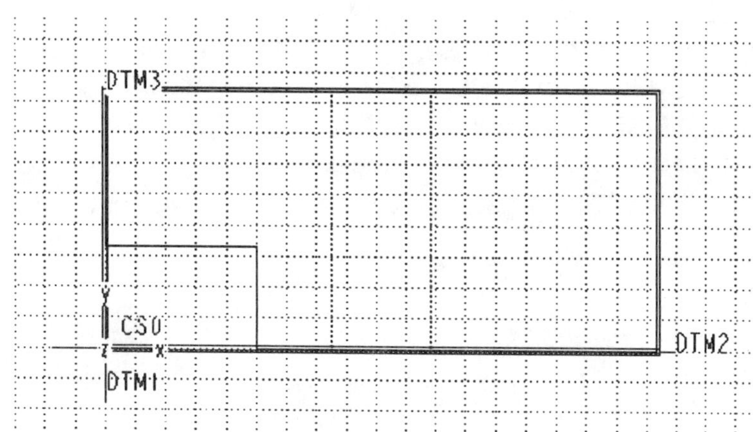

Figure 4-18 Orientation of a Sketch Plane for Modeling the Corner Slot Feature

Choose **Sketch** from the **SKETCHER** and **Line** from **Geometry** menu. Sketch a slope line at the upper right corner to define the feature with a 30° slope and a sharp point, as illustrated in Figure 4-19.

Choose **Alignment** from **SKETCHER**, and double click the upper end-point of the slope line and double click the right side end-point of the slope line. The message "— ALIGNED —" appears in the message window.

Choose **Dimension** from **SKETCHER**. Use the left mouse button to click the upper end-point of the slope line and the left sides of the block. Move the cursor to a suitable location for the dimension. Select the slope line and the lower side of the block, and then move the cursor to position the dimension defining the angle.

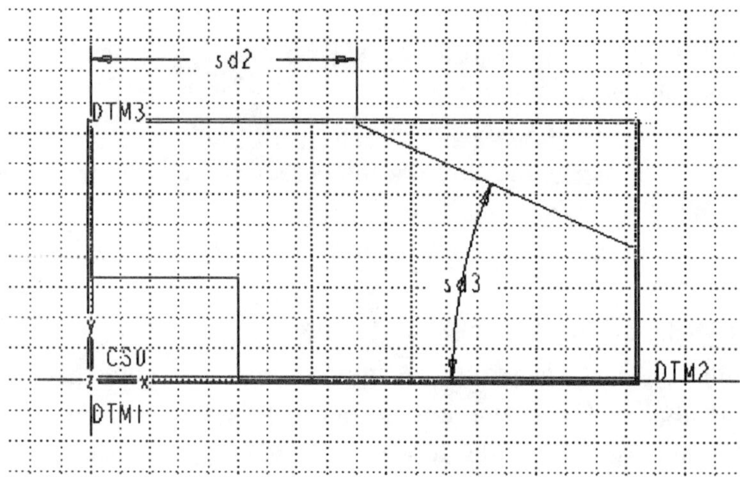

Figure 4-19 Sketch a Rectangle in the 2D Screen

Now choose **Regenerate** from the **SKETCHER** menu and select **Modify** from **SKETCHER.** On the main drawing window, select the distance dimension, type 55/2 in the message window, and hit the **Enter** key. Select the angle dimension, type 30 in the message window, and hit the **Enter** key. Choose **Regenerate** from the **SKETCHER** menu for the second time. The 2D sketch that appears on the screen is illustrated in Figure 4-20.

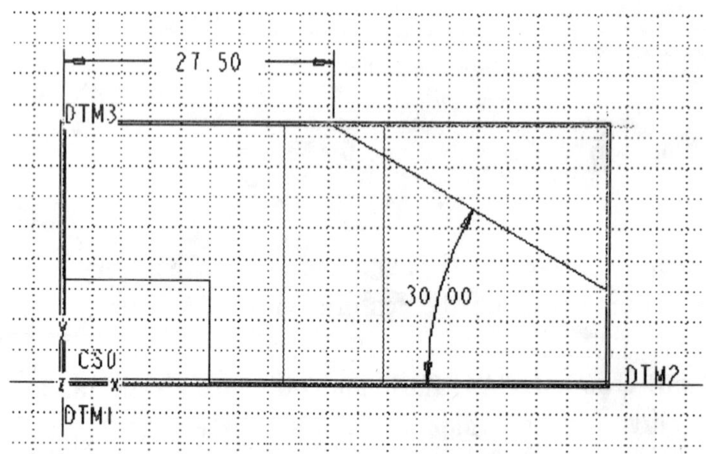

Figure 4-20 Completion of a 2D Sketch of the Block Feature

Choose **Done** from **Sketcher**, the message window will indicate that a 2D sketch of the slope feature has been successfully completed. Now extend the 2D-sketch to a solid model in the third dimension by employing the through-all command.

Choose **Thru All** option from the **SPEC TO** menu and then **Done**. This choice means that the slope line is cut completely through the block. Choose **OK** from the Element Information Window, to complete the construction of a slope feature with the size of 27.5 x 30°. To view the 3D feature, type dv and hit the **Enter** key (see Figure 4-21).

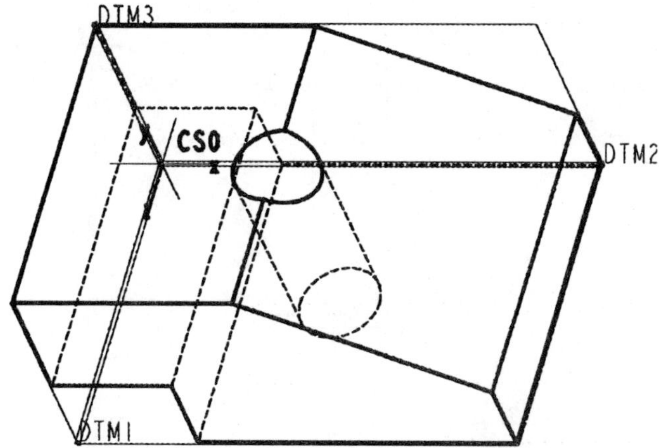

Figure 4-21 Completion of a Solid Model for the Hole Feature

To continue our modeling process, we cut the slot on the right side of the block. The size of the cross section is 16 mm in the z direction and 1 mm below the slope edge in the y direction. Following the procedure previously used for creating the slot feature at the corner, we first select the command sequence starting from the **FEAT** menu:

Create, Solid, Cut, Extrude, Solid, Done.

Now choose the **One Side** option from the **SIDES** menu. The feature will be extruded to only one side of the sketching plane.

Choose **Plane** from **SETUP PLANE** menu and select the right side of the block feature as your **Sketching Plane**. This selection indicates that you will cut the slot starting from the right side which is a plane parallel to DTM1, namely the y-z plane. Select the right side plane for the 2D-sketch. When selecting this plane, an arrow towards the inner side is shown, indicating that the direction of milling cuter is towards the block material. Since this is the correct direction, we choose **Okay**.

To permit the screen to show a sketch plane, select **Top** from **SKET VIEW** first. Choose the top surface of the block feature. The 2D layout presented in Figure 4-22 is shown on the screen of the monitor. Note that the origin of the coordinate system is located at the lower and right corner. This layout gives you a clear understanding of the orientation because it is similar to holding a sketch pad prior to making a drawing of the side slot.

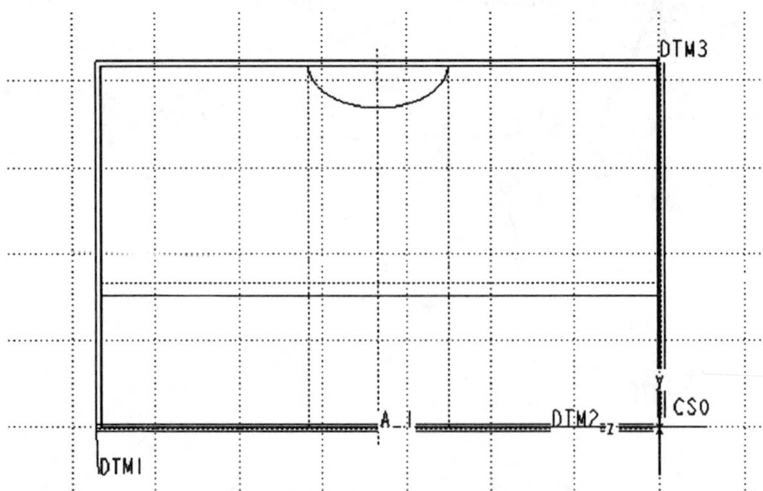

Figure 4-22 Orientation of a Sketch Plane for Modeling the Slide Slot Feature

Choose **Sketch** from the **SKETCHER** and **Rectangle** from **Geometry** menu. Select two points on the 2D screen to indicate the diagonal of the rectangular slot, thus defining the width and height dimensions of the slot as illustrated in Figure 4-23.

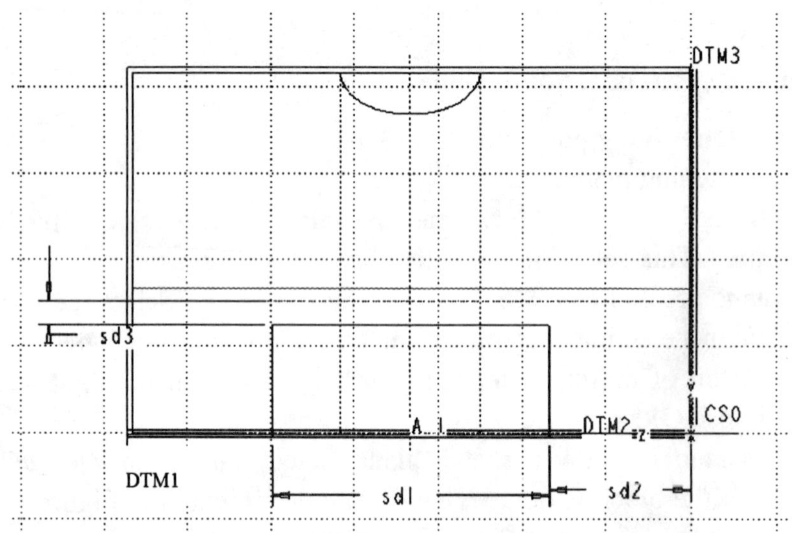

Figure 4-23 Sketch a Rectangle in the 2D Screen

Choose **Alignment** from SKETCHER, and double click the bottom of the slot. The message "—ALIGNED —" appears in the message window. The slot is now positioned along the bottom side of the block feature.

Choose **Dimension** from **SKETCHER**. Employ the left mouse button to click the left and right sides of the slot. Move the cursor to a locate the position the dimension. Select the right side of the slot and the right side of the block, and move the cursor to position the dimension. Select the top line of the slot and the slope edge and move the cursor to position the thickness dimension.

Now choose **Regenerate** from the **SKETCHER** menu and select **Modify** from **SKETCHER.** On the main drawing window, select the dimension of the slot width, type 16 in the message window, and hit the **Enter** key. Select the distance dimension, type 12.5 in the message window, and hit the **Enter** key. Select the height dimension, type 1 in the message window, and hit the **Enter** key. Choose **Regenerate** from the **SKETCHER** menu again. The 2D sketch that appears on the monitor screen is illustrated in Figure 4-24.

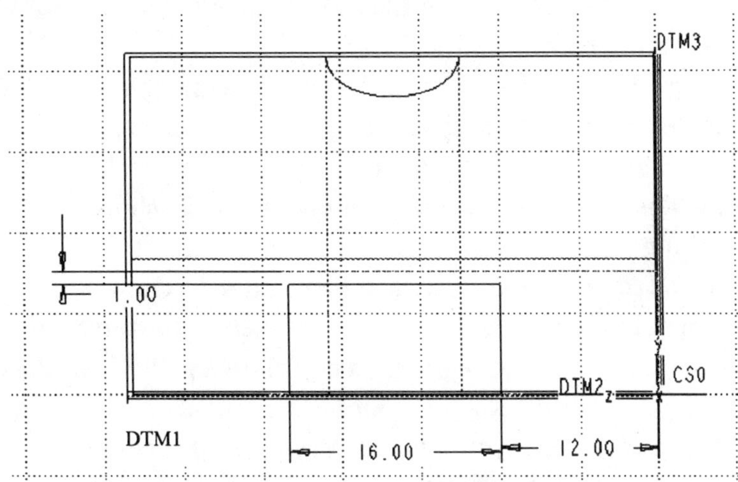

Figure 4-24 Completion of a 2D Sketch of the Side Slot Feature

Choose **Done** from **Sketcher**, indicating that the 2D sketch of the side slot feature has been successfully completed. Now to extend the 2D-sketch to a solid model in the third dimension by selecting the through-all command.

Choose **Thru All** option from the **SPEC TO** menu and then **Done**. This choice cuts the slot completely through the block. Choose **OK** from the Element Information Window, to complete the construction of the side slot feature. To view the 3D feature, type dv and hit the **Enter** key (see Figure 4-25).

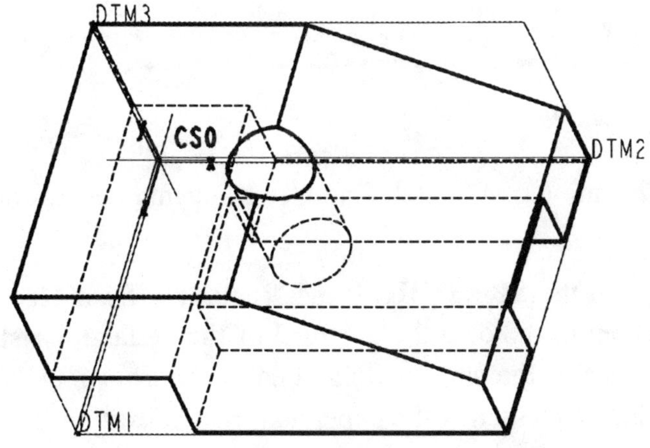

Figure 4-25 Completion of a Solid Model for the Side Slot Feature

To continue our modeling process, we cut the remaining slot from the bottom of the block. The size of the cross section of this bottom slot is 20 x 20 mm in the x direction and the z direction. Following the procedure previously used for creating the slot feature at the corner, we first select the command sequence, starting from the **FEAT** menu:

Create, Solid, Cut, Extrude, Solid, Done.

Now choose the **One Side** option from the **SIDES** menu. The feature will be extruded to only one side of the sketching plane.

Choose **Plane** from **SETUP PLANE** menu and select the bottom of the block as your **Sketching Plane**. This selection implies that the slot will be cut starting from the bottom surface which is the DTM2 plane (the x-z plane). Therefore, we select DTM2 for our 2D-sketch. When selecting this plane, an arrow pointing down is shown, indicating that the direction of the milling cuter is not towards the block material. This is not correct so we choose **Flip** to reverse the direction of the cutter and then and then **Okay**.

To permit the screen to show a sketch plane, first select **Bottom** from **SKET VIEW**. Choose the bottom surface of the block feature. The following 2D layout is shown on the computer screen. Note that the origin of the coordinate system is located at the lower left corner, as shown in Figure 4-26. This layout gives a very clear idea about the orientation of the block and is similar to holding a sketch pad before making a drawing of the bottom slot.

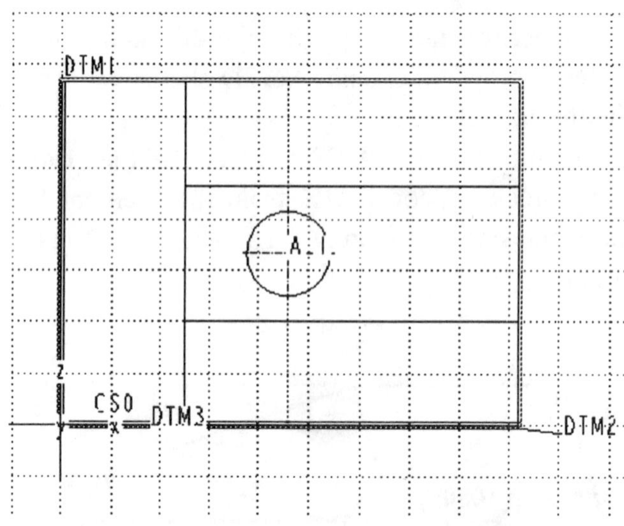

Figure 4-26 Orientation of a Sketch Plane for Modeling the Bottom Slot Feature

Choose **Sketch** from the **SKETCHER** and **Rectangle** from **Geometry** menu. Select two points on the 2D screen to indicate the diagonal of the rectangular slot, thus defining the width and height dimensions of the slot, as illustrated in Figure 4-27. It is very important to make sure that the rectangle is located at the upper and right corner.

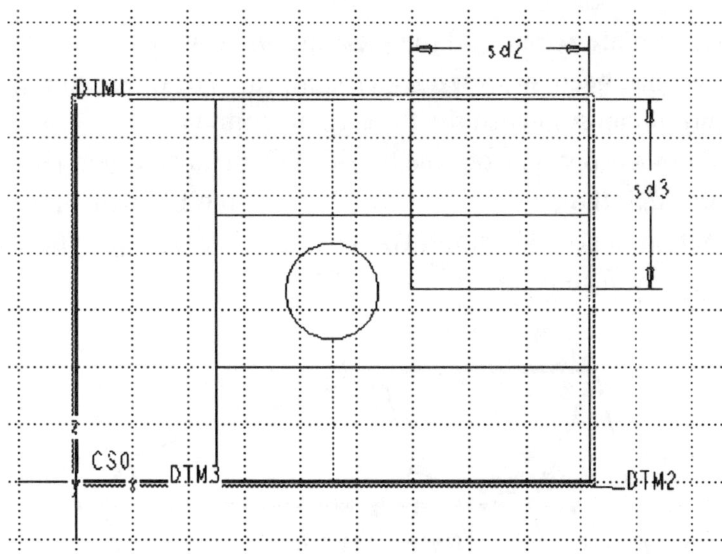

Figure 4-27 Sketch a Rectangle in the 2D Screen

Choose **Alignment** from **SKETCHER**, and double click the top line of the rectangle and double click the right side of the rectangle. The message "—ALIGNED —" appears in the message window. The bottom of the slot is now located along the two sides of the block feature.

Choose **Dimension** from **SKETCHER**. Use the left mouse button to click the left and right sides of the slot. Move the cursor to locate a suitable position for the dimension. Select the upper and the lower sides of the rectangle, and move the cursor to position the dimension.

Now choose **Regenerate** from the **SKETCHER** menu and select **Modify** from **SKETCHER.** On the main drawing window, select the dimension of the slot width, type 20 in the message window, and hit the **Enter** key. Select the dimension of the slot height, type 20 in the message window, and hit the **Enter** key. Choose **Regenerate** from the **SKETCHER** menu for the second time. The 2D-sketch appearing on the screen of the monitor is illustrated in Figure 4-28.

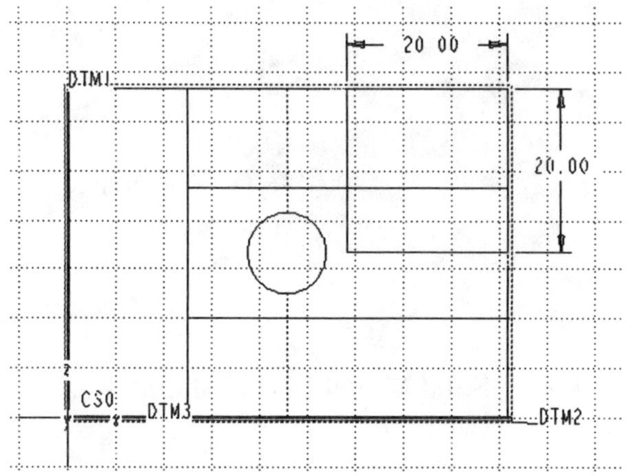

Figure 4-28 Completion of a 2D Sketch of the Bottom Slot Feature

Choose **Done** from **Sketcher**, and the message window will indicate that a 2D-sketch of the bottom slot feature has been successfully completed. Now we extend the 2D-sketch to a solid model in the third dimension with the through-all command.

Choose the **Thru All** option from the **SPEC TO** menu and then **Done**. This choice cuts the slot completely through the block. Choose **OK** from the Element Information Window, to complete the construction of the side slot. To view the 3D drawing, type dv and hit the **Enter** key to display the illustration shown in Figure 4-29.

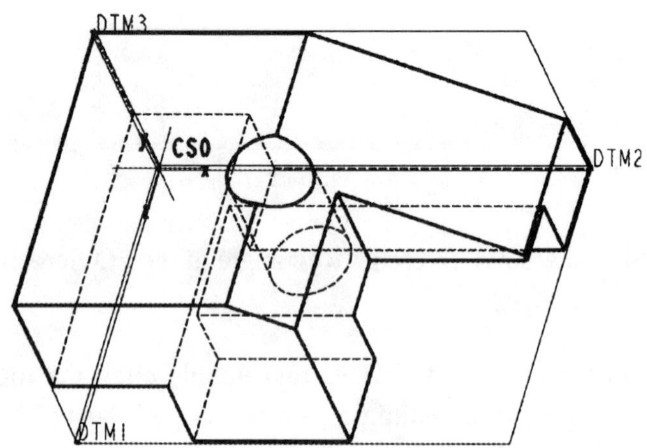

Figure 4-29 Completion of a Solid Model for the Bottom Slot Feature

We have successfully completed the process of modeling the sliding basket. A shaded drawing of the solid model for the sliding basket that we have constructed is presented in Figure 4-30.

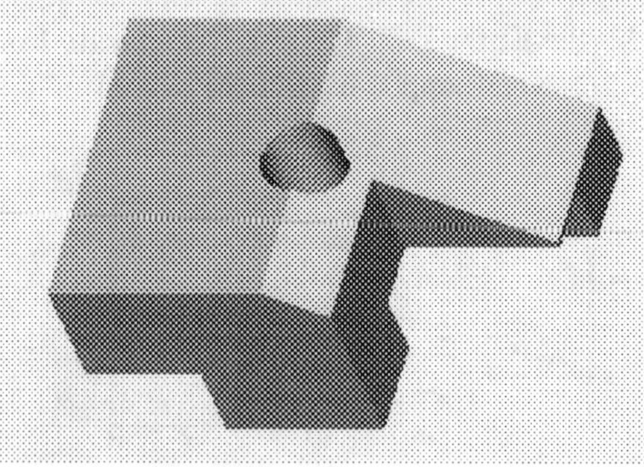

Figure 4-30 Solid Model for the Sliding Basket with Shading

Under the **Pro/ENGINEER** design environment, the routine procedure to shade a 3D visualized object on the computer screen is listed below:

(1) From the **MAIN** menu, select **Environment**; and
(2) From the **ENVIRONMENT** menu, select **Shading** and **Done-Return**.

If you convert from a shaded 3D object to a non-shaded version, select **Environment** from the **MAIN** menu, and then select **Wireframe** and **Done-Return** from the **ENVIRONMENT** menu.

Remember to save all your work. The procedures to save a file when you are using a PC or Unix station are:

Using a PC:

(1) Insert a 3 ½ " disk in Drive A;
(2) From the **MAIN** menu, select **Misc, Change Dir**, in the message window, type in *a:* then *return*.
(3) From the **MAIN** menu, select **DBMS** (Data Base Management System);
(4) From the **DBMS** menu, select **Save**;
(5) On the message window, you will read the statement:
(6) Enter object to save [PRT0001]:
(7) You should type in *filename.prt*
(8) Hit the Enter key and observe Drive A on your PC. The light should be on for a while indicating the process of storing the data to the disk.

Using Unix Station

(1) From the **MAIN** menu, select **Dbms** (Data Base Management System);
(2) From the DBMS menu, select **Save**;
(3) On the message window, you will read the statement:
(4) Enter object to save [PRT0001]:
(5) You should type in *filename.prt*
(6) Hit the Enter key, the file that you have just created will be saved to your own account and next time you can retrieve it directly from **PART** menu.

It is important to know the procedure for ending the Pro/ENGINEER session. Choose **Exit** from the MAIN menu. Confirm the selection by choosing **Yes** when the confirmation window appears. An inappropriate exit may cause the current license for the site to be occupied. As a result, other users may experience difficulty in using the Pro/ENGINEER design system.

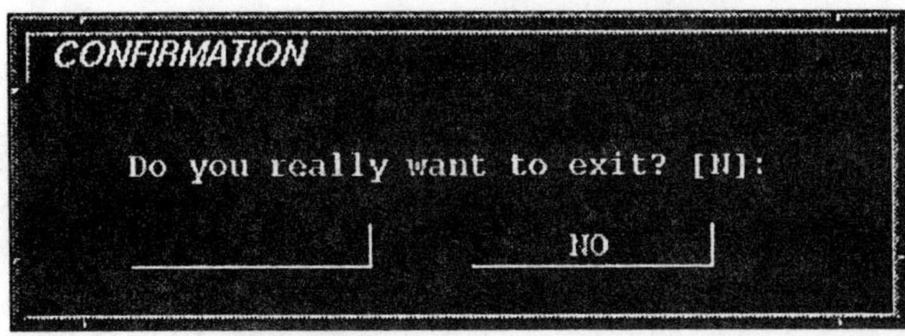

4.2.2 Example 4-2: Mouse Functions

In Pro/ENGINEER, the mouse is the most important element of the user interface. Efficient use of the mouse will increase the speed of the design process. This next example covers the most common uses of the mouse in Pro/ENGINEER. For simplicity, we will use the abbreviations LMB, MMB, RMB to stand for the left, middle and right mouse buttons as identified in Fig. 4-31. The functions of the mouse include:

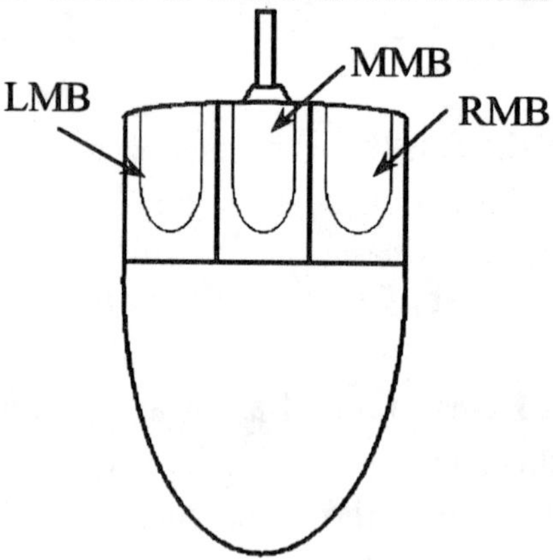

Figure 4-31 The Three Mouse Buttons

1. **Activating Windows**

   Since both Unix workstation and Windows 95/NT support multiple windows, the mouse can be used to switch from Pro/ENGINEER to other applications. This capability is important when answering the prompt in the message window. Always make sure that Pro/ENGINEER is active by clicking inside the window and leaving the mouse cursor within a Pro/ENGINEER window. To switch windows within Pro/ENGINEER, choose **Change Window** from the MAIN menu and click inside the window to which you wish to change. The menu will also change automatically according to the mode in the new active window.

2. **Selecting a Menu Item**

   Pro/ENGINEER is a menu-driven system. Menu options are selected by moving the cursor to the desired position, highlighting the item, and clicking the LMB. As you move through the menu options, a one-line description about the corresponding item will show in the message window.

3. **Seeking on-line Help**

   On-line help can be easily accessed by using the right mouse button. Place the cursor on the menu option for which you require information, hold on the right mouse button, and highlight the message "? Get Help". This action will open the user guide to the page where the menu option is described in detail.

4. Selecting a Geometric Feature

Many commands in Pro/ENGINEER require the selection a specific geometric feature that you are attempting to construct. Move the cursor to the feature you want to choose and click the left mouse button. This feature will be highlighted when it is selected. Use **Query Sel** to choose an item from several adjacent items. If the first item you choose is the correct one, click the middle mouse button (equivalent to choose **Done Sel).** Otherwise, click the right mouse button to pick the next item.

5 View Function

One advantage of Pro/ENGINEER is that the designer can check the model by using the zoom, rotate and pan features. To zoom, depress the Ctrl key and the left mouse button while moving the mouse left or right. The geometry will zoom in or zoom out according to the movement of the mouse.

To rotate, depress the Ctrl key and the middle mouse button while moving the mouse around to rotate the model. Pay attention to the track of the mouse that determines the rotation direction. To pan, depress the Ctrl key and the right mouse button while moving the mouse; the geometry will pan accordingly.

6 Sketching and Dimensioning

Another common use of the mouse is to sketch lines and other geometric features. Choose **Sketch** from the SKETCHER menu and select **Mouse Sketch**.

6.1 Sketching a line

Click the left mouse button at the location where you want to start the line. A red "rubber band" line appears, attached to the cursor. Click the left mouse button at the location where you want to terminate the line. Pro/ENGINEER creates a line between these two points and begins a second rubber band line. Move the mouse to another position where the line is to be terminated and click the left mouse button. Another new line is drawn. Repeat the above steps until all the lines have been drawn as illustrated in Figure 4-32. Click the middle mouse button to end the line drawing operation.

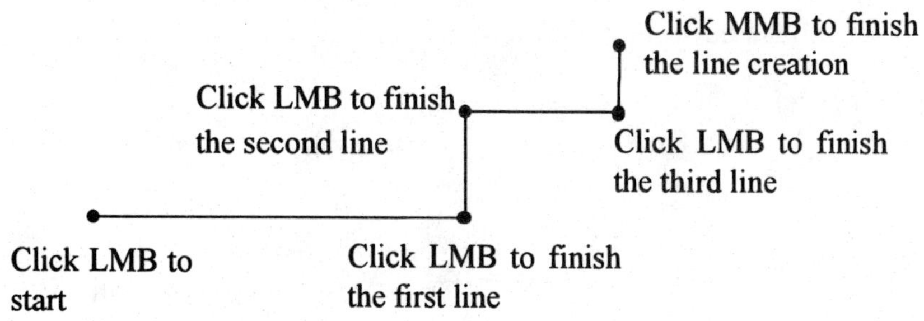

Figure 4-32 Sketching a Line

6.2 Sketching a Circle

Click the middle mouse button at the location of the center point. A red rubber band circle appears that is centered on this point. As indicated in Figure 4-33, move the mouse away from the center to increase the diameter of the circle. Click the middle mouse button again to complete the circle or click the left mouse button to abort the operation.

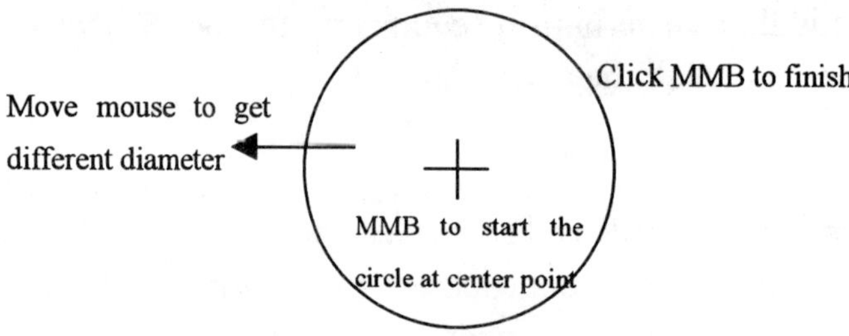

Figure 4-33 Sketching a Circle

6.3 Sketching an Arc

Click the right mouse button on the endpoint of an existing entity such as a line. A red rubber band arc appears. The arc will be tangent to the existing entity. Move the mouse away from the start point to change the radius of the arc as shown in Figure 4-34. Click the right mouse button to finish the arc or click the middle mouse button to abort the operation.

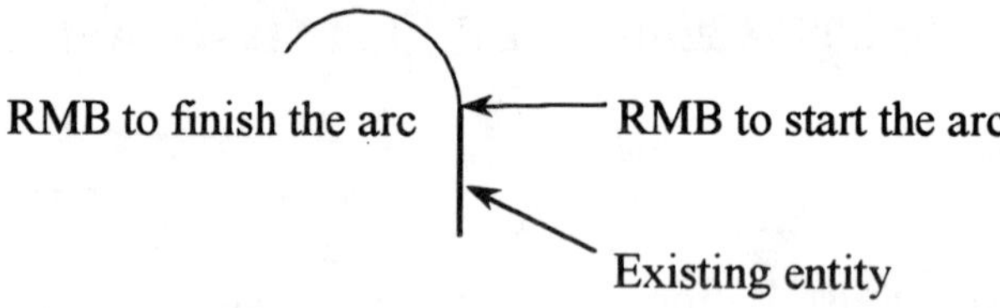

Figure 4-34 Sketching an Arc

6.4 Dimensioning

After choosing **Dimension**, pick the feature to be dimensioned by clicking the left mouse button. The geometry should be highlighted. Move the mouse away from the geometry and place it at the location of the dimension as indicated in Figure 4-35. Click the middle mouse button to add the dimension.

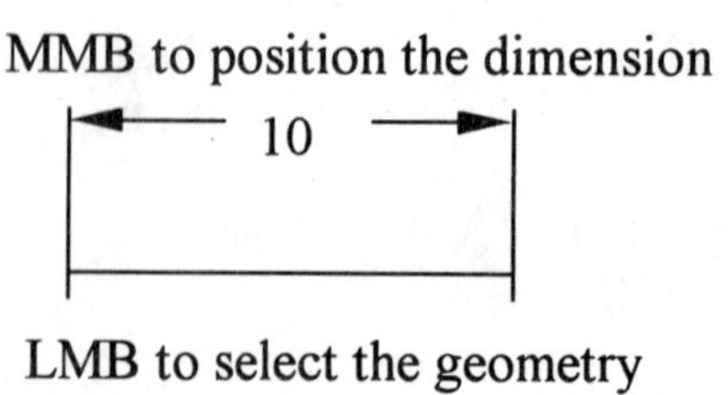

Figure 4-35 Dimensioning

4.2.3 Example 4-3: Solid Modeling of a Base Plate

There are two procedures for constructing geometric features in Pro/ENGINEER.

(1) Feature construction by sketching.
(2) Feature construction by picking and placing.

All of the features in Example 4-1 were constructed by sketching. In Example 4-3, we introduce the second procedure for constructing geometric features by picking and placing. Examine the geometry of the base plate shown in Figure 4-36. We will use "pick and place" to construct the hole features and the round corner features. Let's start the modeling process by selecting the command sequence, starting from the **FEAT** menu:

Create, Solid, Cut, Extrude, Solid, Done

Now choose the **One Side** option from the **SIDES** menu. The feature will be extruded from only one side of the sketching plane.

Choose **Plane** from **SETUP PLANE** menu and pick DTM3 as your **Sketching Plane**. If the arrow direction is towards you, or is in the direction along the positive z direction, choose **Okay.** Otherwise, choose **Flip**, then **Okay**.

From the SKET VIEW, choose **Top**, then select DTM2 on the main drawing window. The 2D layout that appears on the monitor screen is presented in Figure 4-37.

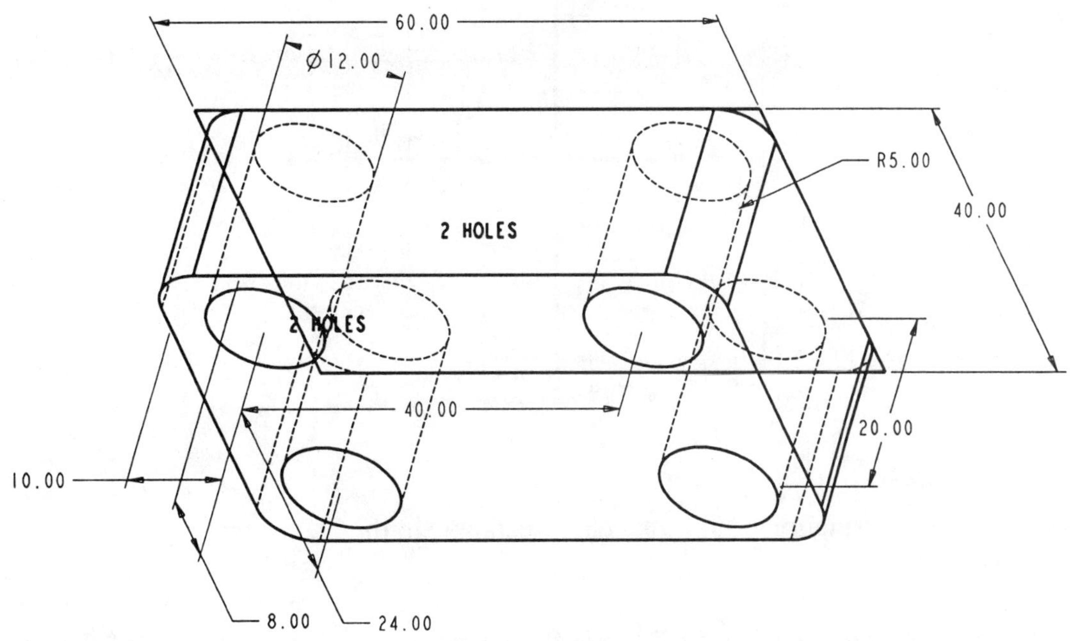

Figure 4-36 Geometry and Dimensions of a Base Plate

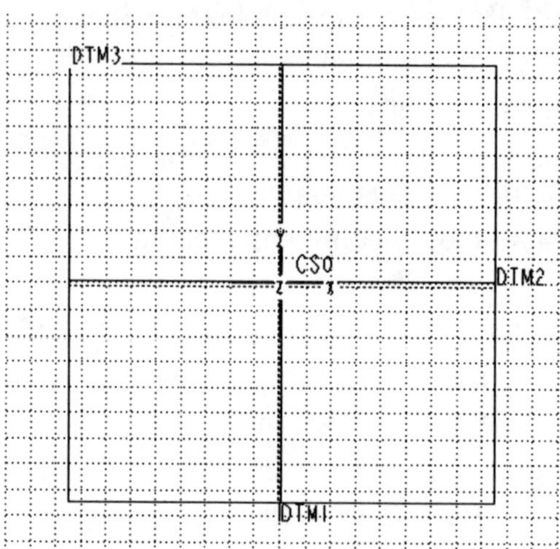

Figure 4-37   Define the orientation of a Sketch Plane

Choose **Sketch** from the **SKETCHER** and **Rectangle** from **Geometry** menu. Select two points on the 2D screen to indicate the diagonal of the rectangle defining the outline of the base plate. As illustrated in Figure 4-38, one point is selected on the positive y axis and the other point is selected on the positive x axis.

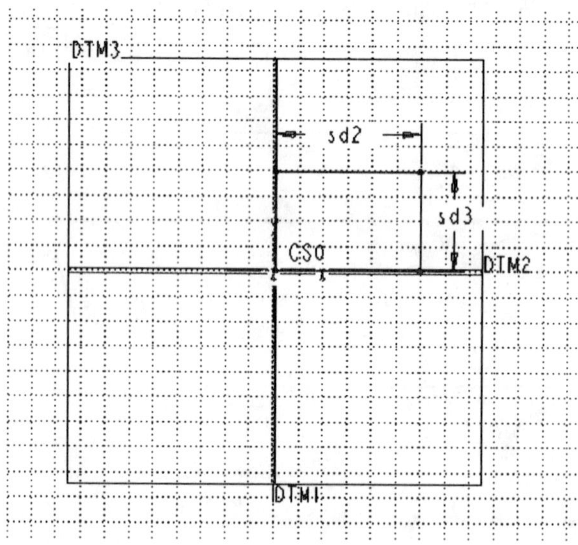

Figure 4-38   Sketch a Rectangle in the 2D Screen

Choose **Alignment** from **SKETCHER**, and double click the left side of the rectangle and double click the bottom side of the rectangle and read the message in the message window to ensure to ensure "—ALIGNED —" is displayed.

Choose **Dimension** from **SKETCHER**. Use the left mouse button to click the top edge of the rectangle. Move the cursor to a suitable location and press the middle mouse button to show the width dimension, sd2. Repeat the process by clicking on either the right or left edge of the rectangle, moving the cursor to a suitable location, and pressing the middle mouse button to show the sd3 (height) dimension.

Choose **Regenerate** from the **SKETCHER** menu and select **Modify** from **SKETCHER.** On the main drawing window, select the width dimension, type 60 in the message window, and hit the **Enter** key. Select the height dimension, type 40 in the message window, and hit the **Enter** key. Choose **Regenerate** from the **SKETCHER** menu for the second time. The 2D sketch of the rectangle is presented in Figure 4-39.

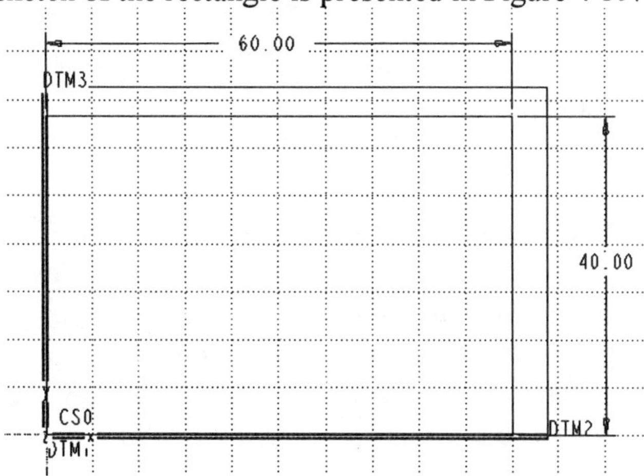

Figure 4-39 Completion of a 2D Sketch of the Rectangular Block.

Choose **Done** from **Sketcher**, and choose **Blind** option from the **SPEC TO** menu and then **Done**. In the message window, you must provide the third dimension by typing 20 in the message window and hitting the **Enter** key. Choose **Okay** from the Element Information Window, thus completing the construction of the block with the three dimensions equal to 60, 40 and 20 mm, respectively. To view the 3D drawing, type dv and hit the Enter key. Recall that dv stands for three-dimensional view. The block illustrated in Figure 4-40 is displayed on your monitor screen.

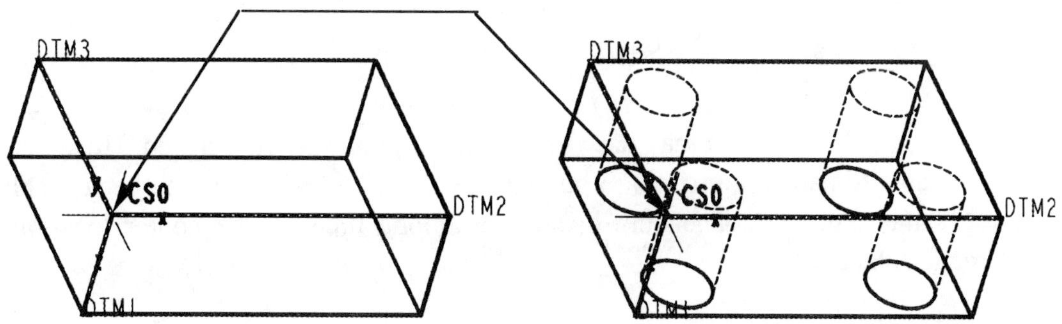

Figure 4-40 Completion of a Solid Model for the Block Feature

Let us continue our modeling process by drilling 4 holes each with a diameter of 12 mm. Because holes belong to a well defined feature, the command sequence from the **FEAT** menu is changed to: **Create, Solid, Hole, Done**

Note that the design of the holes is specified in the **HOLE OPTS** menu. Select **Straight** and then **Done** because all the holes are straight. Now the user should select a plane on which the drilling operation is initiated. From the **PLACEMENT** menu select **Linear** and then **Done** because the boundary of a straight hole is a straight line and linear. Note the message displayed in the message window:

"Select the placement plane"

Use your cursor to select the front side of the block as the placement plane and note the message in the message window:

"Select two edges, …, for DIMENSIONS"

Use your cursor to select the left edge and the message displayed in the message window is asking you to input the dimension from the edge. Type 10 and hit the **Enter** key. Use your cursor to select the top edge and type 8 in the message window and hit the **Enter** key. In the **SIDES** menu, select **One Side** and then Done. Choose **Thru All** from the **SPEC TO** menu and then **Done**. At this moment, the message window is prompting the user to input the dimension for the hole diameter. Type 12 and hit the **Enter** key. Choose **OK** from the Element Information Window. The screen should show the 3D drawing shown in Figure 4-41.

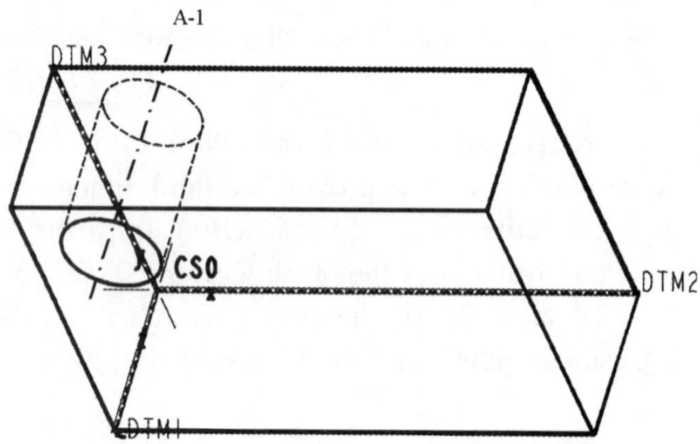

Figure 4-41 Creating a Hole Directly from the **FEAT** Menu

To create the other 3 holes, you may use the procedure previously described. However, it is easier to employ an alternative approach. You may use the "COPY" command because all the holes have the same diameter. The only difference among them is the center location. From the **FEAT** menu select:

Copy, Move, Done

From the **SELECT FEAT** menu, choose **Select** and pick the center line of the existing hole, A-1, and then **Done**. In the **MOVE FEATURE** menu, select **Translate.** In the message window, the user is prompted to indicate the direction for the movement. On the drawing window, select the right side of the block and an arrow appears to indicate the direction. If the arrow direction is to the right, select Okay, type 40 as the moving distance, hit the Enter key.

The copying process is complete so select **Done Move, Done, and OK** from the Element Information Window. The graphic display shown on the screen is illustrated in Figure 4-42.

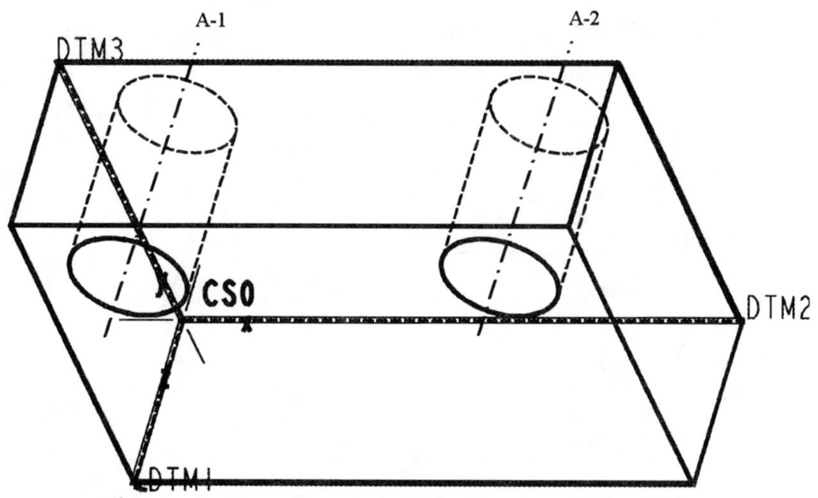

Figure 4-42 Create a Hole Feature Using COPY

Now we can use **Copy** to create the other two holes. From the **FEAT** menu:

Copy, Move, Done

From the **SELECT FEAT** menu, choose **Select** and pick the centerlines of the holes, A-1 and A-2, and then **Done**. In the **MOVE FEATURE** menu, select **Translate.** In the message window, the user is prompted to indicate the direction for the movement. On the drawing window, select the front side of the block and an arrow appears to indicate the direction. If the arrow direction is to the front, select Okay, type 24 as the distance for movement, and hit the Enter key. The copying process is complete. Next select **Done Move, Done** from the **GP VAR DIMS** menu, and **OK** from the Element Information Window. The graphic display appearing on the screen of the monitor is presented in Figure 4-43.

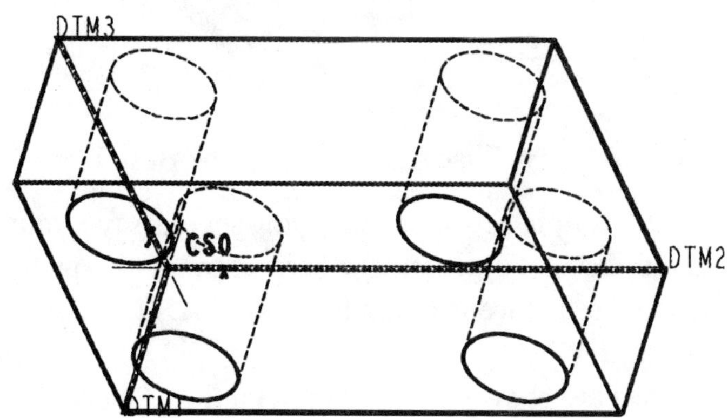

Figure 4-43 Completion of a Solid Model for the Base Plate before Rounding

The next step involves rounding the four corners. From the **FEAT** menu, select **Create, Solid, Round, Done.** Then select **Simple**, **Done** from the **ROUND TYPE** menu. Next choose **Constant**, **Edge Chain** because we intend to round the four corners. Now we select the four edges which define the corners and choose **Done**. Note the prompt in the message window: "RADIUS or <ESC> to select option in the menu:"

Type 5 for the radius. Choose **OK** from the Element Information Window. The display on the monitor screen is presented in Figure 4-44.

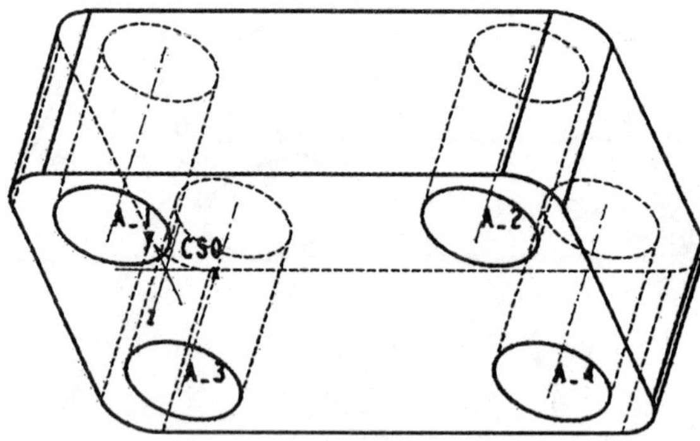

Figure 4-44 The Base Plate after Rounding the Corners

In order to produce a shaded picture, type sd or choose Environment from the MAIN menu and follow the routine procedure discussed previously. A shaded 3D view of the base plate is depicted in Figure 4-45.

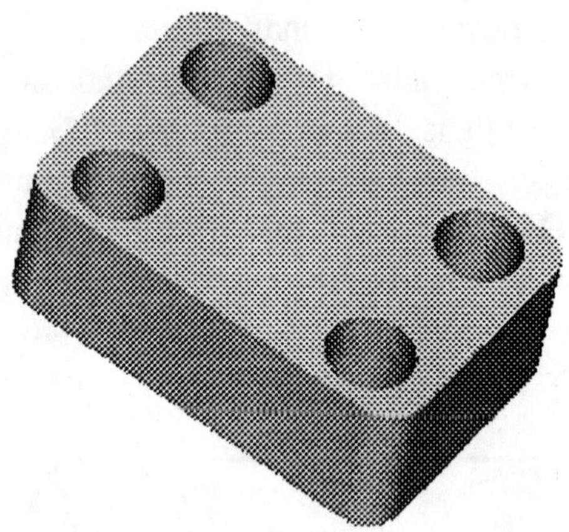

Figure 4-45 A Shaded 3D View of the Base Plate

As the design of the base plate is complete, you should save your work to a file and store the file onto a suitable disk. To generate a hard copy of the picture displayed on the screen, the following command sequence is used. From the **PART** menu, select:

Interface, Export, Plotter

All the print commands are shown on the Plot Window. Select **OK** and **Yes** buttons to print the hard copies required.

4.2.4 Example 4-4: Revolving a 2D Sketch to Construct a Cylindrical Part

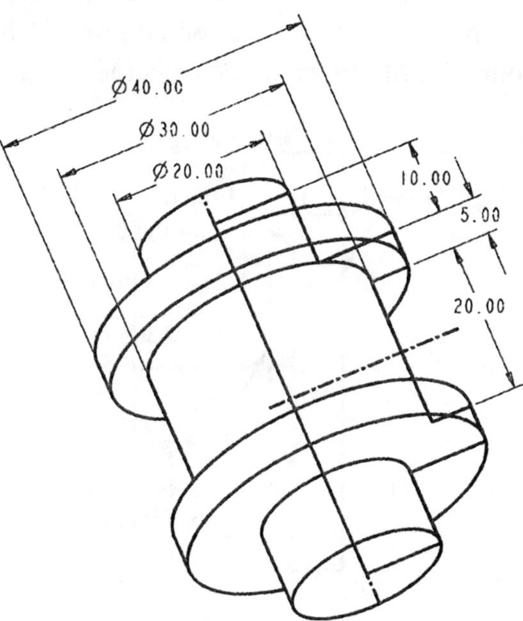

Figure 4-46 Geometry and Dimensions of a Stepped Shaft

In Examples 4-1 and 4-3, we have demonstrated the method used in creating a solid model using the **Extrude** function. In this example, we introduce another function for generating solid models called **Revolve**. The geometry that we use to demonstrate this approach is presented in Figure 4-46. Following the procedure described in the previous examples, we start the drawing from the **FEAT** menu by selecting:

Create, Solid, Protrusion, Revolve, Solid, Done

Now choose the **One Side** option from the **SIDES** menu. Choose **Plane** from **SETUP PLANE** menu and pick DTM3 as your **Sketching Plane**. If the arrow direction is towards you, or is in the direction along the positive z direction, choose **Okay.** Otherwise, choose **Flip**, then **Okay**. From the **SKET VIEW**, choose **Top**, then select DTM2 on the main drawing window.

To use the Revolve function to generate a cylindrical feature, the first requirement is to have a centerline, about which a 2D sketch can be rotated. Choose **Sketch** from the **SKETCHER** and **Line** from **Geometry** menu. Then select **Centerline** and **Vertical**. Use your left mouse button to click on DTM1, or on the y-axis, and a vertical centerline will appear. The second centerline required is a horizontal one, about which a mirrored feature will be created in the next sequence of operations. Select **Centerline** and **Horizontal**, and. use your left mouse button to click on DTM2, or on the x-axis on the current sketch view. A horizontal centerline appears. Choose **Alignment** from **SKETCHER** and double click x and y axes, and read the message in the message window to ensure that "—ALIGNED —" is displayed.

The next step in using the **Revolve** function is to create the geometry of the cross-section of the cylindrical part. Choose **Sketch** from the **SKETCHER** and **Mouse Sketch** from **Geometry**, and begin to draw the shape illustrated in Figure 4-46. You may begin the drawing by sketching a series of lines starting at the origin of the coordinate system. In sketching the lines a, b, c, d, e, f, g, generate a polygon as illustrated in Figure 4-47. The end point for this polygon should be located on the positive x axis. Choose **Alignment** from **SKETCHER**, and double click the vertical line around DTM2 (y-axis), which is the location for line a.

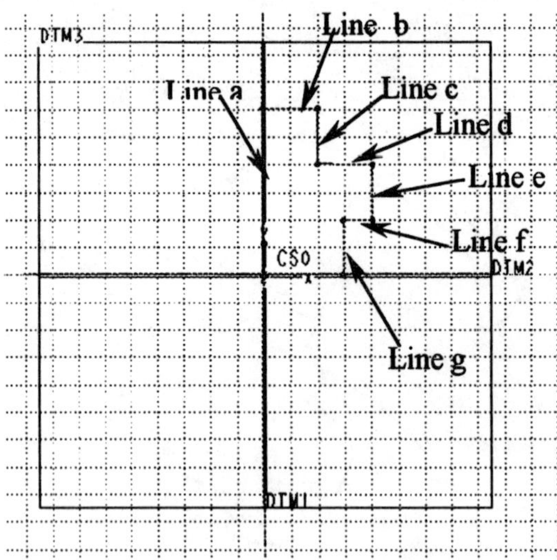

Figure 4-47 Geometry of the Cross-section before Using **Mirror**

As you may have noticed, the geometry of the stepped shaft is symmetric with respect to the x-axis (DTM2), so only the upper portion was drawn in Figure 4-47. The missing symmetric portion can be constructed by using **MIRROR**. This function is used to map a shape from one side to the other side about a centerline. Choose **Geom Tools** from the **SKETCHER** and **Mirror** from the **Geometry** menu. Select the horizontal center line, then pick all those lines you just drew as the entities that are to be mirrored. Select **Done** from **MIRROR**. The mirrored geometry appearing on the screen is depicted in Figure 4-48.

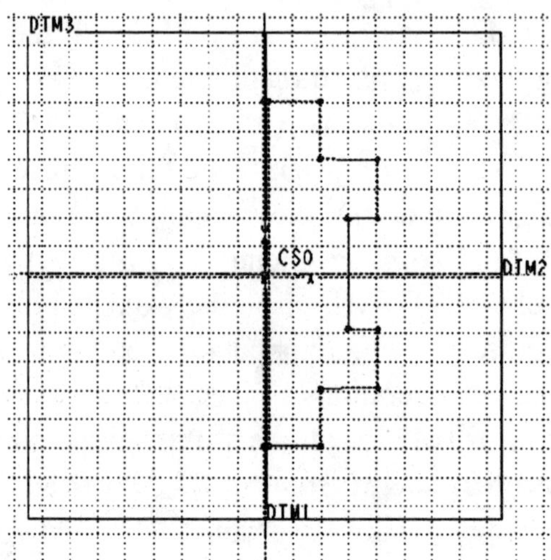

Figure 4-48 Mirrored Geometry of the Cross-section

Choose **Dimension** from **SKETCHER**. We first define dimensions for diameters of the shaft at different sections. Use the left mouse button to click the vertical centerline, or click on the y axis, then click one of the cylindrical edges, say line c, and then click the vertical centerline again. (Note: when you click on the vertical centerline, you must select a location above or below line a). Use the middle mouse button to position the dimension. Repeat this process for line e and line g. On the drawing window, dimension sd3, sd4, and sd5 appear as illustrated in Figure 4-49. After dimensioning the diameters is complete, you must also provide dimensions for the height of each of the 5 sections of the cylinder, i.e., sd6, sd7, and sd8.

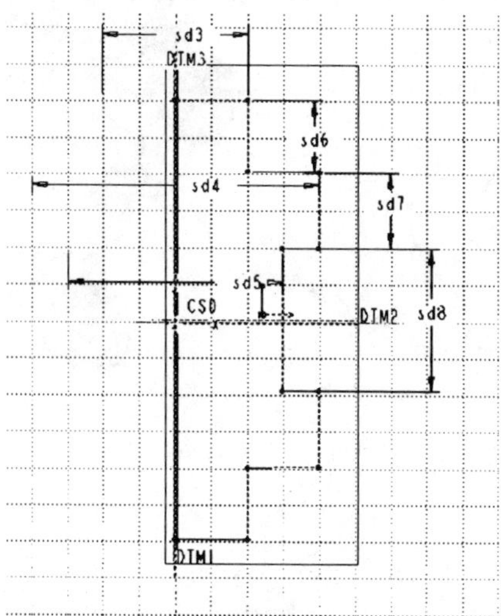

Figure 4-49 Dimensions for the Cross-section: 2D Sketch View

Choose **Regenerate** and then **Modify** from the **SKETCHER** menu. Modify the dimensions of the shaft defined in Figure 4-49. Choose **Regenerate** again. The dimensioned 2D sketch that appears on the screen of the monitor is presented in Figure 4-50.

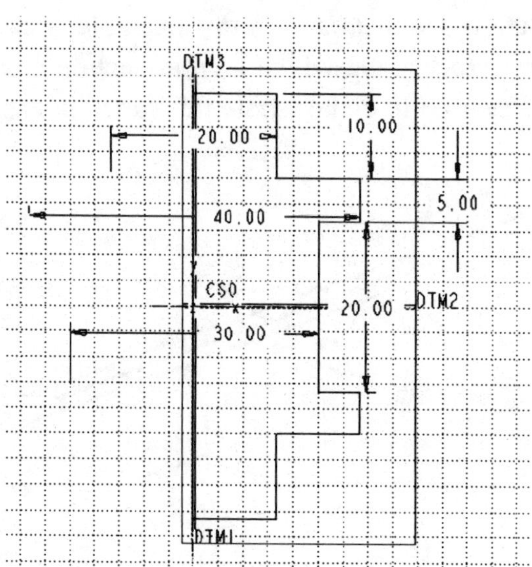

Figure 4-50 Completion of the 2D Sketch of the Shaft Feature

Choose **Done** from the **SKETCH** menu, and select the **360^0** option from the **REV TO** menu and then **Done**. Choose **Okay** from the ***Element Inform*** Window. The completed 3D drawing and the shaded view of the stepped shaft are shown in Figure 4-51.

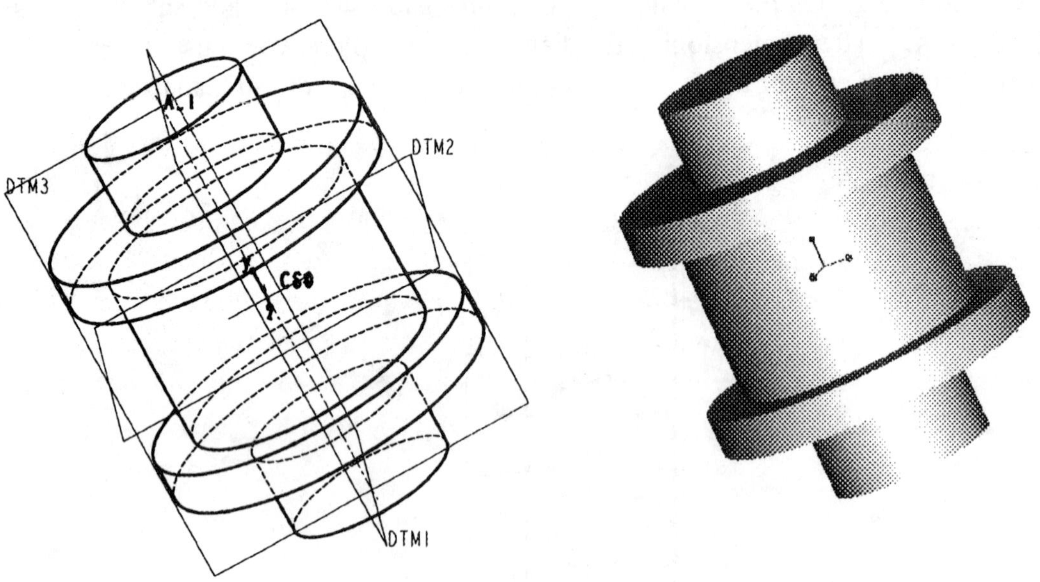

Figure 4-51 The 3D Drawing and Shaded Illustration of the Shouldered Shaft

4.2.5 Example 4-5: Construction Using Sweep along a Defined Trajectory

The solid feature function demonstrated in this example is called **SWEEP**. To employ **SWEEP**, we first define a trajectory along which a cross-section area will be defined. A cane or walking stick is the geometry used to demonstrate this function is shown in Figure 4-52.

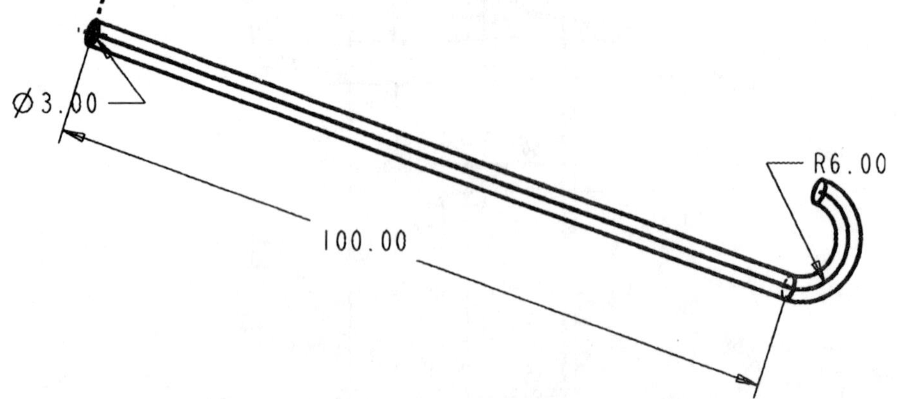

Figure 4-52 Geometry and Dimensions of a Cane.

Following the procedure we have used in the previous examples, begin with the **FEAT** menu:

Create, Solid, Protrusion, Sweep, Solid, Done

To define the trajectory shown in Figure 4-52, choose the **Sketch Traj** option from the **SWEEP TRAJ** menu. Choose **Plane** from **SETUP PLANE** menu and select DTM3 as your Sketching Plane. If the arrow direction is pointed inside, or in the negative z direction, choose **Okay.** Otherwise, choose **Flip**, then **Okay**. From the **SKET VIEW** menu, choose **Top**, then select DTM2 on the main drawing window.

The trajectory consists of a straight line connected to a half circle. Choose **Sketch** from the **SKETCHER** menu and **Line** from the **Geometry** menu. Let's use the y-axis to locate the straight line. Click once on somewhere on the positive y axis to identify the start point of the straight line, and then click on the origin of the coordinate system to locate the end point. Click the middle mouse button to escape from the line-sketching mode. Now use the right mouse button to draw the tangent arc. Click right button at the origin of the coordinate system, then move your mouse. On the drawing window, you will observe an arc extending from the origin. Move the mouse until the end point of the arc meets the x-axis. Click the right mouse button to complete construction of the arc. A wireframe drawing of the cane is illustrated in Figure 4-53.

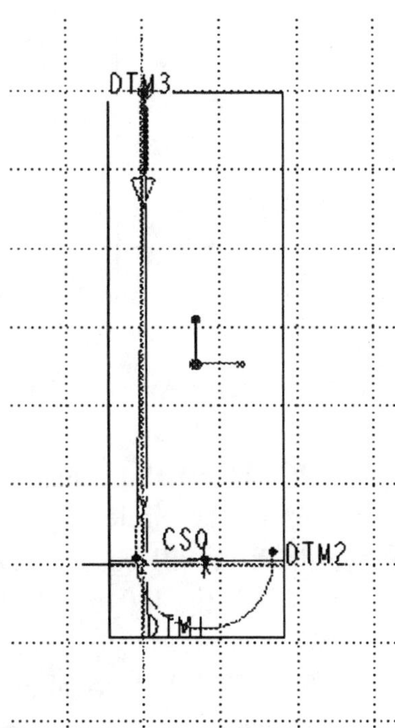

Figure 4-53 A Wireframe View of the Trajectory of a Cane.

Choose **Alignment** from **SKETCHER**, and double click on the y-axis to align the vertical line to DTM1 (which is the y-axis). Also click the end point of the arc to align the point to DTM2 (the x positive axis) to ensure the angle of the arc is 180 degrees.

Choose **Dimension** from **SKETCHER**. Define the dimension for the height of the vertical line and also the radius of the arc. Choose **Regenerate** from the **SKETCHER** menu and select **Modify** from **SKETCHER.** Modify the dimension for the stick, to obtain the size shown in the Figure 4-54. Choose **Regenerate** from the **SKETCHER** menu for the second time. The wireframe sketch in 2D is presented in Figure 4-54.

Choose **Done** from **Sketcher**. Recognizing the fact that the patch, or the trajectory along which the part to be swept is always perpendicular to the trajectory, the Pro/ENGINEER design system automatically switches the current sketch plane to another sketch plane that is used to sketch the cross-section of the sweep part. Obviously this plane is perpendicular to the previous plane which is used for the trajectory sketching.

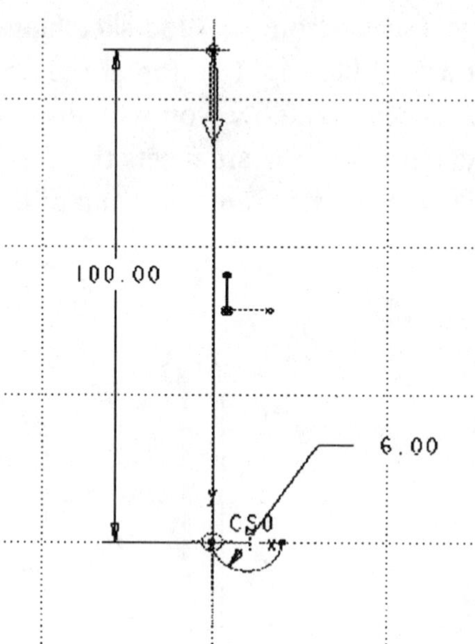

Figure 4-54 Dimensions of the Trajectory

Choose **Sketch** from the **SKETCHER** menu and **Circle** from the **Geometry** menu. Select **Ctr/Point**, to define the center of the circle. Click the origin of the coordinate system, and move the mouse move outward. You will observe a circle appear with the radius varying with the movement of the mouse. Click the left mouse button to complete the sketch of the circle.

Choose **Alignment** from **SKETCHER** to align the circle center to the origin by clicking the circumference of the circle once and then CSO. Choose **Dimension** from **SKETCHER** again to define the diameter of the circle. **Modify** the diameter to give a dimension of 3 and **Regenerate** again. The cross-section shown on the screen of the monitor is presented in Figure 4-55.

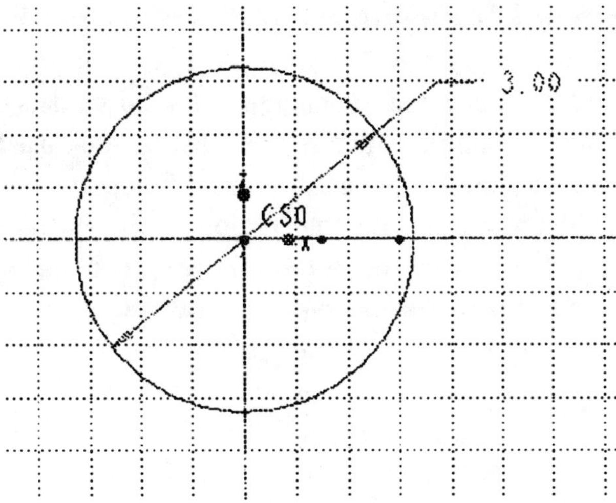

Figure 4-55 Completion of the Cross-section for the Cane

Choose **Okay** from the ***Element Inform*** Window. Rotate the view a little bit and **Zoom in/out** the view. You generate the cane that you created as well as the shaded model, as illustrated in Figure 4-56.

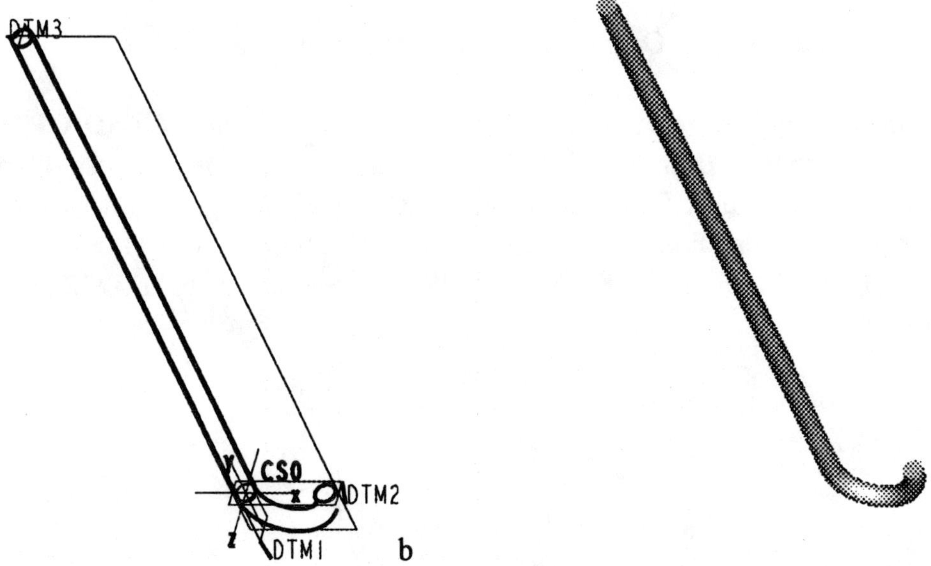

Figure 4-56 Solid Models of the Cane

4.2.6 Example 4-6: Solid Modeling Using the Blending Function

We previously have demonstrated the creation of solid models using **Extrude**, **Revolve** and **Sweep**. There is another function, called **Blend,** that is often used to construct protrusion type solid geometry. In this example, we use the **Blend** function to create the end region of a rod. The geometry and dimensions for the end region of the rod are shown in Figure 4-57. Let's begin the modeling process by selecting the following command sequence, starting from the **FEAT** menu: **Create, Solid, Protrusion, Blend, Solid, Done**

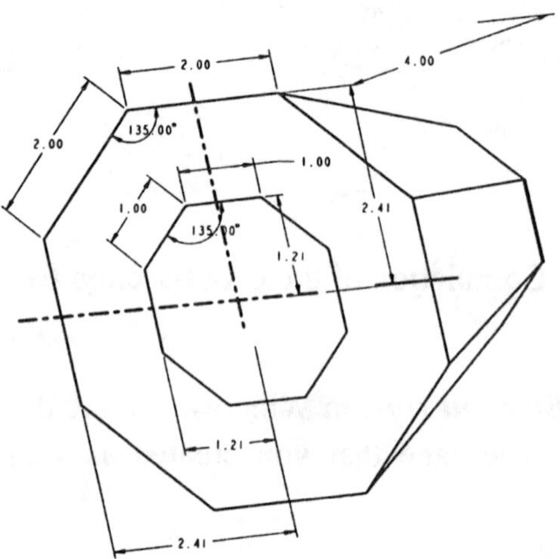

Figure 4-57 Geometry and Dimensions of a Rod End

Choose the **Parallel, Regular Sec, Sketch Sec** option from the **BLEND OPTS** menu, then **Done**. Choose **Smooth** from **ATTRIBUTES** menu, then **Done**. Choose **Plane** from **SETUP PLANE** menu and pick DTM3 as your **Sketching Plane**. If the arrow direction is towards you, or is in the direction along the positive z direction, choose **Okay.** Otherwise, choose **Flip**, then **Okay**. From the **SKET VIEW**, choose **Top**, then select DTM2 on the main drawing window.

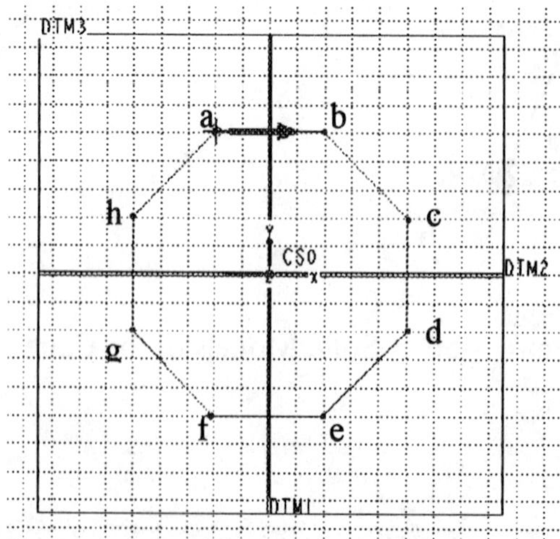

Figure 4-58 Sketched 2D Geometry of the Large Section

First we create an octagon to describe an outline of the cross-section of the larger end of the rod. Choose **Sketch** from the **SKETCHER** and **Line** from **Geometry** menu. We begin to draw the geometry as illustrated in Figure 4-57. The sequence of the line sketching proceeds from: a, b, c, d, e, f, g, and h. Note the lines ab and line ef are both horizontal, and the lines cd and gh are vertical. Choose **Dimension** from **SKETCHER**, to provide dimension lines for the edges of the octagon as shown in Figure 4-59.

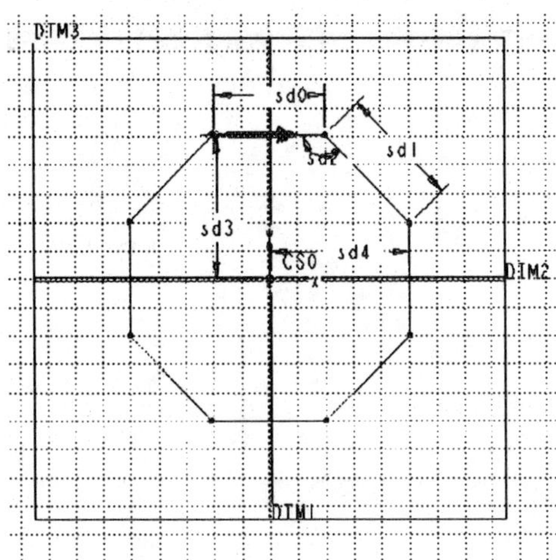

Figure 4-59 Define the Dimensions for the Large Cross-Section

Choose **Regenerate** from the **SKETCHER** menu and select **Modify** from **SKETCHER**. Modify the dimensions for the octagon outline of the larger rod end as shown in the Figure 4-60. Choose **Regenerate** from the **SKETCHER** menu again. The 2D sketch presented in Figure 4-60 is displayed on the screen of the monitor.

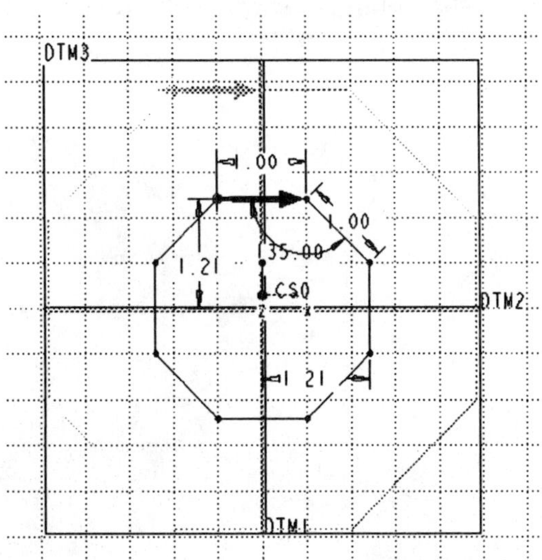

Figure 4-60 2D Sketch and Dimensions of the Cross-Section of the Larger End of the Rod.

In order to construct a blend part, we need to create the cross section for the second end of the rod. Therefore, the second task involves drawing the cross-section for the small end of the rod. Choose **Sec Tools** from **SKETCHER** menu, then **Toggle** from **SEC TOOLS** menu. The first cross section will turn gray and become inactive. Choose **Sketch** and begin to sketch the second cross section. Make certain that the starting point corresponds to the starting point used to draw the first cross section. You may change the starting point by choosing the **Start Point** from **SEC TOOLS** menu. In this example, the geometry of the small end is the same as thc large end, so you can follow the same procedure to draw the second octagon. The only difference between the two 2D sketches is the dimensioning. The dimensions for the small end of the rod are shown in Figure 4-61.

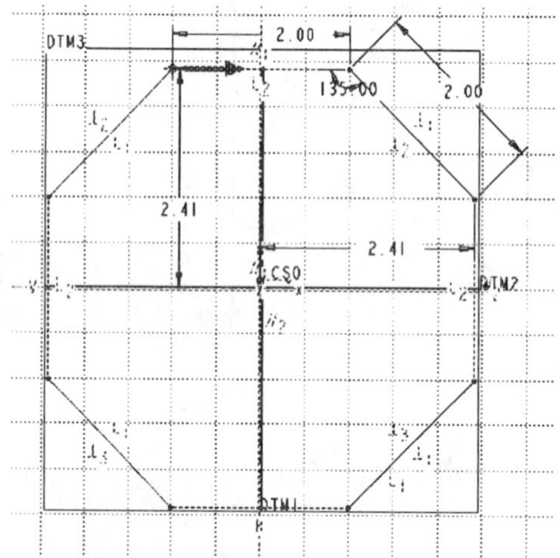

Figure 4-61 Dimensions for the Small Section of the Blend Part

Choose **Done**, then enter the depth, which is the distance between the two ends of the rod. Here we type in 4, and then hit the **Enter key**. Select **OK** from the ***Element Inform*** Window. The complete 3D and shaded illustration of the rod end is displayed in Figure 4-62.

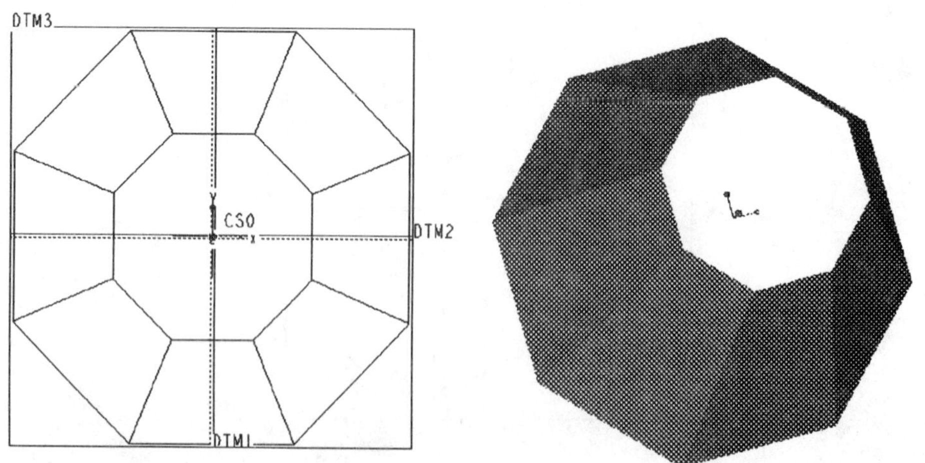

Figure 4-62 Completion of 3D Geometry and Solid Model of a Stick End

4.2.7 Example 4-7: Design of a Walking Stick

Examples 4-1 and 4-3 demonstrated the use of the **Extrude** and **Copy** functions in Pro/ENGINEER. Example 4-4 involved the use of **Revolve**, and Examples 4-5 and 4-6 provide information on employing the **Sweep** and **Blend** functions. In this example, we employ all of these functions to create a more complex walking stick. The geometry and dimensions of this walking stick are illustrated in Figure 4-63. To facilitate the modeling process in this example, we begin with the result for the cane created in Example 4-5.

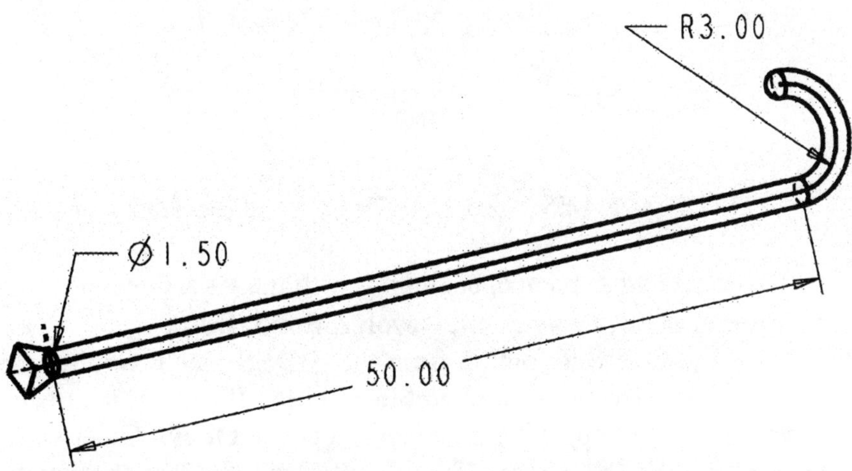

Figure 4-63 Geometry and Dimensions of a Stick

First, the half-spherical feature extended from one end of the stick must be created, with the geometry and dimensions shown in Figure 4-64.

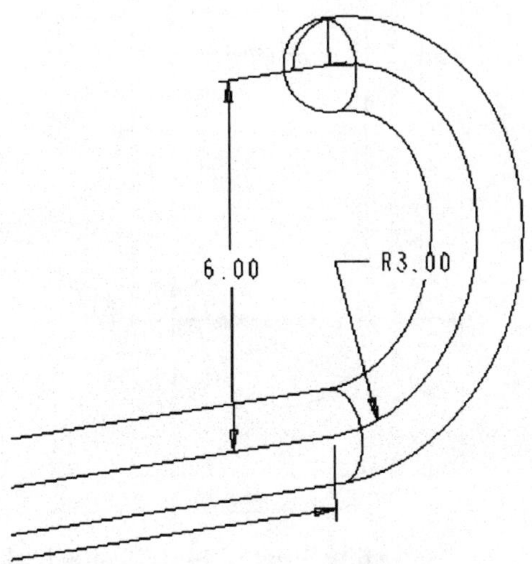

Figure 4-64 Magnified View and Dimensions of the Handle of the Walking Stick

Choose **Pan/Zoom** from **MAIN VIEW** menu and **Zoom-in** from **PAN-ZOOM** menu. Next select two locations to define a box for the zoom area. Click at a preferred location on the screen to define the upper-left corner of the box, and click again at another location to define the bottom-right corner of the box. The image on the screen is presented in Figure 4-65.

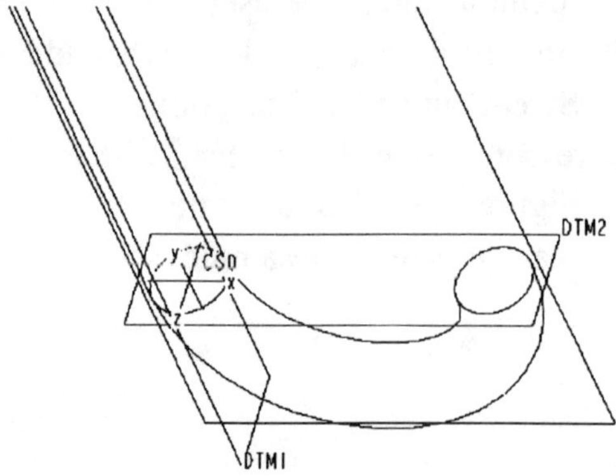

Figure 4-65 Zoom in One End of the Stick Bar

Follow the command sequence, beginning with the **FEAT** menu:

Create, Solid, Protrusion, Revolve, Solid, Done

Now choose the **One Side** option from the **SIDES** menu. Choose **Plane** from **SETUP PLANE** menu and pick DTM3 for the **Sketching Plane**. If the arrow direction is towards you, or is in the direction along the positive z direction, choose **Okay.** Otherwise, choose **Flip**, then **Okay**. From the **SKET VIEW**, choose **Top**, then select DTM2 on the main drawing window.

Next, create a centerline for the revolution feature. Choose **Sketch** from the **SKETCHER** and **Line** from **Geometry** menu. The select **Centerline**, and **Vertical**. The location of the resulting centerline is shown as a dashed line in Figure 4-66.

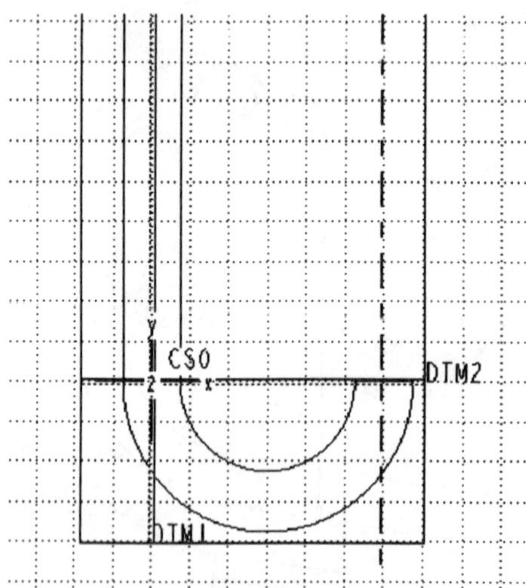

Figure 4-66 Centerline for the Half-Spherical Feature

The second step is to draw the cross-section geometry for the end of the handle, namely a quarter circle. Choose **Sketch** from the **SKETCHER**, **Arc** from **Geometry** menu and **Ctr/Ends** from **ARC TYPE** menu. First select the center point for the arc, which is the intersection of the center line and DTM2. Select point a for the starting location of the arc.

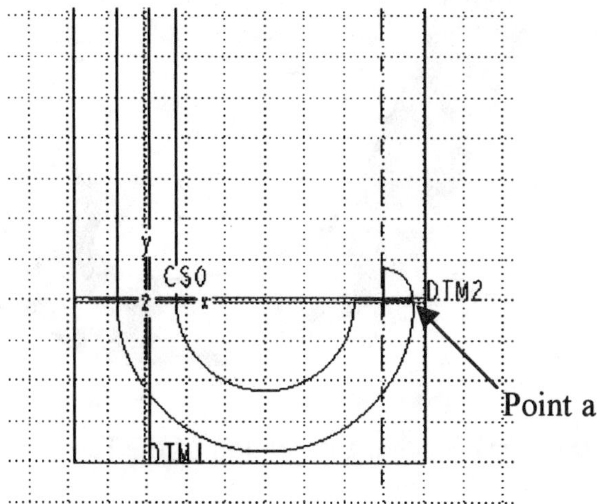

Figure 4-67 Completion of 2D Sketch of the Cross-Section

An arc will appear as you move the mouse. Terminate the arc at a location along the centerline. The next step for the 2D sketch is to close the quarter arc section. To accomplish this sketch two lines to connect the two ends of the arc, one vertical line along the centerline and one horizontal line along DTM2. The completed cross-sectional sketch is shown in Figure 4-67.

Choose **Alignment** from **SKETCHER**, double click the horizontal edge along DTM2 to align the horizontal edge to DTM2, click point *a* then click the outside circle of the handle to align the starting point *a* of the arc to it. Choose **Dimension** from **SKETCHER**. Define the dimension from centerline to DTM1. Choose **Regenerate** and **Modify** from **SKETCH.** Modify the dimension as 12. The 2D sketch on the screen is illustrated in Figure 4-68.

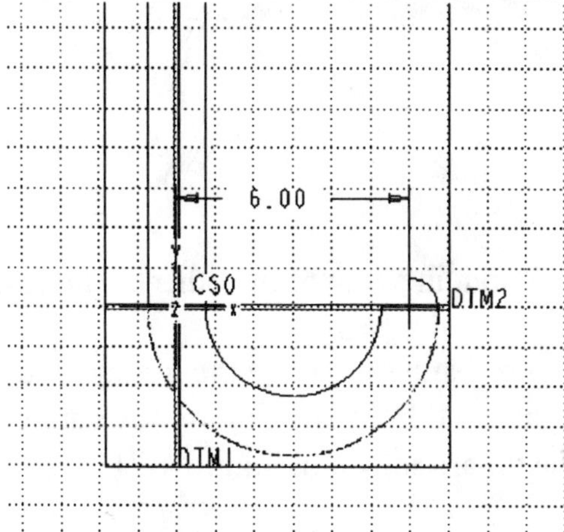

Figure 4-68 Dimensions for the 2D Sketch of the Cross-Section

Choose **Done** from **Sketcher**, and choose the **360** option from the **REV TO** menu and then **Done**. Choose **Okay** from the ***Element Inform*** Window. The handle of the cane appearing on the monitor screen is presented in Figure 4-69.

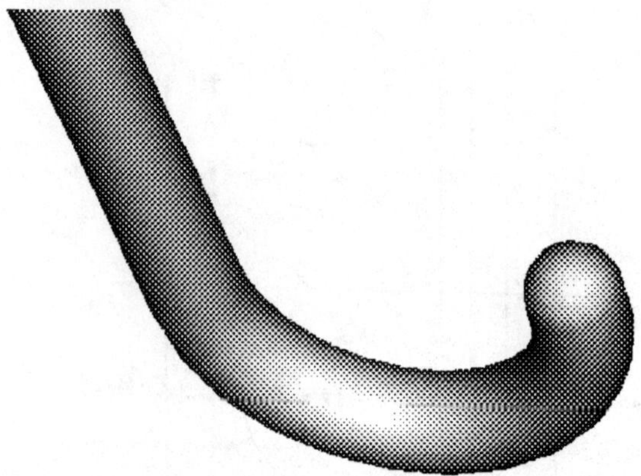

Figure 4-69 3D Shaded Illustration of the Handle of the Cane.

Let's consider the other end of the cane. Use **Zoom in** following the same procedure as we did previously for the handle end of the cane. The image on the screen after zooming in on the ground end of the cane is presented in Figure 4-70.

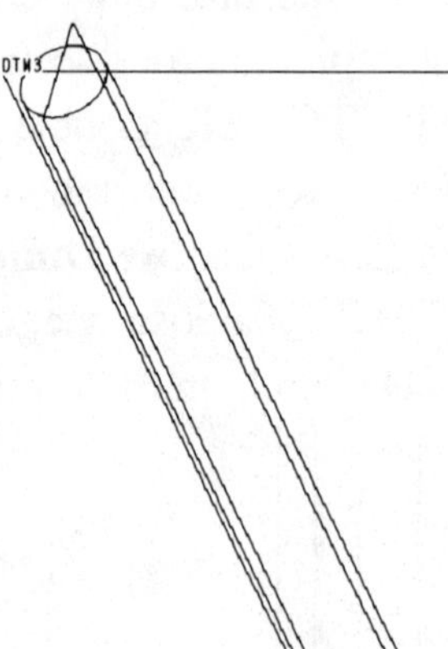

Figure 4-70 Zoom in View of the other End of the Stick Bar

Now we employ the **Blend** feature to create the ground end of the cane. Note that a tip is used on this end to prevent the cane from slipping on the surface of the ground plane. The geometry and dimensions of the tip are shown in Figure 4-71.

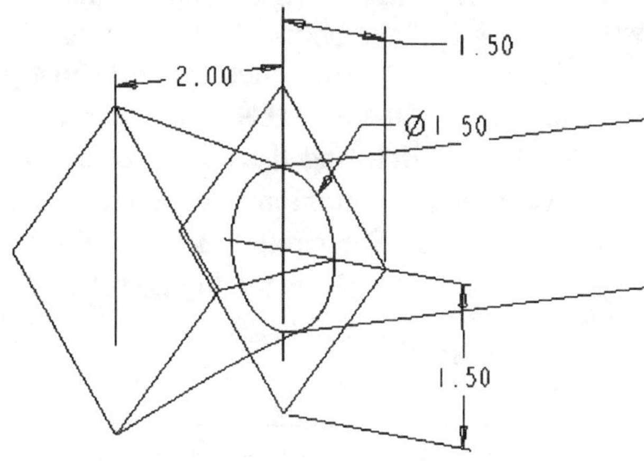

Figure 4-71 Geometry and Dimension of the Tip End of the Cane

Follow the command sequence, beginning with the **FEAT** menu:

Create, Solid, Protrusion, Blend, Solid, Done.

Choose the **Parallel, Regular Sec, Sketch Sec** option from the **BLEND OPTS** menu, then **Done**. Choose **Smooth** from **ATTRIBUTES** menu, then **Done**. Choose **Plane** from **SETUP PLANE** menu and select the end surface of the cane as your **Sketching Plane**. If the arrow direction is pointing out, choose **Okay.** Otherwise, choose **Flip**, then **Okay**. From the **SKET VIEW**, choose **Top**, then select DTM3. The sketch plane appearing on the screen of the monitor is shown in Figure 4-72.

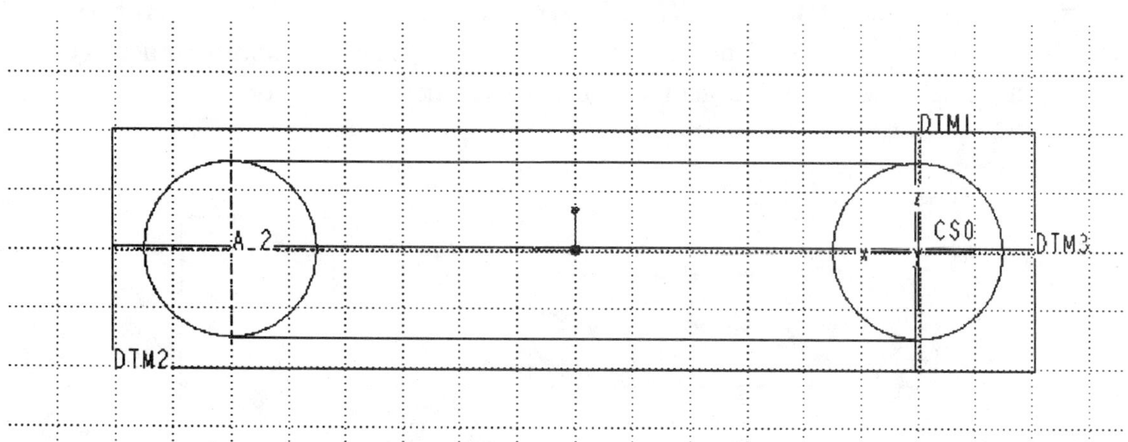

Figure 4-72 Sketch View of the Tip End of the Cane

Choose **Geom Tools** from **SKETCHER** menu and **Use Edge** from the **GEOM TOOLS** menu. Select the upper half and bottom half of the right circle as the geometry for the first cross section for the **Blend** feature. Then click **Done/Return** from the **Use Edge** menu.

Choose **Divide** from the **GEOM TOOLS** menu, and use the left button to click along the upper intersection of the right circle and the DTM1 plane to divide the upper half of the circle into two entities. Also click the bottom intersection of the circle and the DTM1 plane to divide the bottom half into entities. The circle has been divided into four entities. Choose **Regenerate**. The image on the screen is presented in Figure 4-73.

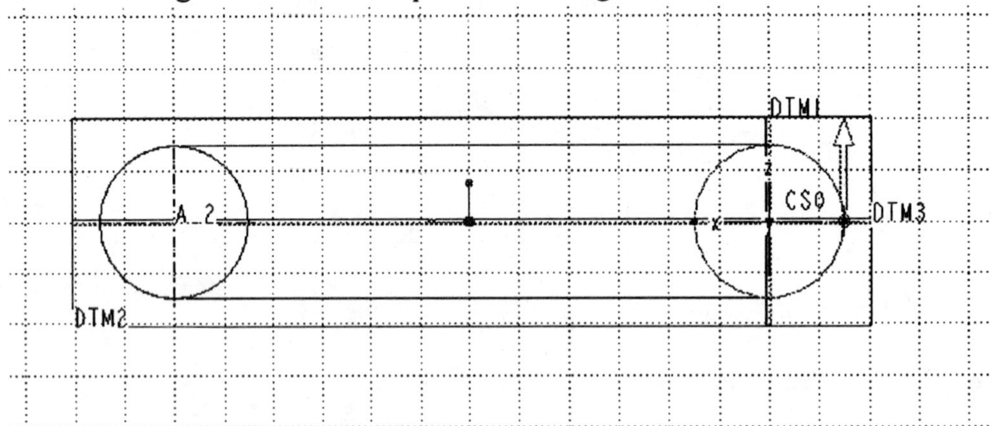

Figure 4-73 First Subsection for the Blend Feature

To create the second cross section for the tip of the cane, choose **Sec Tools** from **SKETCHER** menu, then **Toggle** from **SEC TOOLS** menu. The first cross section will become gray and inactive. Choose **Sketch** and begin to sketch the second cross section. Ensure that its starting point corresponds to the starting point used in drawing the first cross section. You may change the starting point by choosing the **Start Point** from **SEC TOOLS** menu. Choose **Alignment** from the **SKETCHER** menu to align the four corners of the rectangle to the appropriate DTM plane, namely DTM1 or DTM3. **Modify** and **Regenerate** the 2D sketch. The dimensions and geometry for the second cross section are shown in Figure 4-74.

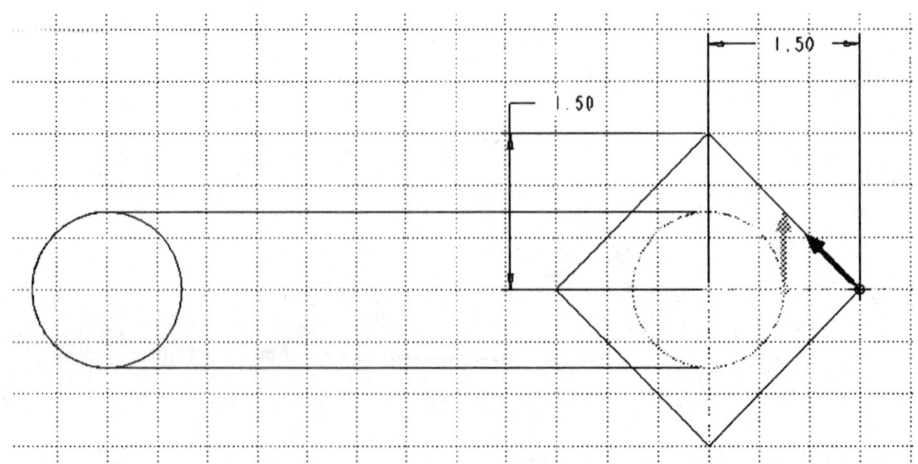

Figure 4-74 Dimensions and Geometry for the Second Cross Section of the Cane Tip.

Choose **Done**, then **Blind** from the **DEPTH** menu. Enter the depth for cross section 2, by typing 4, and **Enter**. Select **OK** for ***Element Inform*** Window. Depress the Ctrl key and use left button to Zoom in and Zoom out. Also you can employ the middle key to rotate the model. A 3D shaded illustration of the complete cane with a rounded end on the handle and a wider tip is presented in Figure 4-75.

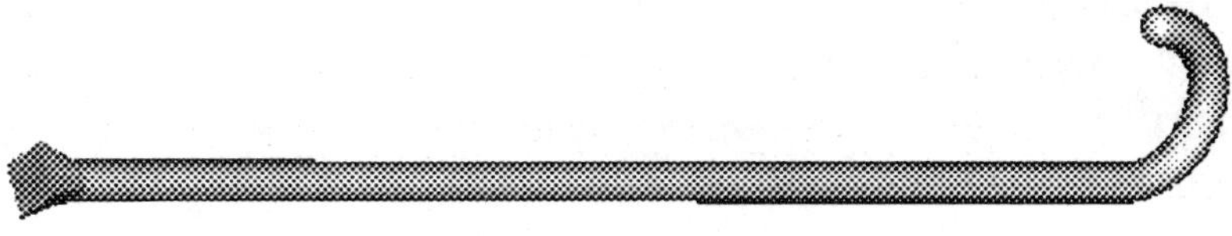

Figure 4-75 Solid Model of the Stick

4.2.8 Example 4-8 : Information Provided by the Pro/ENGINEER System

In addition to the advantages of 3D modeling, Pro/ENGINEER provides the users with a significant amount of useful information for the design and analysis of components. With only a few keystrokes, the user determine information regarding many quantities including mass, volume, center of gravity, the inertia tensor, etc. This is especially useful for designs with irregular or complicated geometry.

In Example 4-8, we demonstrate techniques employed to obtain this information from the data base residing in Pro/ENGINEER. The component considered in this example is the walking stick designed in Example 4-7.

To obtain the information to be used in an engineering analysis, units of length and weight need to be established for the component or the assembly. In this example, we use in. (inch) as the unit for length and lb (pound) as the unit for weight. To define these units, under **PART** menu, choose **Set up, Units, Length, Inch, Done, Same Dims** and **Done**. Select **Mass, Pound** and **Done**. Select **Density,** and enter the density of the material (timber) as 0.023 lb/in^3.

First, let us seek information describing the complete component. From the **MAIN** menu, choose **Info**. From the **INFO** menu, choose **Mass Props, Part MP**. Hit the **Enter** key if you are satisfied with the default value for the relative accuracy, which is 1.0e-5. Since the numerical values determined for many quantities depend on the coordinate system, it should be established and located before beginning to seek information from the data base. In this example, we select CSO as the coordinate system. Under the Pro/ENGINEER design environment, data defining the quantities listed in Table 4-1 are available.

Table 4-1 Information Available about the Components

```
MASS PROPERTIES OF THE PART  ---  Cane
              VOLUME =  1.1198378e+02 INCH^3
        SURFACE AREA =  3.0135657e+02 INCH^2
             DENSITY =  2.3000000e-02 POUND / INCH^3
                MASS =  2.5756270e+00 POUND
    CENTER OF GRAVITY with respect to CS0 coordinate frame:
X  Y  Z       4.9352551e-01  2.2219930e+01 -2.6351238e-05  INCH
    INERTIA with respect to CS0 coordinate frame:  (POUND * INCH^2)
INERTIA TENSOR:
Ixx Ixy Ixz     2.0620848e+03  2.1939176e+00  2.2571187e-05
Iyx Iyy Iyz     2.1939176e+00  6.6882365e+00  1.6774149e-03
Izx Izy Izz     2.2571187e-05  1.6774149e-03  2.0680148e+03
INERTIA at CENTER OF GRAVITY with respect to CS0 coordinate frame: (POUND * INCH^2)
INERTIA TENSOR:
Ixx Ixy Ixz     7.9043255e+02  3.0438507e+01 -1.0924863e-05
Iyx Iyy Iyz     3.0438507e+01  6.0608977e+00  1.6932695e-04
Izx Izy Izz    -1.0924863e-05  1.6932695e-04  7.9573520e+02
    PRINCIPAL MOMENTS OF INERTIA:  (POUND * INCH^2)
I1  I2  I3      4.8814674e+00  7.9161198e+02  7.9573520e+02
    ROTATION MATRIX from CS0 orientation to PRINCIPAL AXES:
```

In conducting an engineering analysis, it is essential that information about the critical cross-section be accessible. Pro/ENGINEER is capable of providing this information.

To determine information concerning a specific cross-section, we must first create this cross section at the location of interest. In this example, a cross-section of interest is located at the intersection of the body of the cane and its tip.

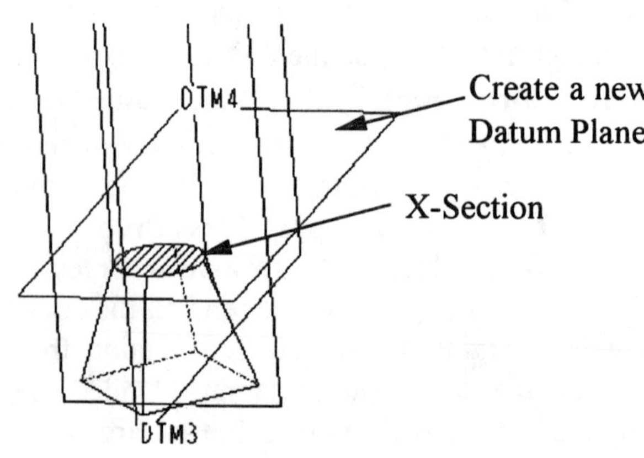

Figure 4-76 Creating the Cross Section Prior to Determining Properties

To create the cross section, we must establish a datum plane (DTM4) which passes through the specified location. Then under **PART** menu, choose **X-section, create, Model, Planar, Single, Done**. Next, name the cross-section --- S. Select DTM4. The cross section will be displayed with crosshatch lines. If you want to change attributes of the crosshatch lines, choose **Modify**, select the name of the cross-section --- S, and choose **Spacing** or **Angle** to modify the crosshatching.

It is now possible to determine the properties of this cross section from the Pro/ENGINEER data base. From the **MAIN** menu, choose **Info, Mass Props, X-section MP** and select the name (S1) for the cross-section created. Hit the **Enter** key keeping the default setting for the accuracy. Select CSO as the coordinate system. The information shown in Table 4-2 is displayed.

Table 4-2 Information about the Cross Section

```
MASS PROPERTIES OF THE CROSS SECTION S
                    AREA =  1.7671382e+00 INCH^2
          CENTER OF GRAVITY with respect to CS0 coordinate frame:
X  Y              0.0000000e+00  5.0000000e+01  INCH
INERTIA with respect to CS0 coordinate frame:  (INCH^4)

INERTIA TENSOR:
Ixx Ixy           4.4179928e+03  0.0000000e+00
Iyx Iyy           0.0000000e+00  2.4850378e-01

POLAR MOMENT OF INERTIA:              4.4182413e+03 INCH^4
     INERTIA at CENTER OF GRAVITY with respect to CS0 coordinate frame: (INCH^4)
INERTIA TENSOR:
angles about x  y  z    0.000      0.000      92.219
Iyx Iyy           0.0000000e+00  2.4850378e-01
           AREA MOMENTS OF INERTIA with respect to PRINCIPAL AXES:  (INCH^4)
I1  I2            1.4726152e-01  2.4850378e-01

POLAR MOMENT OF INERTIA:              3.9576530e-01 INCH^4
          ROTATION MATRIX from CS0 orientation to PRINCIPAL AXES:
                  1.00000      0.00000
                  0.00000      1.00000
       ROTATION ANGLE from CS0 orientation to PRINCIPAL AXES (degrees):
about z axis                 0.000
           RADII OF GYRATION with respect to PRINCIPAL AXES:
R1  R2            2.8867513e-01  3.7499998e-01  INCH
```

In addition to the information provided above, Pro/ENGINEER is capable of performing other design analyses, such as the measuring, checking for interference, surface curvature analysis, and identifying differences in components.

4.2.9 Example 4-9: Adding Color to an Object

In the **Part** and **Assembly** modes the color of shaded parts and surfaces can be easily varied. The use of color is recommended to help identify the different components in an assembly or to produce a more realistic view when using a screen-capture option to create a graphics file. Adding color to engineering drawing represents an integration of art and engineering design. Color also facilitates the communication of the function of a design to others. In this example, we employ the sliding basket created in Example 4-1 to demonstrate the technique used to add color to a component.

To begin the process, we first retrieve the part we created in Example 4-1 by using the command sequence: **Main**, **Mode**, **Part**, **Retrieve**. Input the file name of the part you created in this example in the message window. In the drawing window, the part you created is retrieved and displayed.

Generally, the Pro/ENGINEER design system has only one default color --- white. Choose **Environment** from the **MAIN** menu, choose **Shading** and **Done** from the **Environment** menu. The part appearing on the screen is shown in Figure 4-77. Before adding color to the component, we must define another color.

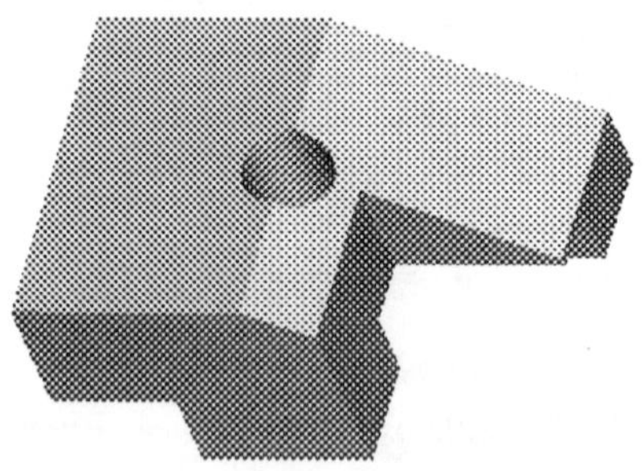

Figure 4-77 Default Shading of the Part

To define a new color, choose **View** from the **MAIN** menu, then choose **Cosmetic** from the **MAIN View** menu. Choose **Appearances** from the **COSM VIEW** menu. Choose **Define** from the **APPEARANCES** menu. Two windows will appear in the screen. The small one called **USER COLOR** shows the current active colors. Before you define your color, recognize that white is the system default color. The larger of the two windows appearing on the screen is the **Appearance Editor** window, which is illustrated in Figure 4-78. It is used to define your new color scheme.

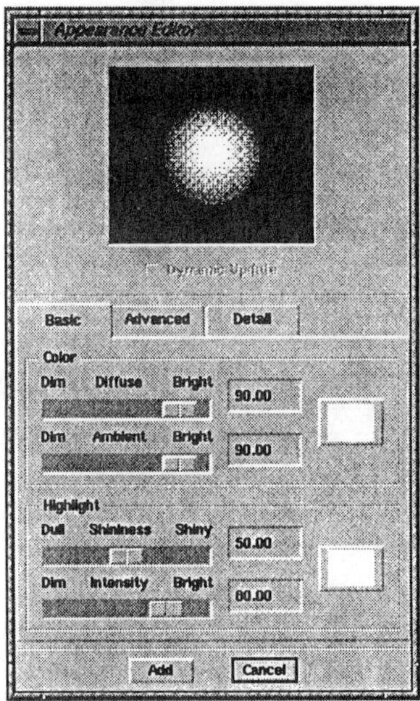

Figure 4-78 Appearance Editor Window

The **Appearance Editor** window is divided into two parts. In the top part, a ball is used to display the new color scheme. The bottom portion of the window is employed to adjust the color and the highlights. Choose the **Basic** tab from the bottom portion of the **Appearance Editor**.

The bottom portion of the **Appearance Editor** window is in turn divided into two areas. The upper portion is used to define color, and the lower region is for specifying the highlight. In this example, we will use **Color** region to define some new colors.

Examining Figure 4-78 indicates two white blocks in the bottom portion of the **Appearance** window. The upper block is for defining the color, the lower one for defining the highlighting. Click on the upper white box, and the **Color Editor** shown in Figure 4-79 appears. The white block in the **Color Editor** window permits you to preview the color being specified. There are two methods for defining color, namely **RGB** and **HSV**. To use the RGB method, click the **RGB** tab in the **Color Editor** to choose this method.

There are three sliding blocks in the **Color Editor** window, namely **R**, **G** and **B**, to adjust the levels of red, green and blue. We know each color can be produced by combinations of red, green and blue. If you use the mouse to move the three sliding blocks to different positions, the color displayed in the block at the top of the **Color Editor** window changes. Adjusting the position of the three sliding blocks, until you achieve the most suitable color for your application.

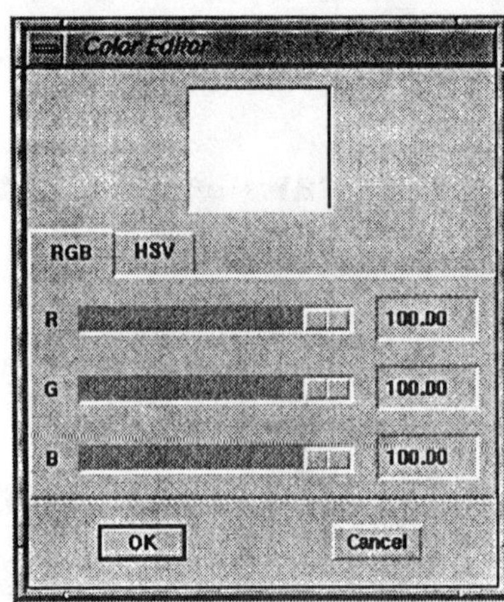

Figure 4-79 Color Editor Window

After complete the adjustments giving the most appropriate color, click **OK** to return to the **Appearance Editor** window. Click **Add**, and the color you have defined is added to the **USER COLOR** window. In this example, let's define green as a color. After you have defined all the colors you want, click **Cancel** from the **Appearance Editor** window to return to the assembly window.

Now, we may add color to our part. Choose **Set** from the **APPEARANCES** menu, and the **USER COLOR** window appears. Choose the green color from the **USER COLOR** window, and a new menu named **OBJECT** appears under the **APPEARANCE** menu. Choose **Part** from the **OBJECT** menu, and the color of the part turns to green. A colored 3D illustration is presented in Figure 4-80[1].

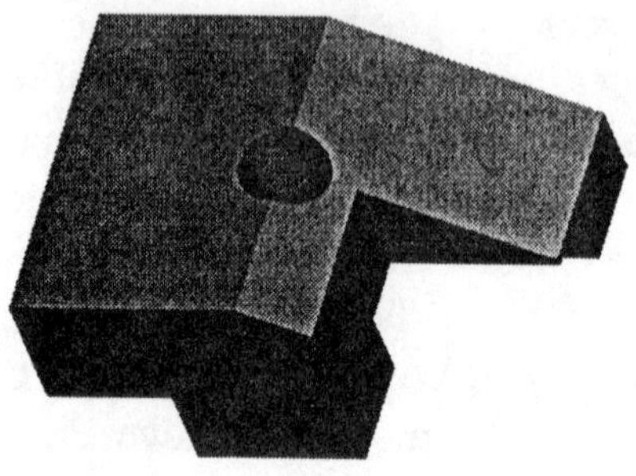

Figure 4-80 A Part with Color

If you want to revoke the color you selected for the component, choose **Unset** from the **APPEARANCES** menu. The **OBJECT** menu appears, and you choose **Part** from the **OBJECT** menu, and the color of the part returns to its original color ---white.

---

[1] The illustration in Figure 80 is not in color. Unfortunately the use of color in a low-volume, low-cost textbook is not currently possible.

4.3 Preparation of Engineering Drawings

4.3.1. Example 4-10: Engineering Drawings of a Housing

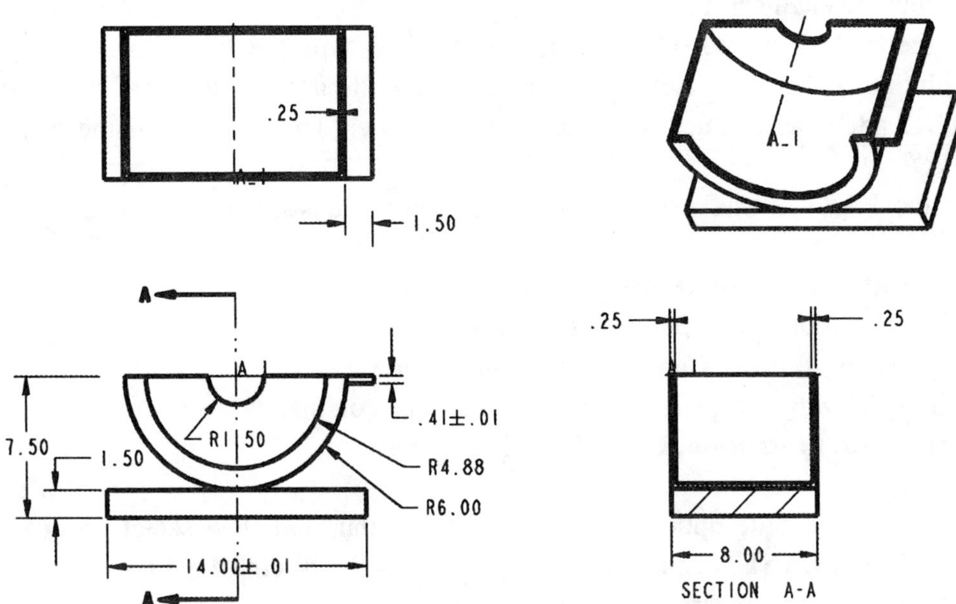

Figure 4-81 Dimensions and Geometry of a Housing (Three Views Shown)

A document representative of information exchange and transmittal in engineering is presented in Figure 4-81. The drawing has three orthogonal projections (views) of a housing. These three orthogonal projections include the front view, the top view and the right side view. To aid in visualizing the geometrical characteristics of the housing, an isometric view of it is also presented. Traditionally a design engineer prepares this type of drawing by:

(1) Make sketches (manually) in 2D space using a pencil on a piece of paper.
(2) Sketch an isometric view of the component in 3D space using a pencil and paper to confirm his or her design intentions.
(3) Make modifications to the original design, by adding dimensions.
(4) Return to the office and use drawing instruments to prepare the engineering document, as illustrated in Figure 4-81.

The Pro/ENGINEER design system follows this traditional approach in the preparation of engineering documents. Recall the procedure used in the feature modeling process with the four basic steps:

(1) Sketch a part geometry in 2D space.
(2) Dimension the 2D sketch by aligning the sketch with the axes and adding dimensions.
(3) Modify the dimensions and regenerate a new 2D sketch.

(4) Transform the 2D sketch to a solid model by adding the third dimension.

In this example, we will present the procedure to prepare the engineering drawings using Pro/ENGINEER. We begin by creating a file, then modeling the part geometry, and finally generating the engineering drawings.

From the **MODE** menu, select **Part,** then **Create**, type the file name in the message window, and hit the **Enter** key. Let's use *housing* as the file name. Note that the file name is saved with .PRT added automatically to the assigned name. The complete name of the file is *HOUSING.PRT*.

From the PART menu, use the following command sequence to create a reference system.

Feature, Create, Datum, Plane, Default

To create the coordinate system of the drawing, use the command sequence:

Create, Datum, Coord Sys, Default, Done

To begin drawing the part, select the following commands starting from the **FEAT** menu: **Create, Solid, Protrusion, Extrude, Solid, Done**

Choose the **One Side** option from the **SIDES** menu, and then select **Done**. Choose **Plane** from **SETUP PLANE** menu and select DTM 3 as the **Sketching Plane**. Select **Okay** when the arrow direction is towards you.

From the **SKET VIEW**, select **Top**, then choose DTM2 on the main drawing window. The 2D drawing plane and axes appear on the monitor screen as indicated in Figure 4-82.

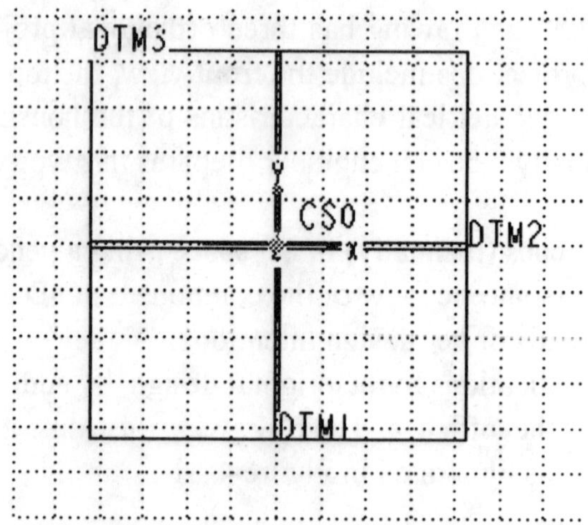

Figure 4-82 Define the Orientation of a Sketch Plane

In the process of modeling the housing, the first feature to create is a solid semi-circular cylinder with a diameter of 12 and a depth of 8. Choose **Sketch** from the **SKETCHER** menu and **Line** from the **Geometry** menu. Select two points on the X axis and align the line with the X axis. Choose **Arc** from the **Geometry** menu, then **Ctr/Ends** to create an arc about the origin CSO with its two ends terminating on the x axis. Align the arc center with CSO and the two ends with the x axis. Dimension the arc. The display on the monitor screen appears as shown in Figure 4-83(a). Modify the dimension of the diameter by typing 6 in the message window and hit the **Enter** key. After regeneration, assign 8 inches to the depth dimension under the **Blind** command. The model of the semi-cylindrical feature is complete, as illustrated in Figure 4-83(b).

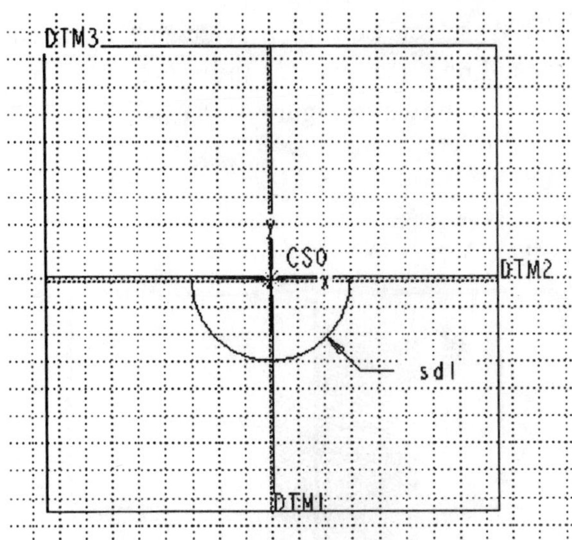

(a) Sketch a Semi-Circle in the 2D Screen

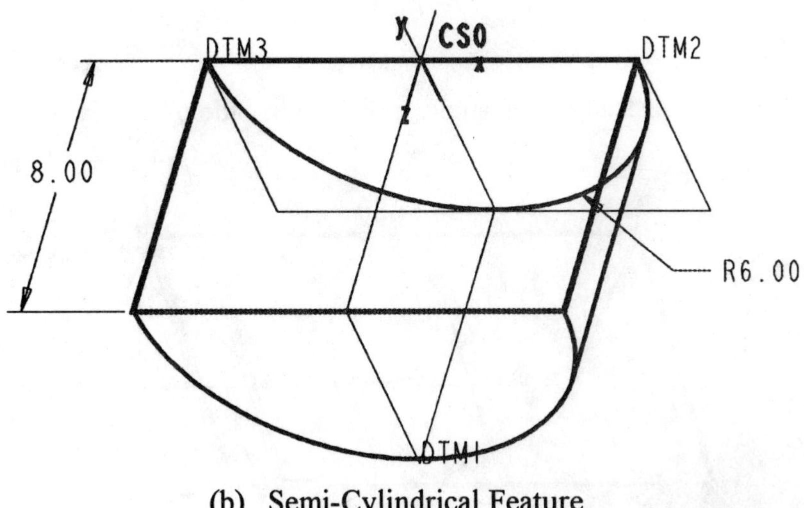

(b) Semi-Cylindrical Feature

Figure 4-83 Creation of a Semi-Cylindrical Feature through Sketching an Arc

The second step in the modeling the housing is to create a cubic volume with dimensions: 1.5 x 8 x 0.41. We will use the top surface of the semi-cylindrical feature as the sketching plane. From the **FEAT** menu select:

Create, Solid, Protrusion, Extrude, Solid, Done

Choose the **One Side** option from the **SIDES** menu, and then select **Done**. Choose **Plane** from **SETUP PLANE** menu and select the top plane of the semi-cylindrical solid as the **Sketching Plane**. Note that the arrow direction is upward, which is opposite to the direction for adding the cubic solid. Therefore, select **Flip** and then **Okay** to ensure that the arrow direction is toward the inside of the semi-cylindrical solid.

From the **SKET VIEW** select **Top**, then select the front end of the semi-cylindrical feature. The 2D layout of the cylinder is presented in Figure 4-84. You are to add the rectangle to the right side of the cylinder as shown in Figure 4-84a.

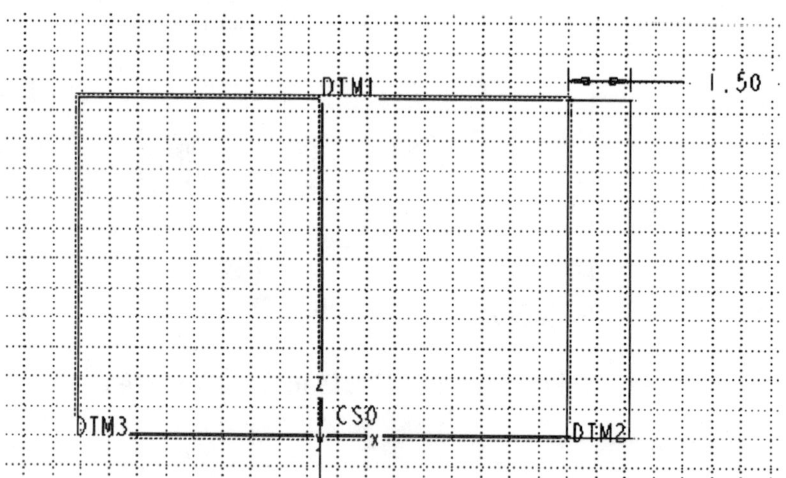

(a) Construct a Rectangle on the Right Side of the Cylinder

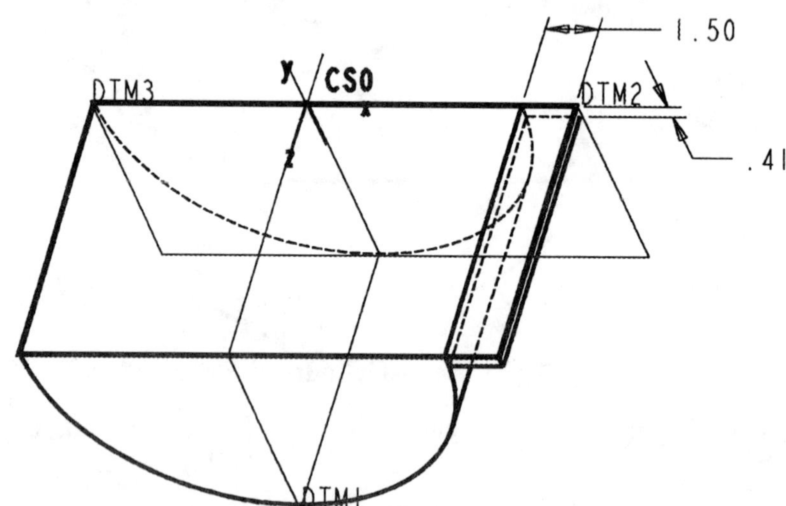

(b) The Cubic Solid Is Positioned on the Right Side

Figure 4-84 Creation of a Cubic Solid on the Right Side of the Semi-Cylindrical Solid

Choose **Sketch** from the **SKETCHER** and **Rectangle** from **Geometry** menu. Select two points on the 2D screen to indicate the diagonal of the rectangle. Make three alignments, with the top, bottom and right side of the rectangle.

Choose **Dimension** from **SKETCHER** to define the width, then **Regenerate** and **Modify** the dimension using 1.5, and **Regenerate** again. Under the **Blind** command, input a depth of 0.41 to complete the construction, as shown in Figure 4-84b.

The third step in modeling the housing is the construction of the base plate. It is a block with dimensions of 14 x 8 x 1.5. From the **FEAT** menu, exercise the command sequence:

Create, Solid, Protrusion, Extrude, Solid, Done

Choose the **One Side** option from the **SIDES** menu, then select **Done**. Choose **Plane** from **SETUP PLANE** menu and select the front face of the semi-cylindrical feature as the **Sketching Plane**. Note that the arrow direction is towards you or out of the computer screen, which is opposite to the direction for adding the base plate feature. Therefore, select **Flip** and then **Okay** to ensure that the arrow direction is towards the inside of the semi-cylindrical solid.

From the **SKET VIEW** select **Top**, then select the top (flat) surface of the semi-cylindrical feature. The 2D layout shown in Figure 4-85a appears on the monitor screen. Add a rectangle to the bottom of the cylindrical solid as shown below.

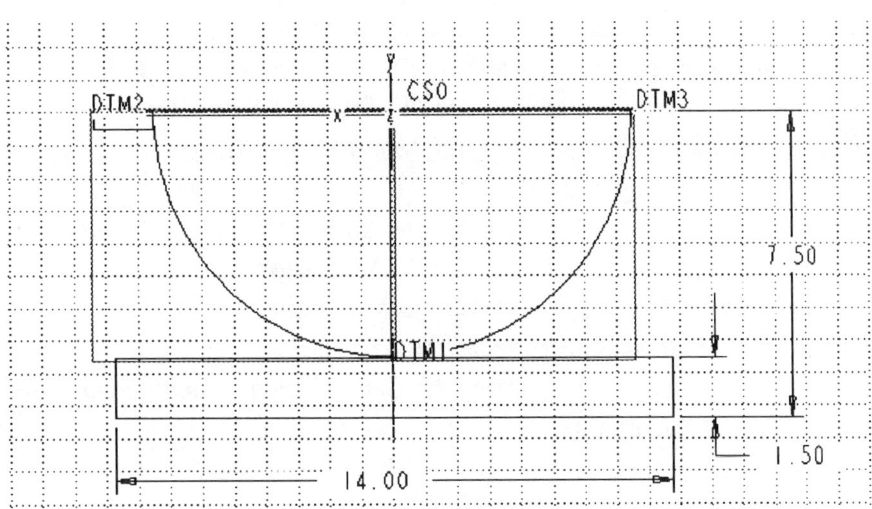

(a) Constructing a Rectangular Solid Attached to the Bottom of the Semi-Cylindrical Solid

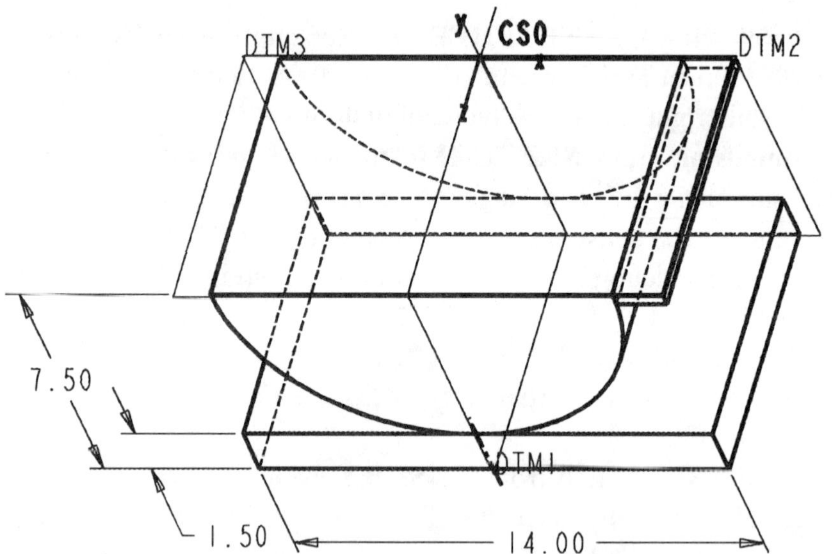

(b) The Solid Base Plate Added to the Semi-cylindrical Solid

Figure 4-85 Creation of the Base Plate Feature

Choose **Sketch** from the **SKETCHER** and **Rectangle** from **Geometry** menu. Select two points on the 2D screen to indicate the diagonal of the rectangle. Choose **Dimension** from **SKETCHER** to define the width and height of the rectangle. Two dimensions for positioning the rectangle are also needed, as illustrated in Figure 4-85a. Select **Regenerate** and **Modify** the dimensions. They are 7.5 and 14 as indicated in Figure 4-85a. **Regenerate** again, and with the **Blind** command, input the depth of 8 to complete the construction, as shown in Figure 4-85b.

The fourth step in modeling the housing is to remove the center of the semi-cylindrical solid to produce a semi-ring with a wall thickness of 0.25 inch. The central region is removed by using **Cut**. In this example, we use REVOLVE to form the cross section of the area to be removed. From the **FEAT** menu, use the command sequence:

Create, Solid, Cut, Revolve, Solid, Done

Choose the **One Side** option from the **SIDES** menu, then select **Done**. Choose **Plane** from the **SETUP PLANE** menu and select the top plane of the semi-cylindrical solid as the **Sketching Plane**. Make sure that the arrow direction is towards the inside of the material, and select **Okay**.

From the **SKET VIEW** menu, select **Bottom**, then select the front surface of the semi-cylindrical feature. The 2D layout, illustrated in Figure 4-86, appears on the monitor screen. To use the **Revolve** command, a centerline is required. From the **SKETCER** menu, select **Line** and **Center Line**. Use the cursor to draw the line along the vertical axis or Z-axis. (Click on the top and bottom ends of the Z axis). Select **Line** and begin your drawing from one end of the Z axis, and terminate it at the other end. The first line is horizontal, the second line is vertical, and the third line is horizontal again ending at the Z-axis (see Figure 4-86a). Hit the middle button of your mouse to escape from the line mode. Use **Alignment** to ensure that the two ends of the lines are positioned on the Z-axis.

Choose **Dimension** from **SKETCHER** to define the thickness at each of the three locations as illustrated in Figure 4-86a. Now choose **Regenerate** from the **SKETCHER** menu and select **Modify** from **SKETCHER.** On the main drawing window, enter the thickness value of 0.25 for each of the three dimensions and **Regenerate** again. The resulting 2D sketch presented in Figure 4-86a appears on the monitor screen.

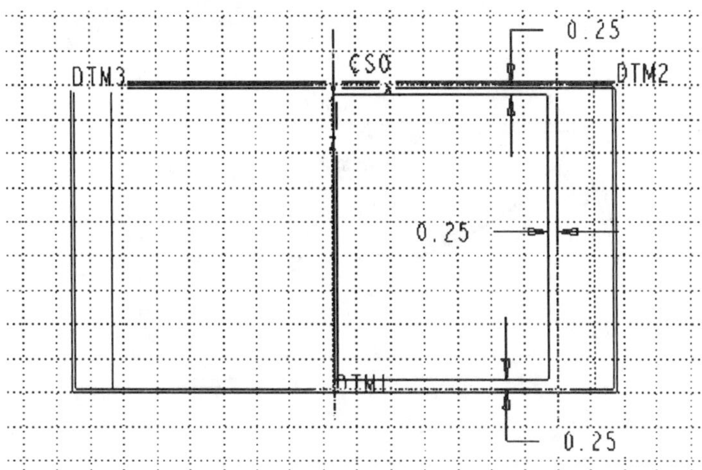

(a) A 2D Sketch of a Rectangle Prior to Performing the Revolving Operation

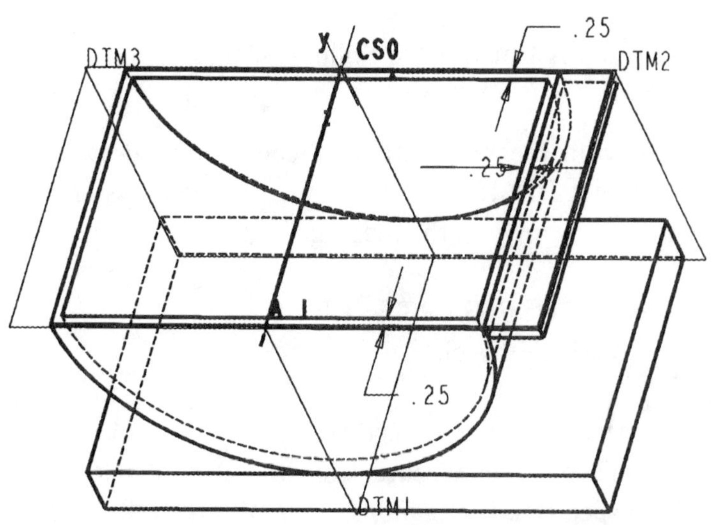

(b) Constructed Hollowed Portion

Figure 4-86 Creation of the Thin Solid Cylindrical Ring

Next, we select **Done** from the **SKETCHER** menu, and confirm the direction for removing material from the semi-cylindrical solid. Because we are in the **Revolve** mode, a new menu called **REV TO** is displayed. Select **180 degrees** and **Done** from the menu. Select **OK** from the ***Element Inform*** window to complete the removal of the center region from the semi cylindrical solid. The 3D- display that appears on the screen is shown in Figure 4-86b.

The fifth step in the solid modeling process involves removal of a semi circular region of material from the front of the housing. The diameter of this semi-circle area is 3. The sixth step is similar to the fifth one except that a annular ring with an inside diameter of 9.75 is attached to the rear of the housing. You are to complete these two tasks independently. The complete solid model of the housing is presented in Figure 4-87.

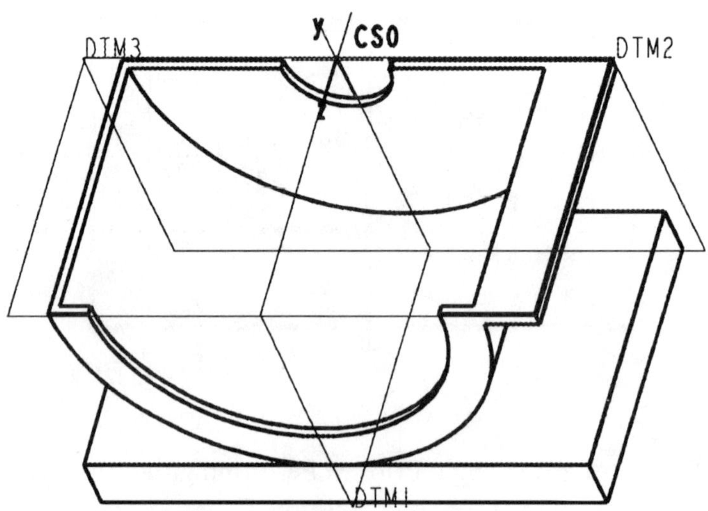

Figure 4-87 A Constructed Solid Model of the Housing Part

Before proceeding with the preparation of engineering drawings, let's save the file. From the **MODE** menu, select **DBMS** (Data base management system) and select **Save** from the **DBMS** menu. A prompt is shown in the message window:

Enter object to save [HOUSING.PRT]: <CR>
Hit the **Enter** key to accept this choice.

To prepare engineering drawings, select **Mode** from the **MAIN** menu. Select **Drawing** from the **MODE** menu. Then select **Create** from the **DRAWING** menu. In the message window, the user needs to specify a file name, say *DRWHOUSING*, and hit the **Enter** key. A new drawing window is shown for preparing the engineering drawings. The user should be aware of the fact that the drawing document is created under a new file name that is different from the original file containing the solid modeling information.

On the **GET FORMAT** menu, select **Retr Format**. In the message window, type *B*, or hit the **Enter** key to accept the default format, which is shown in Figure 4-88.

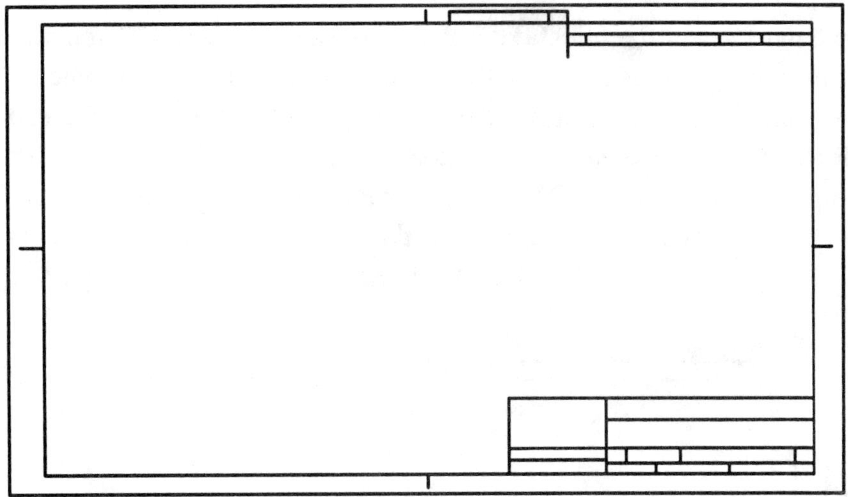

Figure 4-88 The Default Format Used for Preparing Engineering Drawings

To create the engineering drawings, select **View** from the **DRAWING** menu. The user must specify the solid model from which engineering drawings will be made. Type the file name in the message window. In this example, we type *housing*, or hit the Enter key if *HOUSING.PRT* is shown in the message window.

In the **VIEWS** menu, select the following command sequence:

Add Views, General, Full View, No Xsec, Scale, Done

In the message window, "Select **CENTER POINT** for drawing view" is shown. Use the cursor to select a point within the drawing frame to identify the location that you prefer in positioning the front view. After doing so, another prompt is shown in the message window, "Enter scale for view [0.200]." Hit the **Enter** key to accept the scale shown. Upon acceptance, an isometric view of the model is shown in the drawing window and a new menu called "**ORIENTATION**" appears. Figure 4-89 illustrates the display shown on the monitor screen.

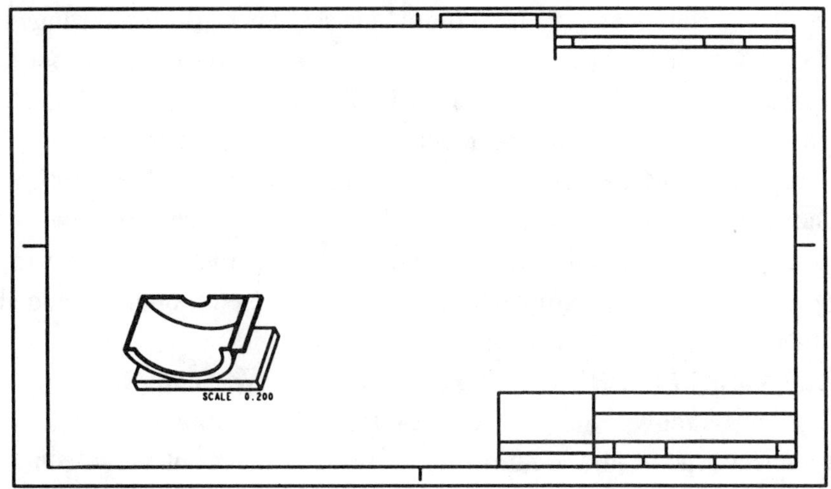

Figure 4-89 Position the Front View on the Drawing Paper

Next, we must select the orientation of the housing to use in he preparation of the front view. In this example, we assume that the front view in the orthographic representation is identical to the front view shown in Figure 4-81. To generate this view, select **Front** from the **ORIENTATION** menu, and then use the cursor to select the front side of the solid model. Select **Top** from the **ORIENTATION** menu. The user is prompted to select DTM 2. This two-step procedure orients the housing so that the front view is generated. Select **Done/Return**, the monitor screen shows the front view presented in Figure 4-90.

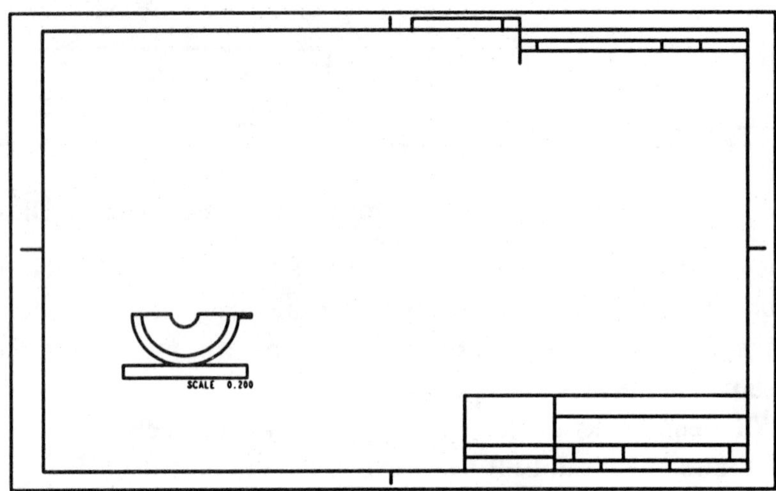

Figure 4-90 The Front View of the Part Model

Let's next prepare the top view and the right-side view. Go back to the **VIEWS** menu, and select commands in the following sequence:

Add Views, *Projection*, Full View, No Xsec, *No Scale*, Done

Note the two differences in the command sequence if comparing the command sequence used in creating the first (front) view. They are ***Projection*** and ***No Scale***.

In the message window, "Select CENTER POINT for drawing view" is shown. Use the cursor to select a point located above the front view because this is the correct location for placement of the top view. Make certain that the distance between this location and the front view currently shown on the screen is appropriate for displaying the top view. Otherwise, you may have to move the top view after it is constructed. On the screen, the top view of the engineering drawing is displayed. Go back to the VIEWS menu again to add the right-sided view following the same procedure of adding the top view:

Add Views, *Projection*, Full View, No Xsec, *No Scale*, Done

In the message window, "Select CENTER POINT for drawing view" is shown again. Use the cursor to select a point at a location to the right of the front view. On the screen, the right-sided view of the engineering drawing is displayed to complete the three orthographic drawings presented in Figure 4-91.

The reader may want to add an isometric view at the top and right corner. Go back to the **VIEWS** menu, select commands in the following sequence:

Add Views, *General*, Full View, No Xsec, *Scale*, Done

Note that we specified ***General***, instead of ***Projection*** in the command sequence.

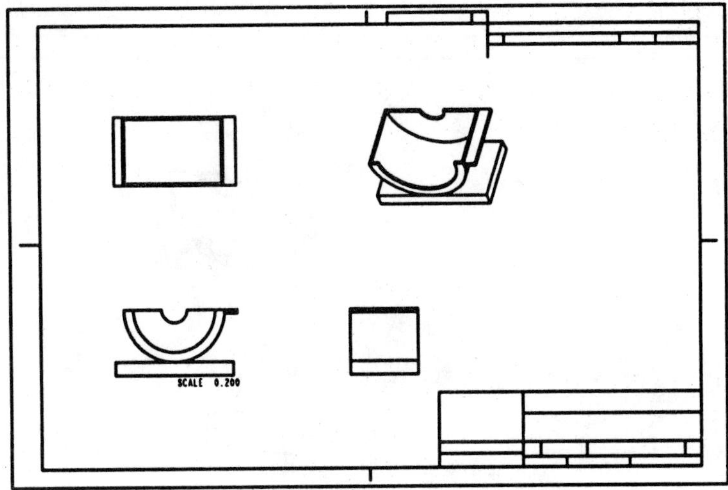

Figure 4-91 Add the Top and Right-Sided View

Figure 4-91 illustrates the three projections of the housing. It is very likely that the three projections initially generated may not be located at the most suitable locations. We usually move the projections within the frame to make the layout more appealing. In the **VIEWS** menu, select **Move Views** and use the cursor to relocate the positions of these projections. You may notice that when moving the front view the other two views move accordingly. When you move the top view or the right-sided view, only the view selected is moved. This action is due to the creation of the front view first. It then became instrumental in locating the positions of the top and right side views as they were created.

In preparing engineering drawings, cross-sectional views are often used to clearly indicate those features that are hidden from the outlined surfaces. Cross-sectional views eliminate the need to use dashed lines and clarify interpretation of the drawing. To create a cross sectional view, we must modify the current projections. Select **Modify View** from the **VIEWS** menu, choose **View Type**, and use the cursor to select the projection to be modified. In this example, let's select the right-sided view. In the **PROJECTION** menu, choose **Section** and **Done**. In the **XSEC TYPE** menu, select **Full**, **Total Xsec**, and **Done**. In the **XSEC ENTER** menu, select **Create**, **Planar**, and **Done**. In the message window, answer the question for labeling the cross section, by typing in the letter A. A prompt in the message window requests the identification of a datum plane and a location to intersect the view in generating the cross section. In this example, let's select the DTM 1 plane and the centerline location in the front view. The prompt in the message window seeks the direction of viewing for the cross-section. Just use the cursor and click the front view, and observe that the right-side view automatically gets cross-sectioned, as illustrated in Figure 4-92.

In the **DRAWING** mode you may not want to observe, or print datum-planes, the coordinate system, etc. To turn off these items, go to **ENVIRONMENT** and remove the checks to those items that apply. To maintain the display on the drawing window updated, use **Repaint** in the **VIEW** menu.

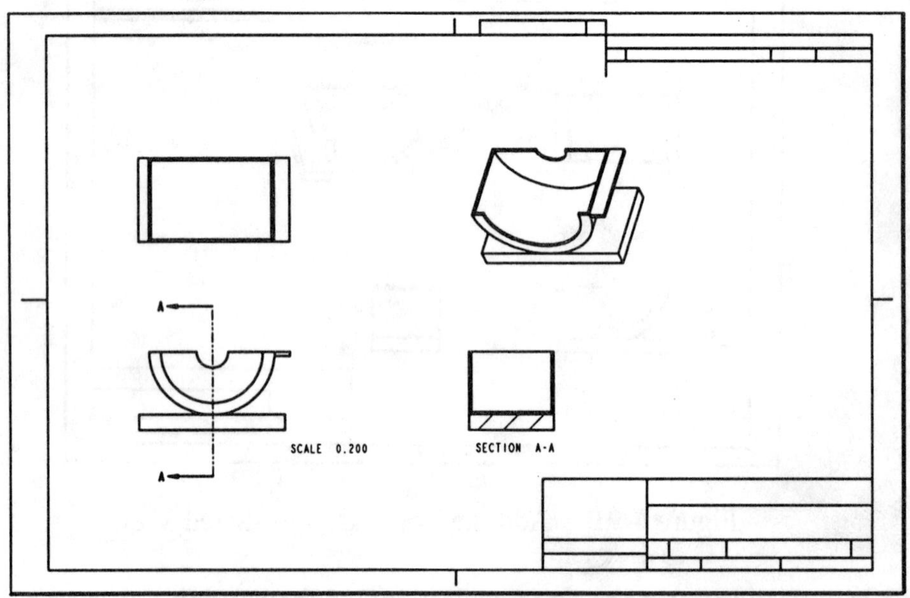

Figure 4-92 Modify the Right Side View with Cross Section

Adding Dimensions to the engineering drawings is a critical operation. The dimensions should be positioned at proper locations to best characterize the feature details. This is extremely for the product realization process, which relies heavily on the preparation of tooling and scheduling of manufacturing facilities. Under the Pro/ENGINEER design environment, the process of adding dimensions is facilitated by using the data base that exists in the file for the solid model.

To add dimensions into the projections, go to the **DRAWING** menu, and exercise the following command sequence:

Detail, Show/Erase, Show, Dimension, Show All, Confirm (Yes), Close

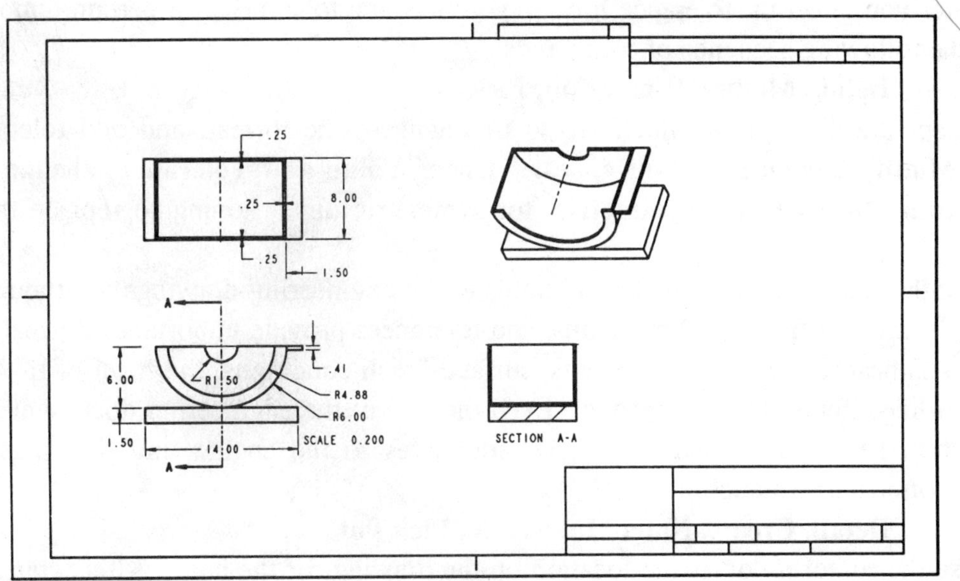

Figure 4-93 Adding Dimensions to the Projections

The dimensions added to the three engineering projections, namely, the front, top and right side views are illustrated in Figure 4-93. All of the dimensions are displayed on your drawing window. It is very likely that the dimensions displayed on your screen may not be displayed as nicely as those shown in Figure 4-93. The **DETAIL** menu has all the options you need to make your drawings look professional. It is very important that you utilize the options available to you to make modifications to improve the appearance of your engineering drawings. To move the dimensions to more suitable locations on the various views, select **Move** from the **DETAIL** menu and use the cursor to select the dimension you want to modify.

In this example, we also present several case studies to demonstrate the use of these options. The first case study involves adding centerlines to the engineering drawings. For bodies that are symmetric, center lines are required. To display centerlines, go to the **DRAWING** menu and exercise the following command sequence:

Detail, Show, Axes, Show All, Confirm (Yes), Done

It is important that you erase some dimensions to avoid redundant dimensioning of engineering drawings. The command sequence to erase dimensions is the same as the command sequence to show dimensions, except that **Erase** is selected, instead of **Show**.

When information on tolerances is needed, the following procedure assists you in adding them to the engineering drawings. From the **DRAWING** menu, select:

Set Up, Drawing Set Up, Retrieve, Set Up Dir

Under the setup direction, select ***iso.dtl***, and then **Quit**. You can select different tolerance types for dimensions in three formats:

Limits: $\left|\begin{matrix}15.01\\14.99\end{matrix}\right|$ Symmetric: 15±0.01 Plus-Minus: $15^{+0.01}_{-0.01}$

After you select the tolerance format, you may add tolerances for certain dimensions by utilizing the following sequence of commands:

Detail, Modify, Dimension, Pick

Select the dimensions which are to be changed, the format, and add tolerances. A window, **Modify Dimensions**, will appear. Under **Value and Tolerance**, change from the default setting: **Normal** to **+-Symmetric**, for symmetric dimensioning to replace the normal format.

Another important issue is to add notes to the engineering drawings. Although the 2D drawings, isometric drawings, dimensions, and tolerances provide important information, other facts, such as heat treatment requirements, surface finish conditions, the form of raw material, and standardized items, are also needed. Even the title of the engineering document should be specified together with the date, etc. To add notes to the engineering drawings, use thc following command sequence:

Detail, Create, Note, Make Note, Pick Pnt

Use the cursor to locate the location on the drawing for the note. After typing the text, hit the **Enter** key twice. Repeat the command sequence given above for additional notes.

The final step in preparing engineering document is to print or plot the drawings. From the **DRAWING** menu exercise the following command sequence:

Interface, Export, Plotter, Size

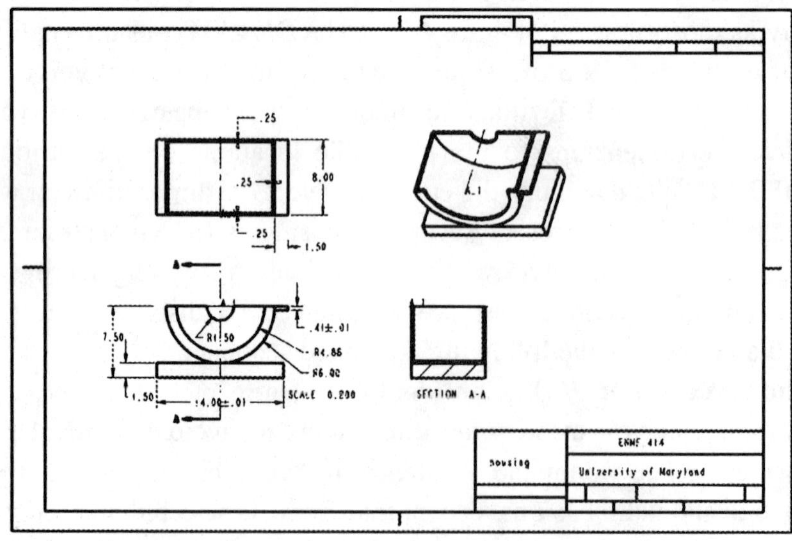

Figure 4-94 An Engineering Drawing with Center Lines and Tolerances Specified

A plotting file is created, and a plotting window appears with the prompt inquiring if it is **OK** to print. Click **OK** and the file is sent to the print spool, and emerges from the printer. The final version of the drawing is illustrated in Figure4-94.

4.3.2. Example 4-11: Engineering Drawings of an Adjusting Bracket

To provide an additional opportunity to explore the capabilities of Pro/ENGINEER IN the preparation of engineering documents, another example is presented. The definition of the adjusting bracket which is the subject of Example 4-11 is presented in Figure 4-95.

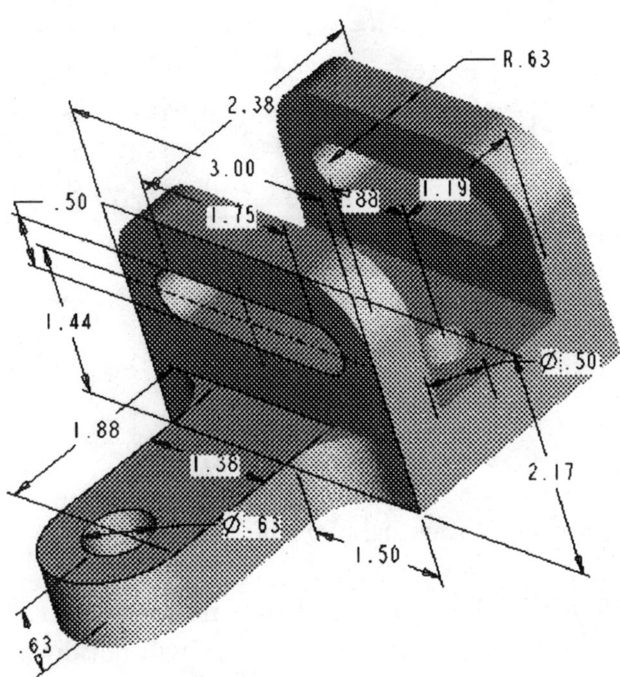

Figure 4-95 Geometry and Dimensions of the Adjusting Bracket

From the FEAT menu, exercise the following command sequence:

Feature, Create, Datum, Plane, Default

To create the coordinate system of the drawing, use the following command sequence:

Create, Datum, Coord Sys, Default, Done

To begin drawing the bracket, select the following commands from the **FEAT** menu:

Create, Solid, Protrusion, Extrude, Solid, Done

Choose the **One Side** option from the **SIDES** menu. Choose **Plane** from the **SETUP PLANE** menu and pick DTM3 as the **Sketching Plane**. Choose the direction so that the protrusion arrow points in the positive Z direction. Then choose **Okay** for the direction. Note that the solid modeling procedure to be demonstrated in this example is an expansion of the coverage presented in the previous examples.

Choose DTM2 as the horizontal referencing plane by picking **Top** in **SKET VIEW** menu and clicking on DTM2. Now you are ready to sketch the shape of the bracket.

Let's first draw the vertical and horizontal centerlines. Choose **Line** from **GEOMETRY** menu and then **Centerline** from ***Line Type***. Draw the lines along DTM1 and DTM2. Click on the top end of DTM1 and click again on the other end of DTM1. Repeat this process with DTM2 by clicking at both ends of the datum.

Choose **Alignment** from **SKETCHER**, select the centerlines (one at a time), and then click on the CSO (the coordinate system of the drawing). You may also align the centerlines to the DTM1 and DTM2, which were used to sketch them. The sketch in the monitor screen appears as indicated in Figure 4-96:

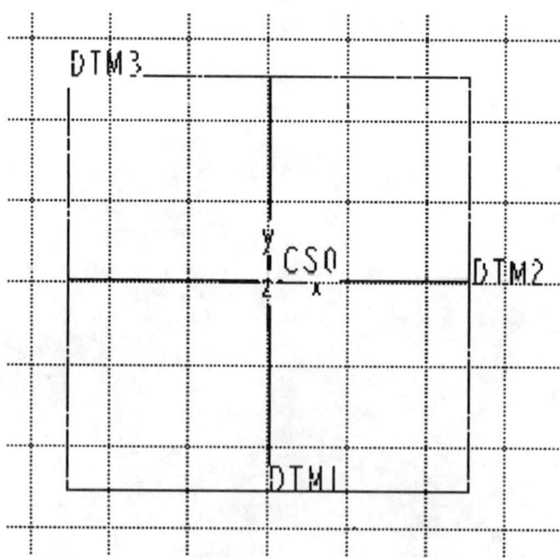

Figure 4-96 Create and Align the Two Center Lines

To create the lip of the bracket, choose **Sketch** from the **SKETCHER** and start to sketch the top view of the front portion of the bracket. We will not draw a circle for the hole since we will drill it later in developing the solid model of the bracket.

Choose **Rectangle** from **GEOMETRY** menu and select the two points defining the diagonal of the box. Pick the points so that the base of the box coincides with the intersection of the DTM1 and DTM2.

Choose **Arc** from **GEOMETRY** and **Ctr/Ends** from **ARC TYPE**. Choose CSO, the intersection of DTM1 and DTM2, as the center of the arc.

Now, select the bottom end of the left vertical line on the box as the starting point for the ARC and the bottom end of the right vertical line as the end point for the ARC. Delete the bottom horizontal line of the box, by selecting **Delete** from **SKETCHER** and clicking on the line.

To provide dimensions for this 2D sketch, **Align** the Arc to DTM1 and to DTM2. (In this case the arc can be **aligned** at the origin of the coordinate system (CSO) since it is centered at that point.). Select **Dimension** from **SKETCHER**, and choose the arc. Note that the dimension of the arc will define the width of the box. Pick one of the vertical lines of the box and locate the dimension at an appropriate position with the MMB. The alignment of the arc establishes the horizontal and vertical lines defining the rectangle.

You now have established all the dimensions necessary to define the geometry and the position of the rectangle and the arc. Click the **Regenerate** function from the **SKETCHER** menu. The prompt in the message window confirms the fact that the regeneration was successful. If this fact is not confirmed, you must add, delete, dimension or align as necessary.

Next modify the dimensions as required. Choose the radius of the arc first and change it to 11/16. Choose the height of the original box and change it to 1+7/8, and regenerate it. The image on the monitor screen at this stage of the development is given in Figure 4-97a. Choose **Done** from **SKETCHER**.

To complete the third dimension, choose the **Blind** option from the **SPEC TO** menu and then **Done**. Enter the depth of 5/8, and choose **Done** from the ***Element Inform*** Window. The monitor screen displays the 3D sketch shown in Figure 4-97b:

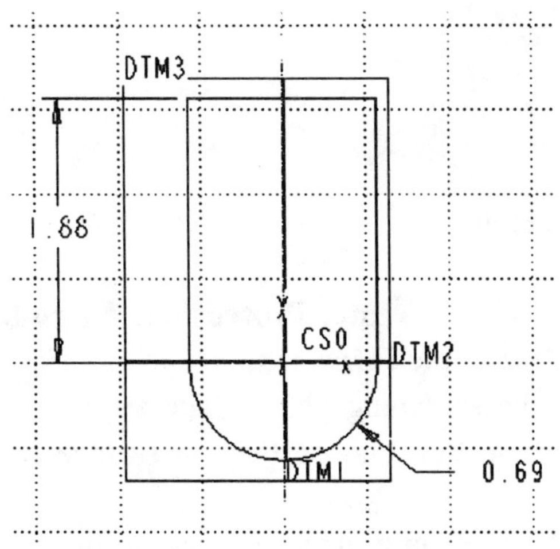

(a) Sketch of the Cross Section

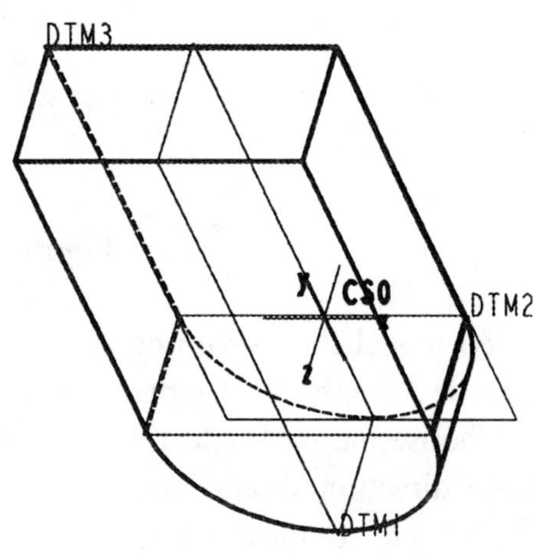

(b) Extruded Cross Section

Figure 4-97 Creation of the Lip for the Bracket

Let's continue the part modeling by drilling a hole at the center of the arc, as illustrated in Figure 4-98. Choose **Create** from the **FEAT** and **Hole** from the **SOLID** menu. Select **Straight**, and **Done**. Choose **Linear** from **PLACEMENT** and then **Done**. Select the top (or DTM3) surface of the model as the placement plane. Choose DTM1 as one of the references for dimensioning, enter 0.00. Choose DTM2 for the second reference and again enter 0.00. These actions will center the hole at the intersection of the two datum planes, completely defining the position of the hole. To drill the hole through the lip, exercise the following sequence of commands.

Choose **One side** from the **SIDE** menu and **Done**.

Choose **Thru All** from **SPEC TO,** and then **Done**. Enter the diameter of 5/8.

Select Preview, and Ok from the ***Element Inform*** window. The hole appears on a 3D sketch of the bracket lip as illustrated in Figure 4-98.

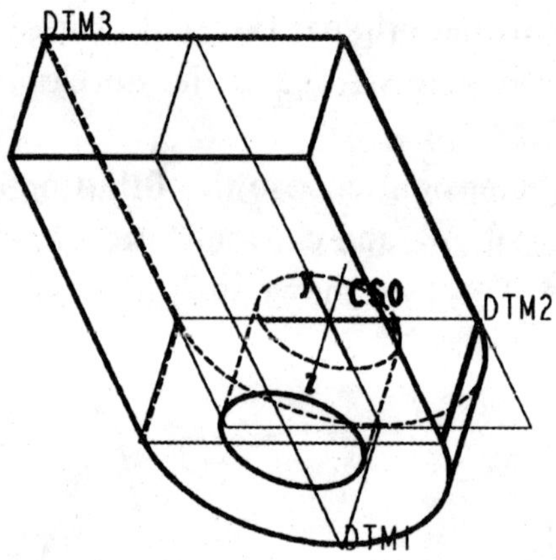

Figure 4-98 Drill the Hole

To model the remaining portion of the bracket, select **Create, Protrusion, Extrude, Solid, Done** from the **FEAT** menu, and from the **SIDES** menu pick **One Side, Done.**

Choose the back (flat) surface of bracket lip as the sketching plane. The arrow head indicating direction should point out of the model. Using **Top** choose the top surface (the DTM3 plane) to complete the sketch.

Select **Sketch**, **Rectangle**, and drag the box in the sketching view. Be certain the base of the new box coincides with the base of the lip. Align the baseline of the box to the base line of the part.

Dimension the height (2+1/6), the width (3.00) and the horizontal position with respect to the DTM1 for the new rectangular solid as 3.00/2 = 1.50.

Regenerate, Modify dimensions, and **Regenerate** again. The display on the monitor screen is given in Figure 4-99a.

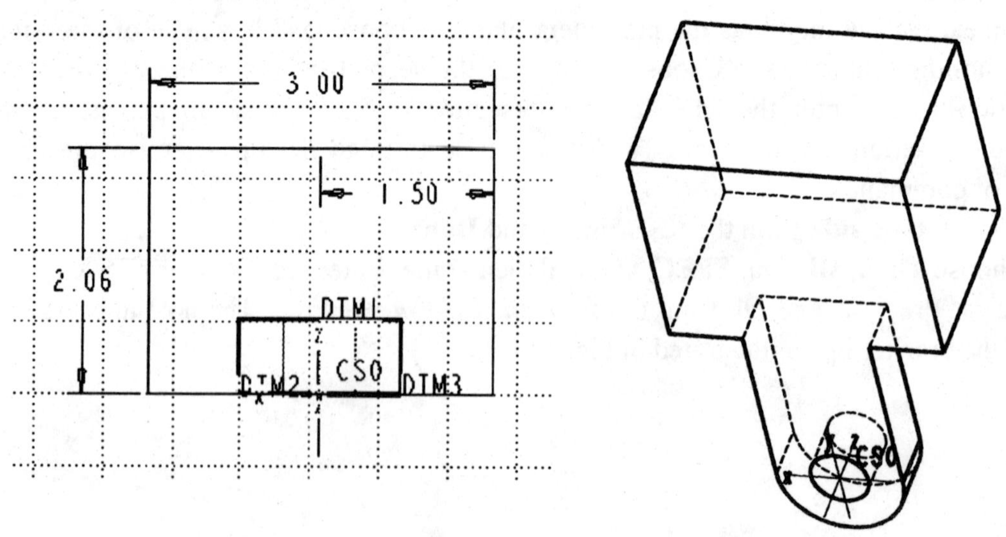

(a) Sketch of the Cross Section (b) Extrude the Cross Section

Figure 4-99 Creation of the Second Half of the Bracket

Choose **Blind,** enter Depth of 2+3/8, and then **Done**. Select **OK** from the ***Element Info*** window. The monitor screen displays the 3D sketch presented in Figure 4-99b.

To create the Rounds on the two edges on the top surface of the bracket, select the following commands from the **FEAT** menu:

Create, Solid, Round, Simple, Done, Constant, Edge Chain, Done

Select the two edges, and then **Done**. Enter the **New Value** of 5/8. Choose **OK** from the ***Element Inform*** window. The rounded corners on the bracket appear as illustrated on the 3D sketch presented in Figure 4-100.

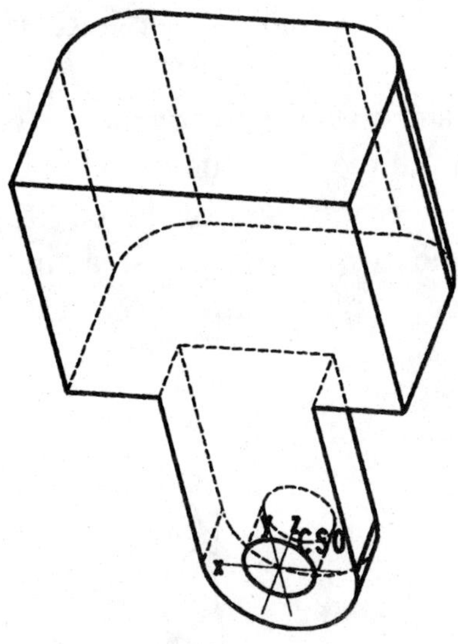

Figure 4-100 Create Round on the Two Edges

To cut away the central portion of the solid model of the second portion of the bracket, we create a new datum plane, DTM4, which is parallel to the base of the bracket. We use the new datum plane to reference the depth of the cut. From the **FEAT** menu select **Create, Datum, Plane, Offset.**

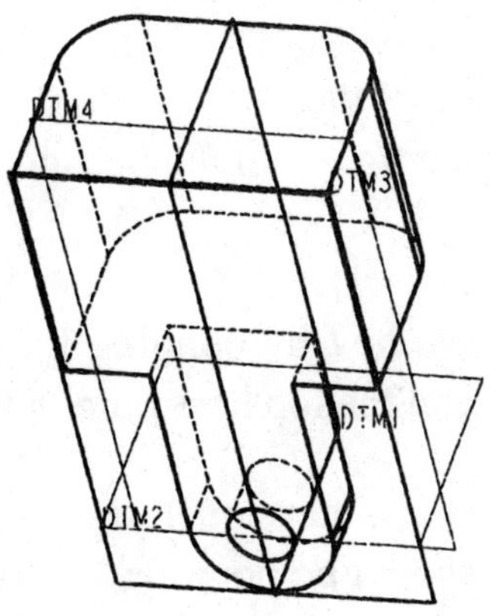

Figure 4-101 Create DTM4

Select the base surface of the part. Choose **Enter Value** from **OFFSET**, and enter the value of 11/16. The sign of your number should be based on the green arrow observed when you choose **Enter Value**. The new datum plane must lie inside the Part. After entering the value of the offset, select **Done**. The appearance of the screen showing the new reference plane is presented in Figure 4-101.

Now that we have defined an additional reference plane, we cut the material from the center of the bracket. From the **FEAT** menu select:

Create, Solid, Cut, Extrude, Solid, Done, One side, Done.

Choose the top surface of the main body of the bracket as the sketching plane. The arrow head indicating the cutting direction should point into the body. Choose either the **Right** or **Left** reference plane for sketching.

Select **Sketch**, **Rectangle** and draw the box on the sketching plane. Make sure the width of the box is the same as the width of the main body and the height of the box is less than that of the main body. Align the two vertical lines of the box to the existing part, by choosing **Alignment**, and then selecting the two lines (one at a time) and aligning them with the appropriate edges. The 2D sketch displayed on the screen is shown in Figure 102a.

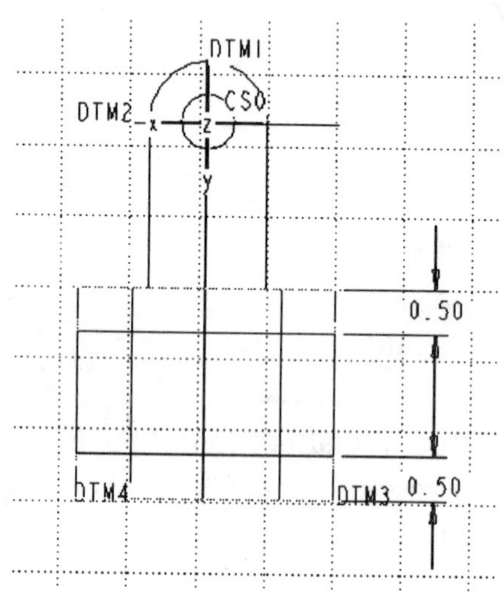

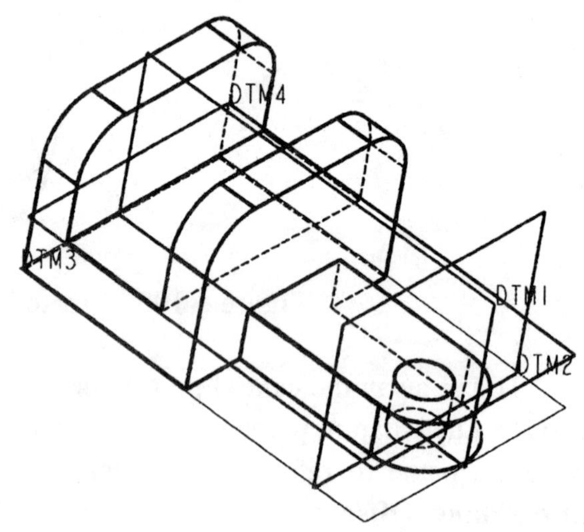

(a) Sketch of the Cross Section of the Cut (b) Extrude the Cross Section

Figure 4-102 Create the Valley in the Second Half of the Bracket.

Now give the dimension of the box by defining the thickness of the walls (1/2) with respect to the base line of the sketch and the top horizontal line of the main body of the bracket.

Regenerate, Modify (the dimensions), **Regenerate, Done**

The Cutting direction arrow should be pointing **INSIDE** the part. Select **Okay**. **Upto Surface, Done**. Select DTM4, and then **OK** from the ***Element Inform*** window. The display on the monitor screen appears as shown in Figure 4-102.

To model the Slots on the sides of the bracket, select:

Feature, Create, Solid, Cut, Extrude, Solid, Done, One Side, Done.

Choose the back plane of the part as your sketching plane, flip the cutting direction arrow as necessary (so that it directed toward the inside the bracket), and then select **Okay**. Pick either **Top**, **Right** or **Left** when selecting your reference plane.

To sketch the slot, select **Sketch**, **Rectangular** and drag a box representing the rectangle. Choose **Arc**, **3 Points** and select the sides near the two corners of one end of the rectangle as the starting and the end point of the arc. Drag the arc until its center is on the vertical lines defining the ends of the box. Repeat this process for the other end of the box, and then delete the two vertical lines.

Dimension the slot using the values given in the 3D model presented in Figure 4-95. (Provide dimensions from the center of one arc to the center of the other; the center of one arc to the top of surface, and the center of an arc to the edge of the side plane)

Select **Regenerate** and **Modify** the dimension as necessary, then **Regenerate,** and **Done**. The arrow head indicating the cutting direction must point toward the inside of the slot. If so, pick **OK,** then **Thru All, Done,** and **OK** from ***Element Inform*** window. The display on the monitor screen for the 2D and 3D drawings appears as shown in Figure 4-103.

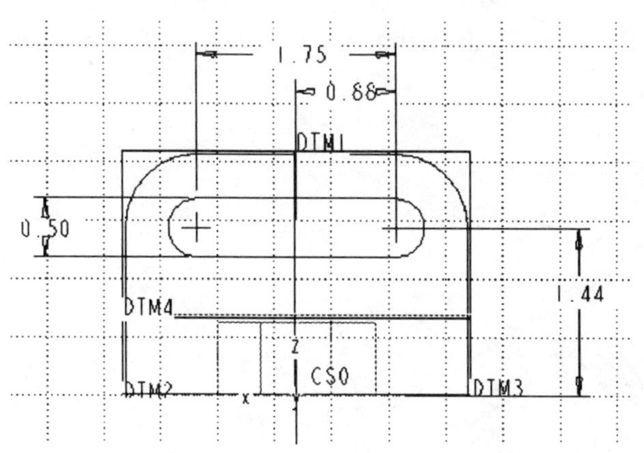

(a) Sketch of the Cross Section

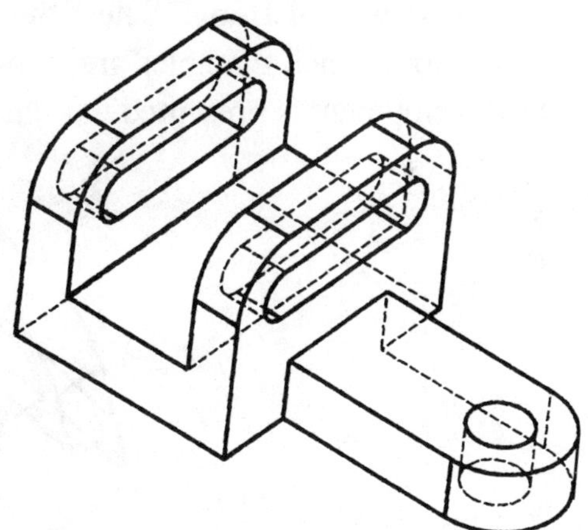

(b) Extrude the Cross Section

Figure 4-103 Cutting the Slot into the Bracket

To drill the two holes located at the base of the cut away region, we use a new method called **COPY & MIRROR**. When using this method, we drill only one hole and use the mirror command to create a second hole on the other side. To drill the first hole, exercise the following sequence:

Create, Hole, Straight, Done, Linear, Done

Select the reference plane DTM4 for the sketching surface. For references, pick DTM 1 and enter the distance of (1 3/4)/2. Pick the back surface of the part, and enter a distance of (2 3/8)/2. Select **One Side, Done, Thru All, Done** and enter a **diameter** of 1/2. **Preview** the hole or just select **OK** from ***Element Inform*** window. A 3D drawing of the bracket showing the first of the two holes is presented in Figure 4-104.

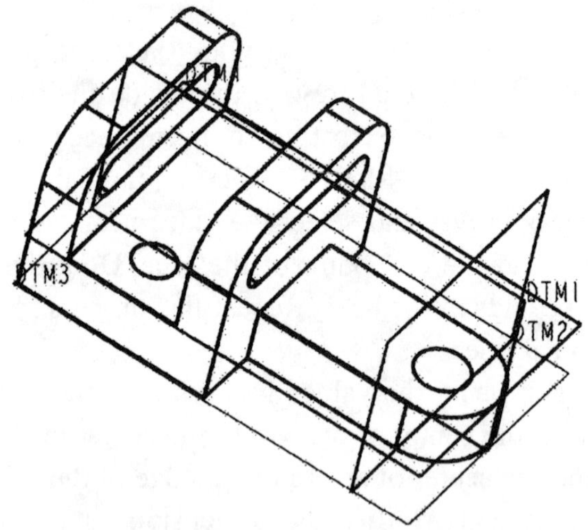

Figure 4-104 Create the First Hole in the Valley

To drill the second hole, choose **Copy** from the **FEAT** menu, and select **Mirror, Select, Independent and Done.** Select the hole just created. **Done Select**, **Done. Pick** DTM 1 as the plane about which the holes are to be mirrored. The 3D drawing of the bracket at this stage of the development is presented in Figure 4-105.

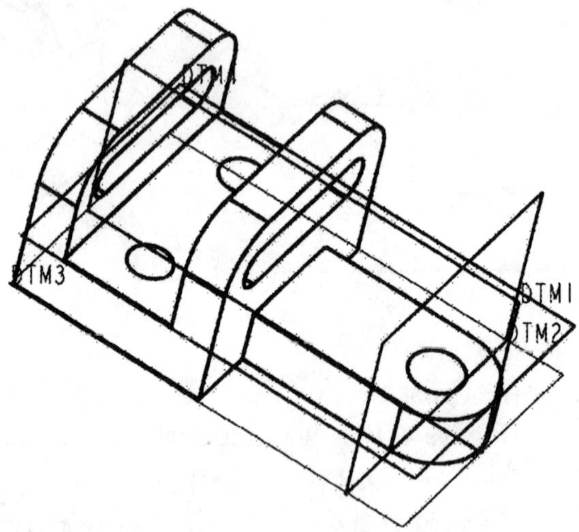

Figure 4-105 Using the Copy Command to Create the Second Hole

To form the fillets where the lip of the bracket blends with the main body, we exercise the following sequence of commands:

Create, Round, Simple, Done; Constant, Surf-Surf, Done.

Pick the front surface of the main body and the right surface of the lip. Enter the **value** of l/2 for the radius of the round. Select **Preview**, and then **OK.** Do the same for the left surface of the lip to complete the solid model of the adjusting bracket. Figure 4-106 illustrates the solid model with all of its features.

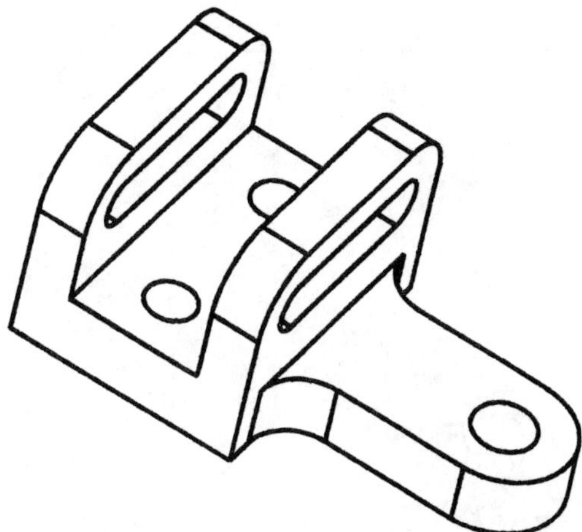

Figure 4-106 An Illustration of the Feature Based Modeling of an Adjusting Bracket

Upon completing and saving the file for the 3D model of the adjusting bracket in the **Part** model, you may continue with the preparation of engineering documentation more commonly employed in design. To begin the preparation of engineering drawings, follow the same procedures described in Example 4-10: Begin with:

Mode, Drawing, Create

In the message window, type in the file name of the part previously created, say *bracket.prt*. The drawing window will appear. First define the drawing format. You may use the default format or if this is not suitable select the **GET FORMAT** menu, and choose **Retr Format**. In the message window, type *B* for the format type.

The following command sequence is used to create the first drawing view:

View, Add Views, General, Full View, No Xsec, Scale, Done

General means a particular view that is independent from other views in the drawing. When you create the first view, the only available selection is **General**. The system first orients the model in the default view orientation that was previously established in the Pro/ENGINEER environment. **Full View** is used to show the shown in its entirety. **No Xsec** means that no cross-sectional views will be displayed. You can select either **Scale** or **No Scale** to specify whether to define scale of the views. We initial will use 0.7 as the scale, but you may change it later by using the **Modify** command under the **VIEW** menu. The message window prompts you to provide the position of the first view or so-called parent view at some location on the drawing frame. Recall this first view defines the relative positions of other projective views. We select the region in the lower left corner of the drawing frame for the position of the first (front) view. The display shown on the screen is illustrated in Figure 4-107.

You may now orient the front view using the command in **ORIENTATION** menu, which is listed in the current command window. Or you may decide to change it later if you have already quit the current menu. Using **Modify** in the **VIEW** menu, select **Reorient** from **MODIFY** menu.

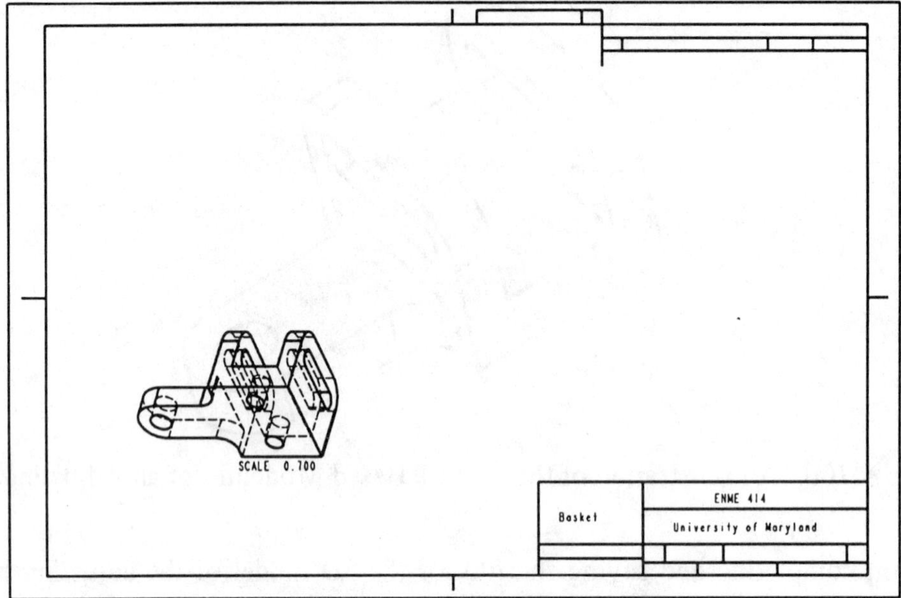

Figure 4-107 Default Orientation of the Front View

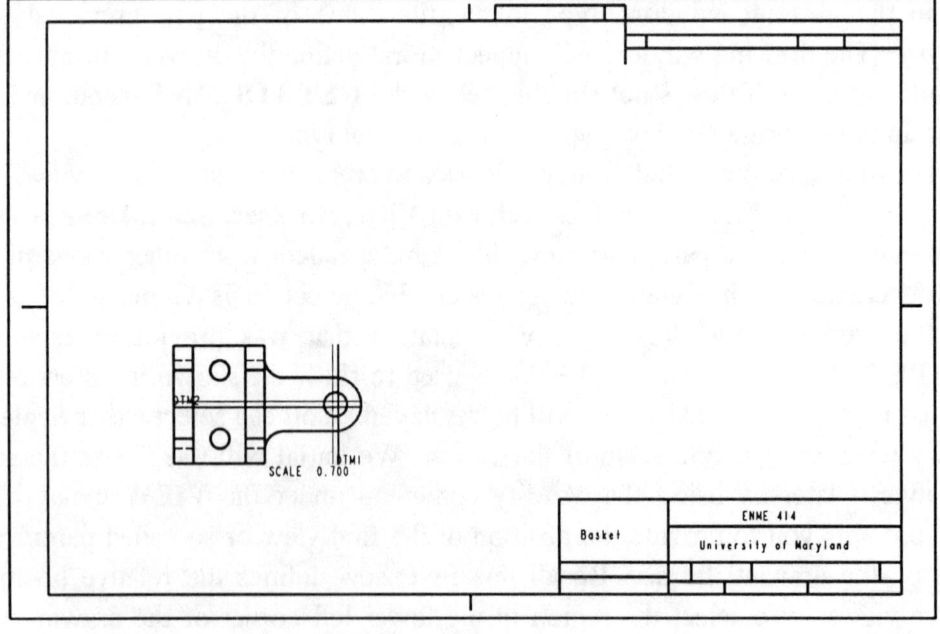

Figure 4-108 The Front View of the Part Model

From the **ORIENTATION** menu, select **Back** and use the cursor to select the bottom side of the bracket. Select **Left** from the **ORIENTATION** menu and use cursor to select the right side of the bracket in the default view orientation window. Pick up **Done/Return**, and the screen depicts an engineering drawing representing the front view (Figure 4-108).

The top-view and right-view are drawn using the following commands: in the **VIEWS** menu, select **Add Views,** ***Projection*****, Full View, No Xsec,** ***No Scale*****, Done**

Projection means a view created from another existed view by projecting the geometry along a horizontal or vertical directions (orthographic projection). Select the center point for the top view somewhere near the top of the of the drawing above the front view. This selection defines the vertical location of the top view. Its horizontal position is already defined by the front view.

Go back to the **VIEWS** menu to add the right side view with **Add Views,** ***Projection*****, Full View, No Xsec,** ***No Scale*****, Done.** Select the center point for the right view somewhere to the right of the front view to define the horizontal position of the right view. Return to the **VIEWS** menu, and select the following commands in sequence to add a general 3D view at the right upper corner:

Add Views, ***General*****, Full View, No Xsec,** ***Scale*****, Done.** The final projection views are illustrated in Figure 4-109.

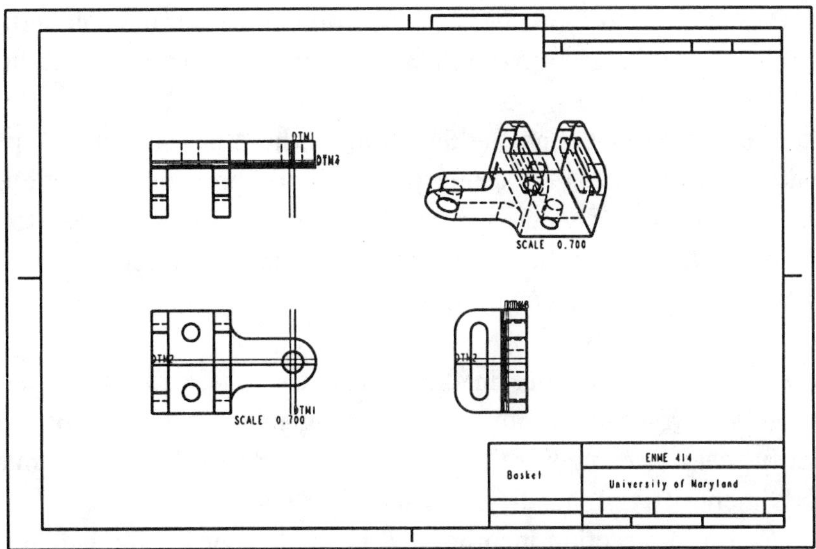

Figure 4-109 Orientation and Arrangement of the Four Engineering Projections

If you want to adjust the relative position of these views, go to the **DRAW** menu, select the commands to move views, and use the cursor to relocate the positions of these views. If you want to change the scale or orientation of the views, go to **VIEWS** menu, select **Modify Views**, and then **Change Scale** or **Reorient**. Please note that **Change Scale** is only for those views for which you have defined scales such as the front view. It is not possible to change the scales of top-view or right-view. Also **Reorient** may only be executed for the general view.

You can create cross section view when you create the top-view or the right-side view. To produce a cross section view, select **Xsec** instead of **No Xsec**. The previous example showed the technique for creating a **total Xsec**. In this example, we will create a Full & Local cross-section view using a different selection under **XSEC ENTER** menu. The following command sequence is employed to modify the one of the three views to generate a cross-sectional view. Starting with the **VIEWS** menu, select **Modify View, View Type, Pick**

Choose the view you want to modify, say the top view. In the **VIEW TYPE** menu, select **Projection, Full View, Section** and **Done** Then in the **XSEC TYPE** menu, select **Full & Local**, **Total Xsec**, and **Done**

In the **XSEC ENTER** menu, select **Create**, **Planar, Single** and **Done**

First type in **A** in the message window to identify (name) the full cross-sectional view. The next step is to select a datum plane to define the position of the cross-section. We select the central plane, which is DTM2 in the front view. Click on the front view to define the direction for viewing the cross-section. The top-view will be automatically changed to cross-section view and shown on the screen.

Now, we will add a local X-section in the top view to show the shape and dimensions of the brackets more clearly. Starting from the **VIEW BNDRY** menu, select **Add Breakout, Show Outer, Choose Xsec, Create**. Next choose the **XSEC ENTER** menu, and select **Create**, **Planar**, **Single** and **Done**.

Type in **B** in the message window to identify (name) the local cross-section. The location of the local cross-section is through the centerline of uppermost hole shown in the front view. Section B-B is made to show the shape and dimensions of the hole more clearly. to prepare Section B-B, we must create a datum plane through the center of the hole, which is plane DTM5.

Click on the front view to define the direction of the cross-section. A prompt in the message window will request: "Select the center position for a breakout to section <B>." Pick the centerline of the hole in the top-view, and a '+' will appear on the screen to confirm this location. Next, you must sketch a spline to define the outline for the local cross-section view. Choose **Done** after completing the drawing of the outline.

Since there are two cross-sections in the same view, one a full cross-section and the other a local cross-section, you must modify the hatching pattern in one cross-section or the other to better identify them. To modify the hatching pattern, follow the command sequence from the **DRAWING** menu: **Detail, Modify, Xhatching, Pick**

Click on the local cross-section incorporated in the top-view. The hatching pattern for the local cross-section will be highlighted. Then click on **Done Sel**, and a new menu, called **MOD HATCH** appears. Select **Spacing**, **Hatch**, **Half**, and then select **Angle, Hatch** to rotate the lines in the hatching pattern by 90°. Finally select **Done.**

To add dimensions to the projections, go to the **DRAWING** menu, and follow the command sequence:

Detail, Show/Erase, Show, Dimension, Show All, Confirm (Yes), Close.

To adjust the dimensions shown in your drawing, select **Move** from the **DETAIL** menu, and use the cursor to select the dimensions you want to modify. Or select **Move Text** from the **DETAIL** menu to move the text for the dimensions to maintain the position of dimension arrows. Or alternatively , select **Switch View** from the **DETAIL** menu to move the dimensions between different views. Use **Erase** instead of **Show** in the **Show/Erase** window to erase the dimensions that are redundant. Use the **Flip Arrows** to reverse the arrows for those dimensions where the arrow heads are too close to each other.

Refer to the previous example about the technique for adding tolerances and notes. The completed drawing for the adjustable bracket is illustrated in Fig. 4-110.

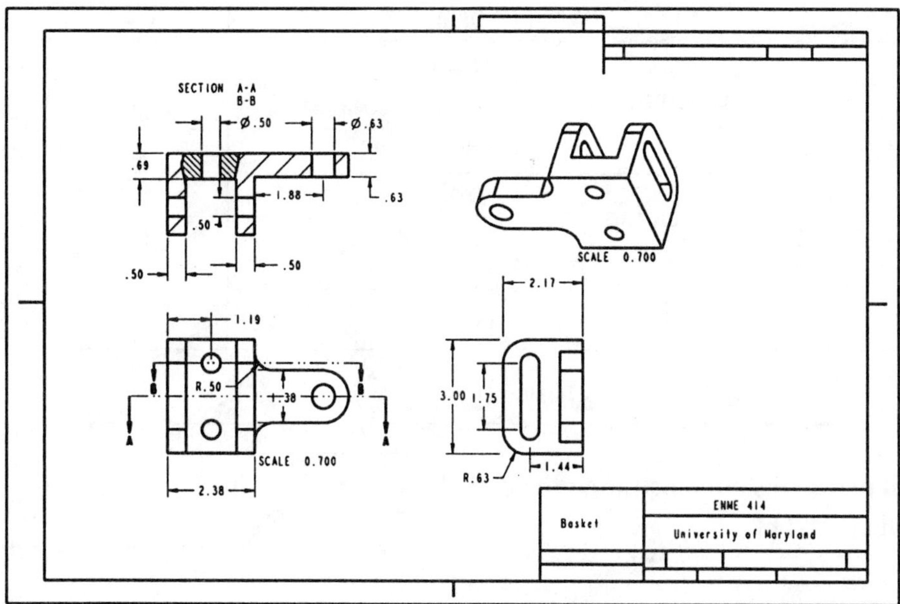

Figure 4-110 Completed Engineering Drawing of an Adjusting Bracket with Dimensions

4.3.3. Example 4-12: Adding Geometric Tolerances to the Sliding Basket

Geometric tolerances state the maximum allowable deviation of a shape or a position from the true geometry and location specified by the designer. Tolerances provide a concise method for expressing manufacturing information such as:

- Where the critical surfaces are located.
- How the surfaces are related to each other.
- How the part is to be inspected in determining its acceptability.

Typical symbols applied to an engineering drawing that convey information in addition to shape and location are presented in Table 4-3.

Table 4-3 Typical Geometric Tolerances

	Tolerance	Characteristic	Symbol
Individual Feature	Form	Flatness	⏥
		Straightness	—
		Circularity	○
		Cylindricity	⌭
Individual or Related Features	Profile	Line	⌒
		Surface	⌓
Related Features	Orientation	Parallelism	//
		Perpendicularity	⊥
		Angularity	∠
	Location	Position	⌖
		Concentricity	◎
	Runout	Circular runout	↗
		Total runout	⌰
Supplementary Symbols	Ø Ⓜ MMC Ⓛ LMC Ⓢ RFS		

Three illustrations demonstrating the use of these symbols in engineering documents are presented in Figure 4-111.

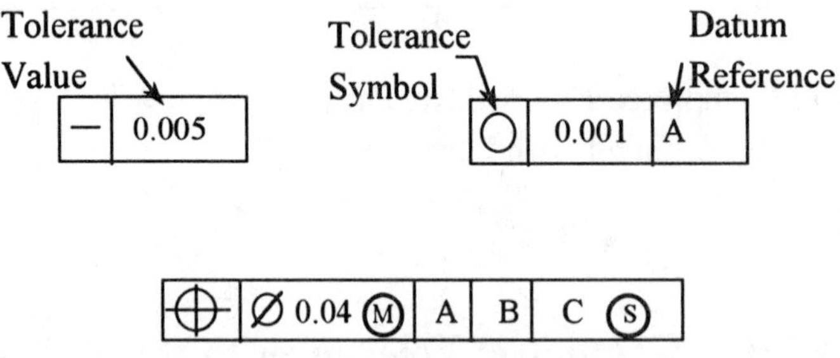

Figure 4-111 Illustrations of Geometric Tolerancing

In Pro/ENGINEER, geometric tolerances can be specified and "basic" dimensions can be created. When geometric tolerances are specified, selected datum entities, such as a plane, an axis, a dimension, or some others, are set as the reference datum. The geometric tolerance (gtol) function can be accessed in Drawing mode, Part mode, and Assembly mode. All the geometric tolerance information will be reflected on the same model regardless of the mode in which you are working. Sometimes it is better to work in the part or assemble mode because the features may be selected more easily. If this is the case, the gtol symbols can be arranged in the drawing mode to create a clear engineering document.

The method to add geometric tolerances is demonstrated using the sliding basket example discussed in Example 4-1. Before starting to specify geometric tolerances, dimensions must be given, and the appropriate reference datum must be established. (The concept of reference datum is employed in the sense of geometric tolerances, as opposed to the concept of datum used in constructing a solid model) To incorporate geometric tolerances, select the **Part** menu, choose **Set up, Geom Tol, Set Datum**, and select DTM1, as illustrated in Figure 4-112.

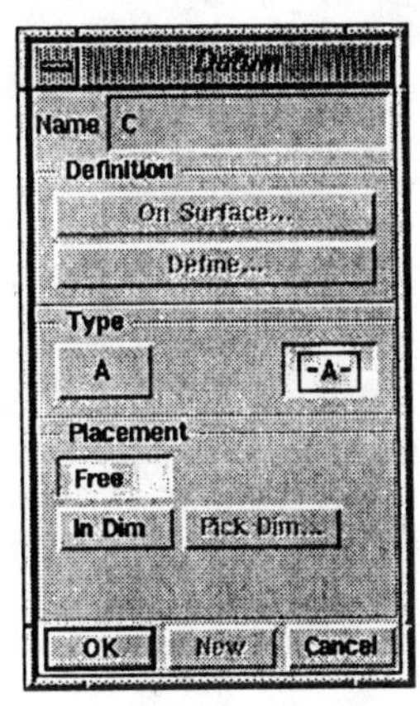

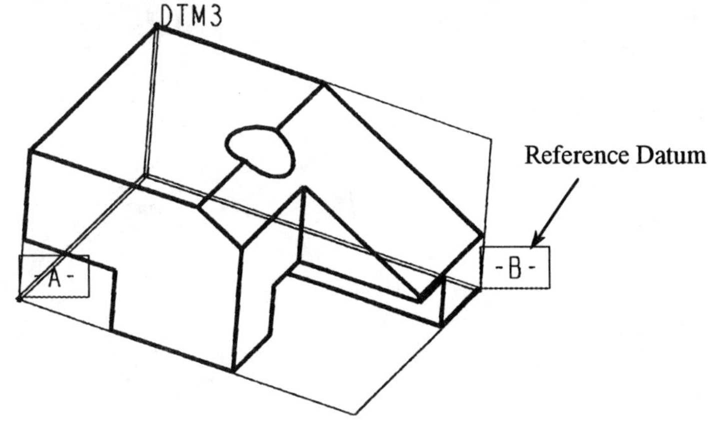

(a) Datum Dialog Box (b) Set Reference Datum

Figure 4-112 Set Reference Datum

In the datum dialog box, set the name of the datum to 'A'. Choose the button on the right of (-A-) to set the type and choose the **Free** button for Placement. Repeat the above steps to set DTM2 as the reference datum as shown in Figure 4-112b.

Now, choose **Specify Tol** from the **GEOM TOL** menu. The Geometric Tolerance dialog box shown in Figure 4-113 is displayed. On the left of the dialog box, are different geometric tolerances symbols that may be selected. The four tabs at the top of the dialog box (**Model Refs, Datum Refs, Tol Value**, and **Symbols**) display different contents when selected. The dialog box shows the Model Refs page when the dialog box is first displayed.

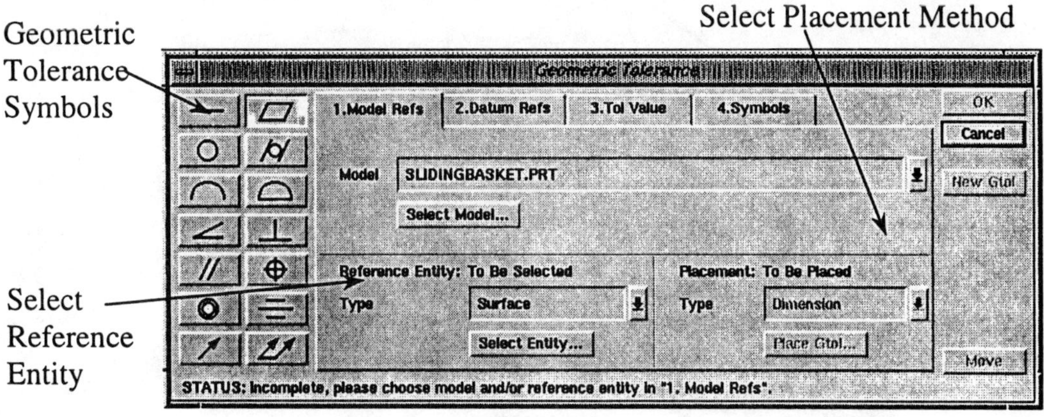

Figure 4-113 Geometric Tolerance Dialog Box

The first geometric tolerance added to Figure 4-21 is to specify the flatness of the top surface. Choose the flatness symbol, and the name of the part will appear in the **Model** text box. Since the flatness is applied to the top surface, set the reference entity to *surface*. Choose the **Select Entity** button from the geometric tolerance dialog box and select the top surface.

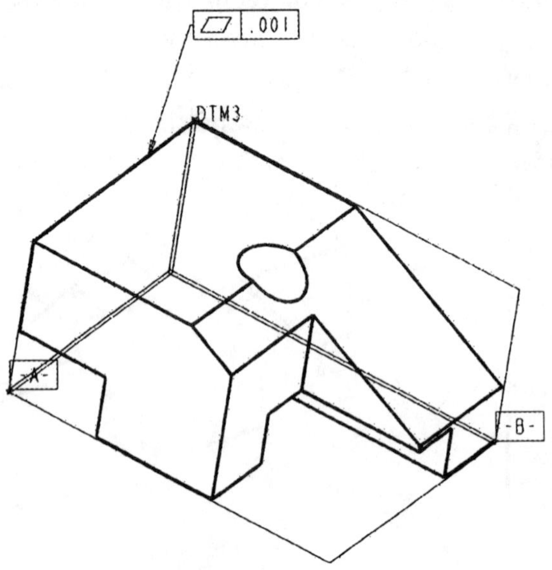

Figure 4-114 Specifying Flatness of the Top Surface Using Geometric Tolerancing

Under the **Placement** option, choose **Leaders**, and select a position to place the symbol. Since flatness does not require datum reference, the **2.Datum Refs** tab need not be selected. Instead, choose the **3.Tol Value** button to set the tolerance value as 0.001. To change the location of the symbol, select the **Move** button on the right side of the dialog box. Finally click on the **OK** button to finish. The surface flatness symbol is shown in Figure 4-114.

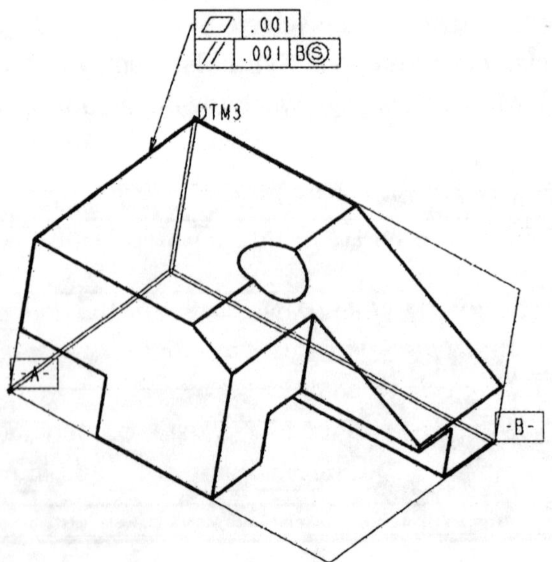

Figure 4-115 Specifying Parallelism of the Top Surface with Geometric Tolerancing

Another geometric tolerance requirement for the top surface is its parallelism with respect to the bottom surface (Reference datum B). To specify this requirement, choose **Specify Tol** again, and select the Parallelism symbol. The reference entity is set to *surface* and the top surface of the part is selected. Under **Placement**, choose **Other Gtol** and select the flatness gtol that was just specified. The new parallelism gtol symbol will attach to the flatness gtol symbol as shown in Figure 4-115. Now choose the **2.Datum Refs** button to specify the reference datum for the parallelism requirement. For basic reference, choose *B* and select **RFS(with symbol)**. Set the tolerance value to 0.001.

Following the same procedure, establish the perpendicularity and parallelism requirements sing geometric tolerances for another surface as illustrated in Figure 4-116.

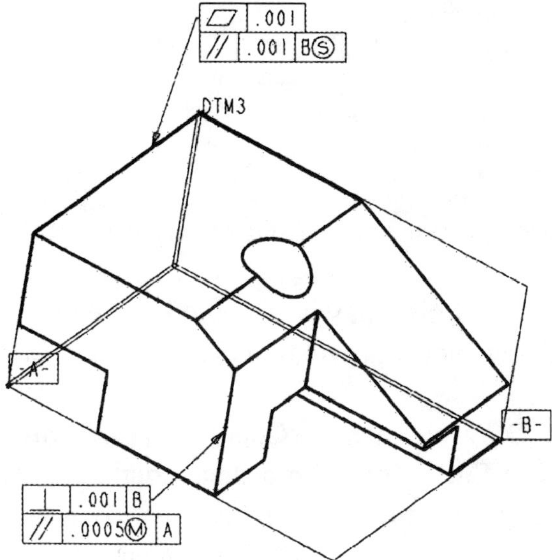

Figure 4-116 Specifying Flatness and Parallelism of the Front Surface

Geometric tolerances may be specified in drawing mode. In drawing mode, the geometric tolerances may be modified and the position may be rearranged to clarify the drawing. In the drawing mode, create the front view and top view as shown in Figure 4-117. Choose **Detail, Show/Erase**, and select the **Geometric Tolerance** button to show the gtol symbols. Use **Move, Move Text**, and **Mod Attach** to adjust the position of the gtol symbols.

Following the process for specifying geometric tolerances in Part mode, let's first add a reference datum. From **Detail**, choose **Geom Tol, Set Datum**, pick DTM3, and identify the datum as C.

We are going to specify the position tolerance for the hole drilled through the center of sliding basket. Follow a similar procedure as was employed in adding **gtol** in part mode. Use the geometric tolerance dialog box to complete the position tolerance for the hole as shown in Figure 4-117. Choose A as the basic datum and C as the second datum.

Under **4.Symbols**, choose **Diameter Symbol** to add the proper symbol before the tolerance of 0.0005.

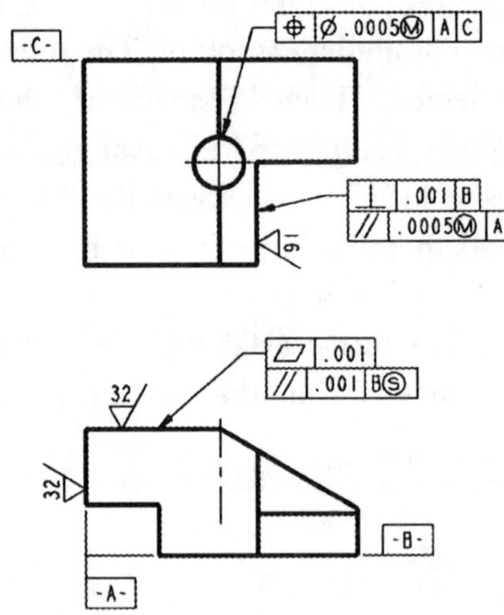

Figure 4-117 Specifying Hole Tolerance and Surface Roughness in the Drawing Mode

In Pro/ENGINEER, the surface finish may be specified while operating in either Part or Drawing mode. However, Drawing mode provides more functions to employ in applying surface finish symbols.

Choose **Detail** from the Drawing menu. Choose **Create, Surf Finish, Retrieve**, ***/machined*****, and** ***standard1***. Choose **Entity** for the attachment option. Select the top surface in the front view and enter 32 for the surface finish value. To create other surface finish symbols in the drawing, choose **Pick Inst** menu from the **Surf Finish** menu, and select the first surface finish symbol that was used. Locate the symbol and enter the surface roughness value. Follow this procedure to provide the additional surface finish symbols illustrated in Figure 4-117. In Pro/ENGINEER, the surface finish symbols can also be customized to display waviness, maximum and minimum average roughness height, lay direction etc.

4.4 Assembly of Components

Almost every product on the market is assembled from components. A pencil used to take notes may consist of three or four components. A car we drive is assembled from thousands of components. Assembly involves the joining together of two or more separate components to form a new entity, called a subassembly or an assembly. The method to accomplish the assembly of these components may be mechanical fastening, welding, adhesive bonding, etc. Assembly of components is one of the critical tasks in product development. In fact, the Pro/ENGINEER design system provides a virtual environment to simulate the physical assembly process observed on the shop floor.

4.4.1 Example 4-13: Assembly of Two Blocks

After creating individual parts in the **PART** mode, you may combine them into assemblies. The process of assembling parts is conducted in the Pro/ASSEMBLY mode. The main purpose of this mode is to arrange components and subassemblies together to form assemblies. There are several other functions included in the Pro/ASSEMBLY mode. They include modifying, adding, or deleting features associated with the components involved in the assembly. A unique function of the Pro/ASSEMBLY mode is the ability to update the various solid models with modifications made during assembly.

In this example, we introduce some basic functions of the Pro/ASSEMBLY mode by considering the assembly of two rectangular plates. The dimensions of these plates are shown in Figure 4-118.

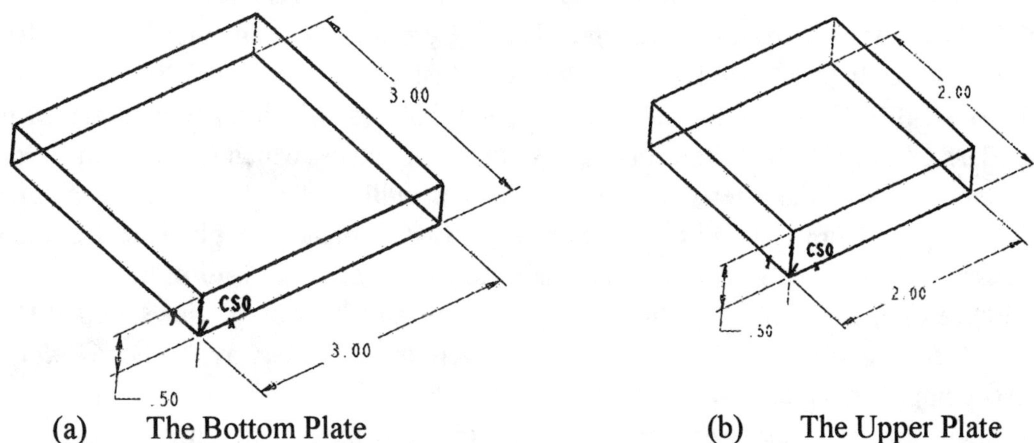

(a) The Bottom Plate (b) The Upper Plate

Figure 4-118 Geometry and Dimensions of Two Plates

The appearance of the plates after assembly is illustrated in Figure 4-119. Under the Pro/ENGINEER design environment, there are 10 placement options available to users, which include mate, mate offset, align, align offset, insert, orient, coordinate system, tangent, edge on surface, and point on surface. In this example, we employ the mate, align and align offset options for the assembly operation.

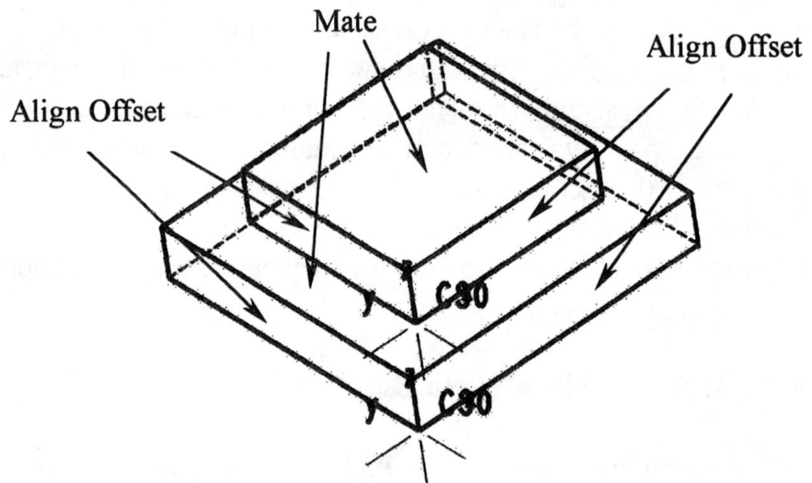

Figure 4-119 Assembly of the Two Rectangular Plates

Starting from the **MODE** menu, utilize the following command sequence:

Assembly, Create, Component, Assemble.

Type in the file name of the bottom plate, *plate1.prt,* in the message window and then hit the **Enter** key. Select **Assemble** again and type in the file name of the upper plate, *plate2.prt*. There are six windows displayed: **Pro/ENGINEER Assembly**, **COMPONENT WINDOW Part, Menu Manage, Model Tree, Component Placement** and **Message Window.** Except for the Message Window, they are illustrated in Figure 4-120.

The **Assembly Window** displays the parts and features, which have been placed in the assembly. The **COMPONENT Window** shows the component, which is about to be placed into the assembly. Note the asterisks in the header bar indicate that the window is currently active. In the **Menu Manager Window**, under the **PLACE** menu, ten placement options are listed as described previously, and we may add constraints. The **Component Placement Window** indicates the current placement status of the part shown in the **COMPONENT WINDOW Part**. In Figure 4-120 no placement activity is underway. The **Model Tree** window lists components included in the current assembly.

To begin the assembly operation, select the **PLACE** menu, and choose **Mate, Select, Query Sel**. Click on the top surface of the bottom plate, and note that it is highlighted. If it is not click on **Next** to change selection entities until the top surface is highlighted. Choose **Accept** to confirm your selection. Choose **Query Sel** and move the cursor to the **COMPONENT WINDOW**, and click on the bottom surface of the small plate. The surface is aligned with x-y plane, by clicking **Next** continuously until the x-y plane is highlighted. Select **Accept** to finish this step.

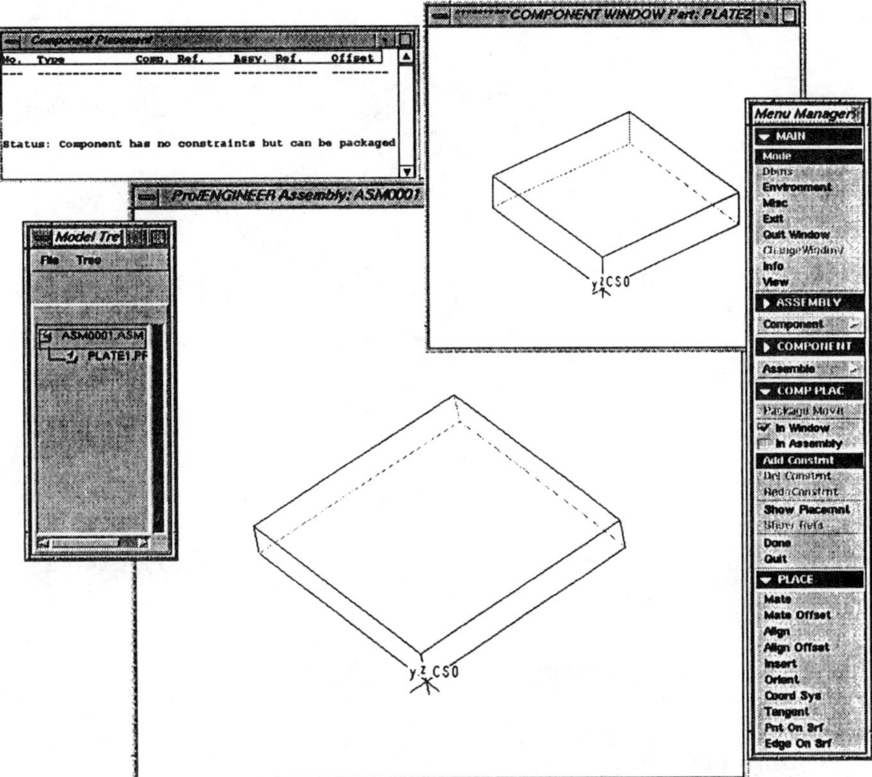

Figure 4-120 The Assembly Windows System

Since the upper plate may slide on the top plane of the bottom plate, we must constrain it. From the **PLACE** menu, choose **Align, Select, Query Sel.** Click the left side surface, which coincides with the y-z plane shown in the **Assembly window**. Click on **Next** to target the highlighted object, and select **Accept** to confirm your selection. Choose **Query Sel** and move the cursor to the **COMPONENT WINDOW Part**, click the left side surface of the small plate, and select it. These selections constrain the movement of the upper plate in the y direction.

From the **PLACE** menu, choose **Align, Select, Query Sel** again. Click the side surface, which coincides with y-z plane in the **Assembly window**. Click on **Next** until the y-z plane is selected, and then click on **Accept**. Choose **Query Sel** and move the cursor to the **COMPONENT WINDOW**, click on the side surface, which coincides with y-z plane of the upper plate, and click on **Accept** to confirm your selection. These selections constrain the movement of the upper plate in the x direction.

A prompt will be displayed in the **Message Window** as well as the **Component Placement Window** that the component may be placed. Click **Done** on the **COMP PLAC** menu to complete the assembly process. The updated assembly window shows the display illustrated in Figure 4-121.

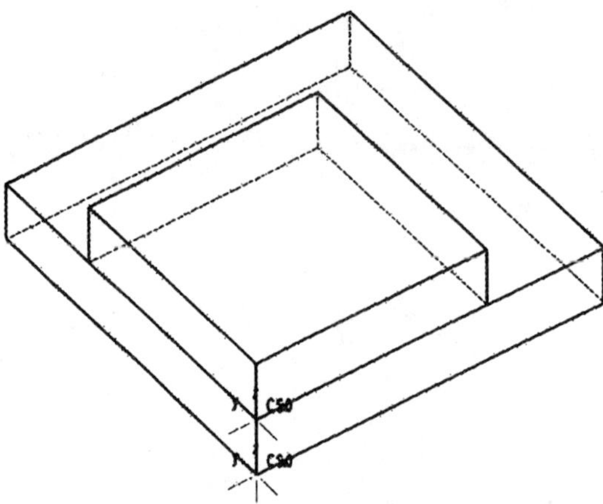

Figure 4-121 Assembly of Two Plates Aligned to a Corner

To move the upper plate to the position shown in Figure 4-119, select the command sequence listed below from the **ASSEMBLY** menu:

Component, Redefine, Select, Pick.

Pick the upper plate in the **Assembly Window**, and the **COMPONENT Placement Window** is displayed as well as **COMP PLAC** in the **Menu Manager Window**.

Select the second placement constraint: **Align** from the **COMPONENT Placement Window**, and note the appearance of a rectangle outside of the line you have selected. Choose **RedoConstrnt** in the **COMP PLAC** menu. In the **CONSTR REDEF** menu, there are three items: **Type, Assembly Ref, Comp Ref**. These items permit the definition of the type of constraint, the reference position in the **assembly window**, and the reference position in the **Component Window Part**. To locate the relative position of the component in the assembly, select **Type** and **Done**.

Select **Align Off** instead of **Align** in the **PLACE CONSTR** menu, and then **Done**. A red arrow head is displayed, which may point to the outside of the bottom plate or the negative x direction. If so, enter -0.5 in the **Message Window** to move the upper plate 0.5 in away from the side surface of the bottom plate. If the arrow points to the x positive direction, enter 0.5, instead of -0.5. The assembly of the two plates after these modifications is illustrated in Figure 4-122.

Following the same procedure, we align the upper plate in the y direction making an offset of 0.5 in from the side surface of the bottom plate. Note the direction of movement as indicated by the arrow head. If the arrow points to the outside of the plate or the negative y direction, enter 0.5 to specify the dimension of the offset. The upper plate is moved to the location indicated in Figure 4-123. Since this position is not correct, we must choose **RedoConstrnt** in the **COMP PLAC** menu again. Then **Type**, **Done**, enter the correct offset of

-0.5 if the arrow is pointing to the outside of the plates. The position of the upper plate will be changed to the correct position as indicated in Figure 4-119.

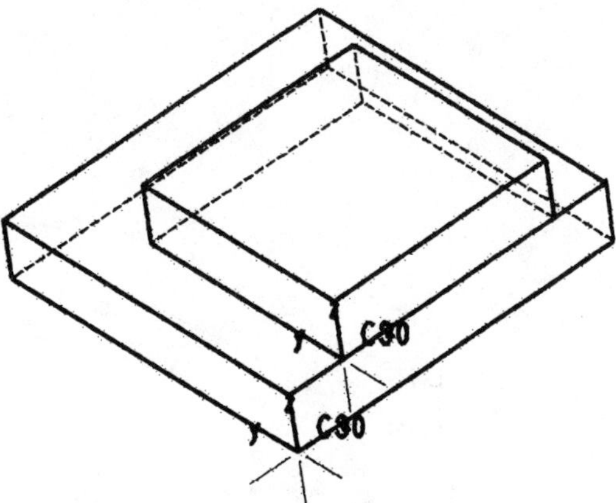

Figure 4-122 Alignment of the Upper Plate along the x Positive Direction

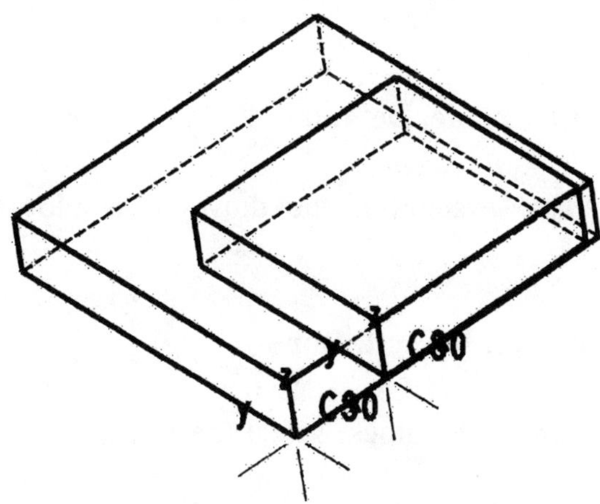

Figure 4-123 Alignment of the Upper Plate along the y Negative Direction

4.4.2 Example 4-14: Assembly of a Block and a Shaft

In this example, we assemble a block and shaft, as shown in Figure 4-124. The block is a simple rectangular solid with a through hole and the shaft is straight with a shoulder at its end.

Let's first create the block by using the following sequence of commands. From the **MODE** menu, select **Part,** then **Create**, and type *block* to identify the file name in the message window, and then hit the **Enter** key.

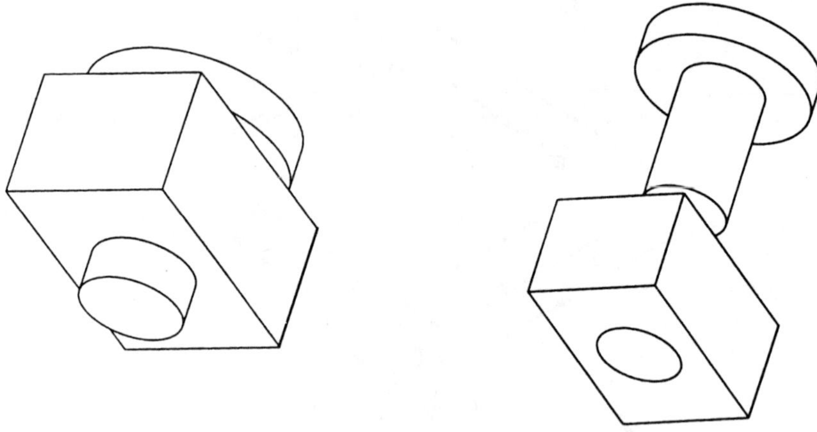

Figure 4-124 An Assembly of a Block and Shaft

From the **PART** menu, exercise the following command sequence to create a reference system.

Feature, Create, Datum, Plane, Default

To create the coordinate system of the drawing, employ the following command sequence:

Create, Datum, Coord Sys, Default, Done

To create the solid model of the block, select the following commands from the **FEAT** menu:

Create, Solid, Protrusion, Extrude, Solid, Done.

Choose the **One Side** option from the **SIDES** menu. Choose **Plane** from the **SETUP PLANE** menu and pick DTM 3 for the **Sketching Plane**.

From the **SKET VIEW**, select **Top**, and then select DTM2 for the sketching plane. The appearance of the drawing of the block at this stage of the development is shown in Figure 4-125.

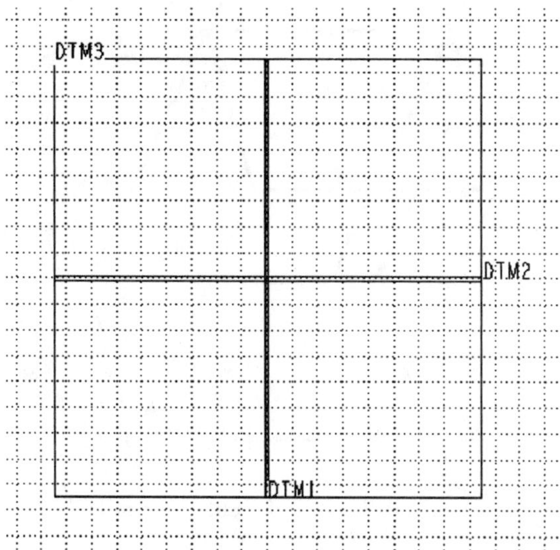

Figure 4-125 Orientation of the Sketching Plane for the Block

Choose **Sketch** from the **SKETCHER** and **Rectangle** from the **Geometry** menu. Select two points on the 2D screen to indicate the diagonal of the rectangle. Align the left side of the rectangle with DTM1 and the bottom side of the rectangle with DTM2. Choose **Dimension** from **SKETCHER**, and define the longer edge of the rectangle as 30 and the shorter edge as 20. Then regenerate it to obtain the display presented in Figure 4-126.

Under the **Blind** command, specify the depth as 15 to complete the drawing of the rectangular solid without the hole as indicated in Figure 4-127.

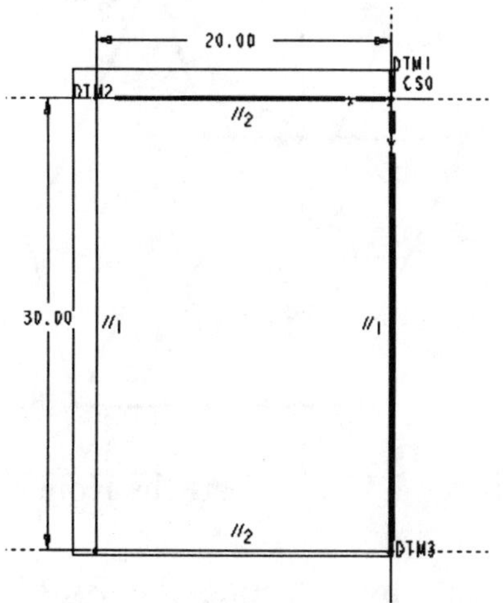

Figure 4-126 The Sketch of the Cross Section of the Rectangle

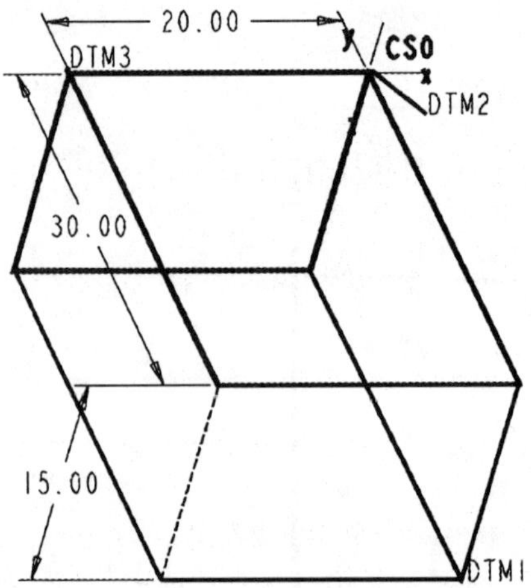

Figure 4-127 Create the Block

Next we cut the hole with diameter of 12. Choose **Create, Solid, Hole, Done.** Select **Straight**, **Done, Linear, Done.** Then chose the front side of the block for the placement plane of the hole. Use the mouse to select the left edge of the block, and input 20/2 in the message window for the distance between the center of the hole and the left edge. Use the mouse to select the top edge of the block and type 30/2 in the message window for the distance between the center of the hole and the top edge. In the **SLIDES** menu, select **One Side** and then **Done**. Choose **Thru All** from the **SPEC TO** menu and then select **Done**. Following the prompt in the message window, input 12 as the diameter of the hole. Choose **OK** from the ***Element Inform*** window. The block is complete as illustrated in Figure 4-128.

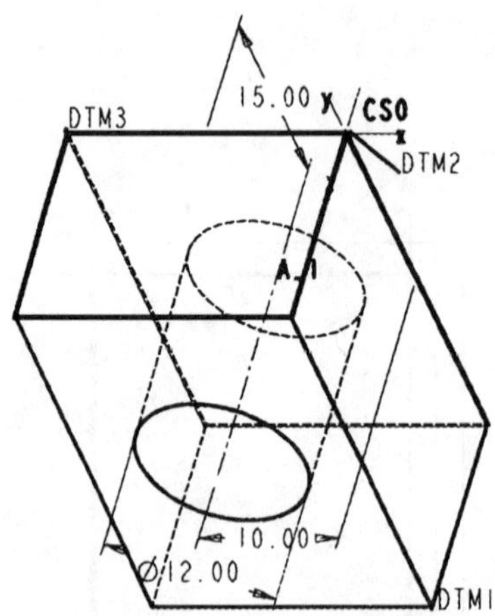

Figure 4-128 Create the Hole Feature

Next, we create the shaft by selecting the PART menu, and using the following command sequence to create a reference system.

Feature, Create, Datum, Plane, Default

To define the coordinate system for the drawing, employ the following command sequence:

Create, Datum, Coord Sys, Default, Done

To begin drawing the shaft, select the following commands starting from the **FEAT** menu:

Create, Solid, Protrusion, Revolve, Solid, Done.

Choose the **One Side** option from the **SIDES** menu. Choose **Plane** from the **SETUP PLANE** menu and pick DTM3 as the sketching plane. From the **SKET VIEW**, select **Top**, then select DTM2 on the main drawing window.

Choose **Sketch** from the **SKETCHER** and **Line, Centerline** from the **Line TYPE** menu. Select two points on DTM1 and align the centerline with DTM1. The centerline will serve as the axis of revolution.

Choose **Line** again from the **Geometry** menu, and then sketch the cross section of the shaft as shown in Figure 4-129. Align the bottom edge of the sketch to DTM2 and the left edge of the sketch to DTM1. Add dimensions to the 2D sketch as indicated in Figure 4-129.

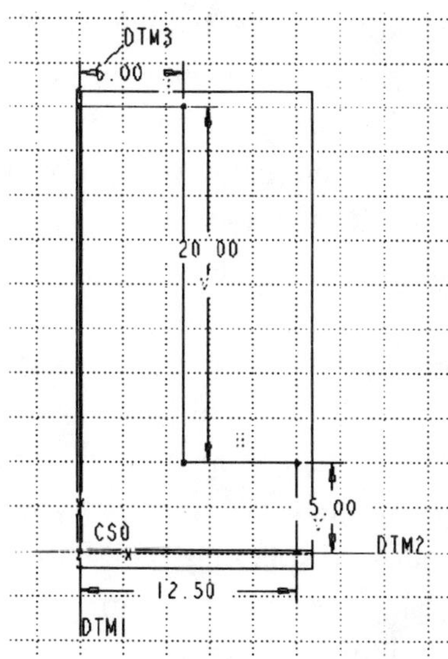

Figure 4-129 Sketch the Cross Section of the Shaft

Use commands **Regenerate, Done** to regenerate the 2D sketch. Choose 360° under the **REV TO** menu, and then select **Done** and **OK** from ***Element Inform*** window. The complete 3D solid model of the shaft is presented in Figure 4-130.

Now that we have prepared the block and the shaft for assembly, let's begin the assembly process.

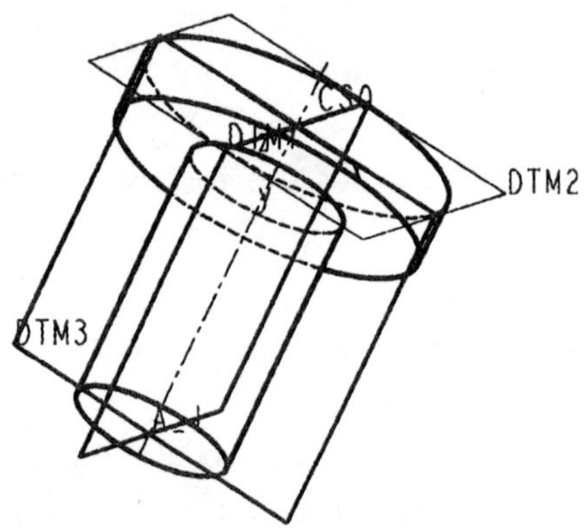

Figure 4-130 Complete 3D Solid Model of the Shaft

From **MAIN** Menu, choose **Mode, Assembly, Create** and type the file name: *assembly1* (default is ASM001) in the message window. Under the **ASSEMBLY** menu, choose **Component, Assemble.** the prompt in the message window, seeks the file name of the first component to be placed in the assembly. From our experience with Example 4-13, we recognize that the first component serves as a reference for positioning the other components in the assembly. In this example, we select the block as the basis for the assembly. Enter *block* in the message window, and hit the **Enter** key. In the drawing window for the assembly, the solid model of the block is retrieved from memory and displayed. Select **Assemble** again, and enter the file name of the second component to be placed in the assembly. Since only two components are involved in this assembly, we enter *shaft* in responding to the prompt in the message window. The model of the shaft is retrieved from memory and displayed in a new window. To aid in visualization of the assembly, the **Component Placement window** is also displayed.

We are now prepared to assemble the block and the shaft. From the **PLACE** menu, choose **Insert**. Note that the assembly of the two components involves inserting the shaft into the hole of the block. Go to the window displaying the reference component (the block) and select the hole. The prompt in the message window, requests you to "select revolved surface to insert in other part". Go to the window displaying the shaft. Inspect the title bar of this window to determine if a line of asterisks is evident indicating the window is active. Select the small cylinder of the shaft, and you will observe that a line is added to the Component Placement window indicating a constraint on the movement of the shaft (i. e. The movement of the shaft is constrained to its centerline of during assembly. Two degrees of freedom remain, namely the shaft may slide along the center line of the hole and rotate in the hole.

Next we will add another constraint to position the shaft with respect to the axis of the hole. Choose **MATE** from the **PLACE** menu, and select the back surface of the block. Next, choose the front surface of the large cylinder on the end of shaft. A prompt in the message window, a message indicates that the component may now be placed. A similar message will also be shown in the Component Placement window. The only remaining degree of freedom permits the rotation of the shaft in the hole. It is not necessary to constrain this motion to complete the assembly process. Finally, choose **Show Placement** under **COMP PLAC** menu. If the image displayed on the screen is similar to that shown in Figure 4-131, choose **Done**.

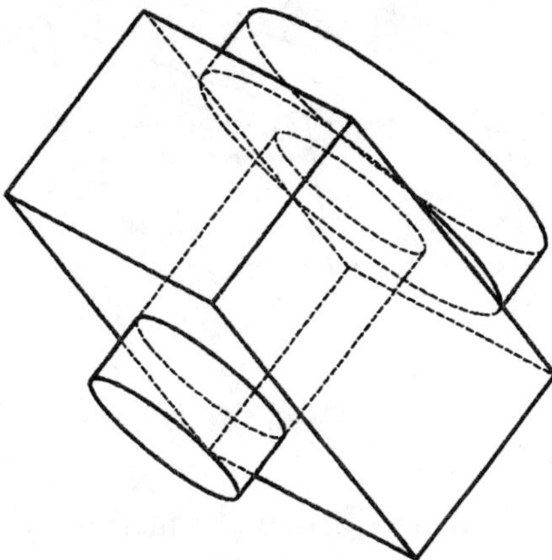

Figure 4-131 Completed Assembly of the Block and Shaft

One advantage of performing an assembly under the Pro/ENGINEER design environment is that it allows the designer to check for clearance and interference between and among components and sub-assemblies.

From the **MAIN** menu, choose **Measure, Clear/Intf** to examine the clearance and/or interference between a pair of mating parts or within the entire assembly. Since the assembly in Example 4-14 is quite simple, let's perform checking for the entire assembly. First, we will check the interference. Choose **Global Intf, Parts Only, Exact Result, Done/Return.** The message window prompt displays " There are no interfering parts". Next, we will check the clearance. Choose, **Global Clr, Parts Only, Done/Return**. The prompt request data for the clearance. Pro/ENGINEER will check to determine if any component in the assembly is within this clearance (interference is included). Type 0.0001 to indicate the clearance. The prompt in the message window indicates that the clearance between the shaft and the hole is 0.0000.

Now, let us modify the radius of the shaft to 6.2, which leads to the interference between the shaft and the hole in the block (support). Check the global interference again. Pro/ENGINEER indicate that the SHAFT and the SUPPORT interfere with a volume of (114.9823 mm^3). The interfering parts, and the corresponding interference volume will be highlighted.

4.4.3 Example 4-15: Assembly of a Block, a Shaft, and a Key

In Example 4-14, we assembled a shaft into a hole in a block. Because the cylindrical nature of the two mating parts, there was no need to constrain the angular position of the shaft when it was inserted into the hole. That degree of freedom was not constrained and the shaft could rotate in the hole. In this example, we add a key between the block and the shaft to constrain the rotation of the shaft in the hole.

In this example, the key is the third component. While the procedure to create the key is not described, all of its dimensions are shown in Figure 4-132.

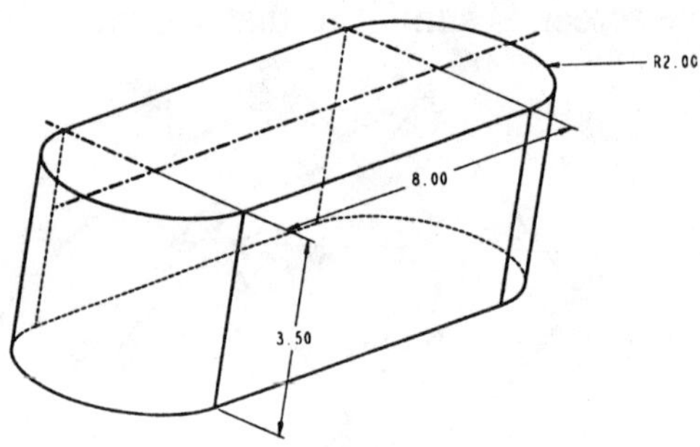

Figure 4-132 Key Used for the Assembly Process

Both the block and the shaft need to be changed to accommodate the key. The sketch and dimensions of the key way cut into the block are shown in Figs. 4-133a and b. The width of the key way is the same as that of the key. A datum plane, DTM4, is created through the center of the hole and parallel to the side surfaces of the key way. It is used later in the assembly of the block and the key to aid in locating the angular position of the shaft.

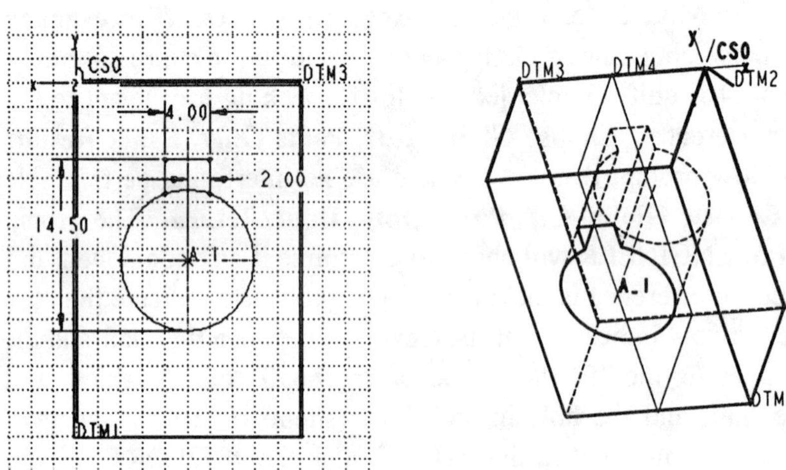

(a) Sketch of Key way (b) Completion of the Key Way on the Block

Figure 4-133 Modification to the Block

Next, we modify the shaft. Consider a perpendicular distance between the bottom of the key way toward the axis of the shaft of 4, and establish a new datum plane, DTM4, at this location as indicated in Figure 4-134b. It is created by using **Offset,** and entering 4 to define the distance between DTM4 and DTM1. Select DTM4 as the sketch plane to create and draw the key way on the shaft, as illustrated in Figure 4-134a.

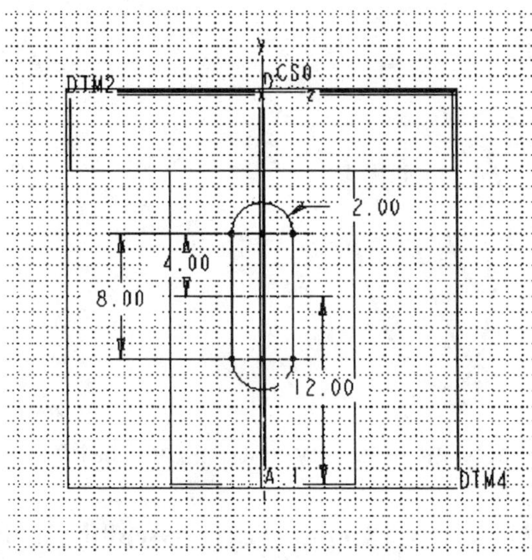

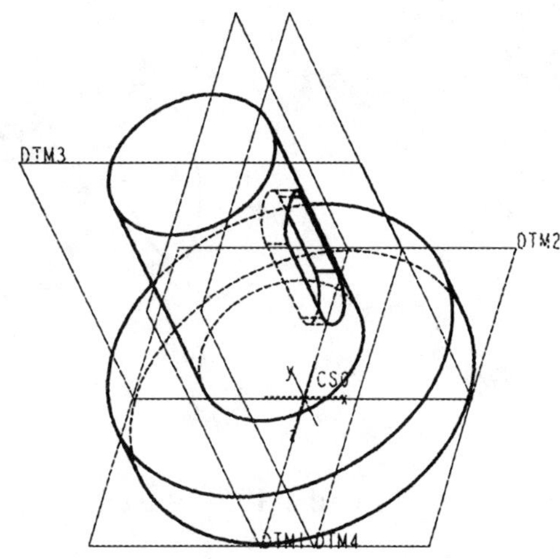

(a) Sketch of the Key Way on the Shaft (b) 3D View of the Key Way on the Shaft

Figure 4-134 Modifying the Shaft

We are now ready to begin the assembly process. However, if we assemble the block and shaft following the same procedure as described in Example 4-14, we will not be able to insert the key. (It is a blind key way). If we follow the assembly procedure that is used on the shop floor, the shaft and key are assembled first. Let's retrieve the shaft from memory to serve as the basis component.

From **MAIN** Menu, choose **Mode, Assembly, Create** and type the file name: *assembly2* (default is ASM001) at the prompt in the message window. Under the **ASSEMLBY** menu, choose **Component, Assemble.** In the message window, enter *shaft* to define the first (basis) part and hit the **Enter** key.

In the drawing window for the assembly, the shaft model is retrieved and displayed. Pick **Assemble** again, and then enter *key*. The key model is retrieved and displayed in a new window. Now we are ready assemble the key into the shaft.

The bottom surface of the key must contact the bottom surface of the key way. From the **PLACE** menu, choose **Mate**. In the assembly window, select the bottom surface of the key way on the shaft, and then select the bottom surface of the key. These actions have introduced a constraint that is listed in the Constraint Placement Window.

Now we add a second constraint. From the **PLACE** menu, choose **Insert**. Go to assembly window and select the cylindrical surface at one end of the key way. Also select the cylindrical surface at one end of the key. Make sure that the ends selected are on the same side of those two parts. These actions produce the second constraint, and the message window indicates that the key may now be inserted into the key way. However, if you choose **Show Placement**, you may find that the key is not yet aligned correctly with the key way. This occurs because the Insert option permits relative rotation between the key and key way. Select

Mate again and constrain the other ends of the two parts. The assembly displayed is presented in Figure 4-135 after you choose **Show Placement**. Then choose **Done** to save your work.

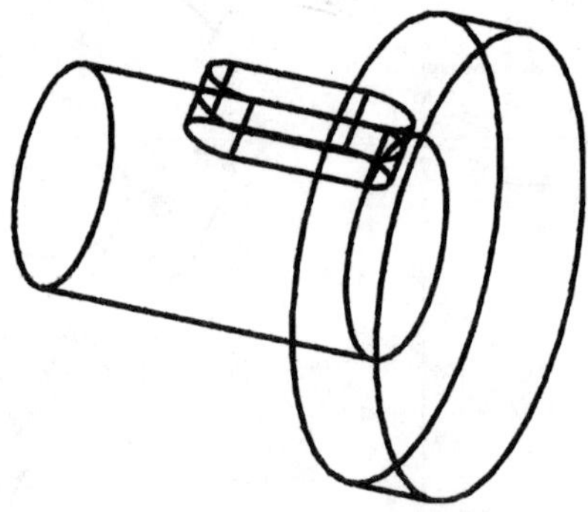

Figure 4-135 Assembly of the Key into the Key Way in the Shaft

Now we bring the block into the assembly. Choose **Component**, **Assemble**, and enter the part name, *block*, and the solid model of the block is displayed in a new window. First we use **Insert** and **Mate** to provide two constraints controlling the assembly of the shaft and block by following the same procedure that we employed in Example 4-14.

The critical issue is determining the angular position of the shaft relative to the hole. A good approach is to require the DTM4 plane on the block to be co-planar with the DTM3 plane on the shaft. We may use the **Align** command to provide this constraint.

From the **PLACE** menu, choose **Align**. In the assembly window, pick DTM3, and an arrow head is displayed. You are prompted to choose the face of the datum plane (red or yellow). The arrow head indicates the yellow side of DTM3. The use of the colors red and yellow for the datum place Aids in adding the final constraint prior to assembly. We recognize that we still have two ways to assemble the components even though we have aligned the DTM4 plane of the block and the DTM3 pane of the shaft. One is correct with the key aligned with the key way of the block. The other assembly geometry shows the shaft rotated about its axis 180° away from the correct position. In this position the key is not in alignment with the key way and the components cannot be assembled. When the datum plane DTM3 of the shaft and the reference plane DTM4 of the block are properly correlated the assembly can be completed as illustrated in Figure 4-136.

Consider the assembly of the shaft and the block, with the orientation defined in Figure 4-133. When you choose the datum plane as DTM3 for the shaft and DTM4 for the block, you must be certain that the direction arrows for both of these planes coincide. The correct direction is obtained if the arrow for DTM3 of the shaft and the arrow for DTM4 of the block both point in the same direction as arrow 1 or arrow 2.

Using the same technique that was employed for aligning the key and key way, add the last constraint to the assembly by selecting **Align**. The prompt in the message window will indicate that the component may be placed. Choose **Show Placement** to make sure the assembly is correct, and **Done** to complete the assembly. A 3D drawing of the assembly is presented in Fig. 4-136.

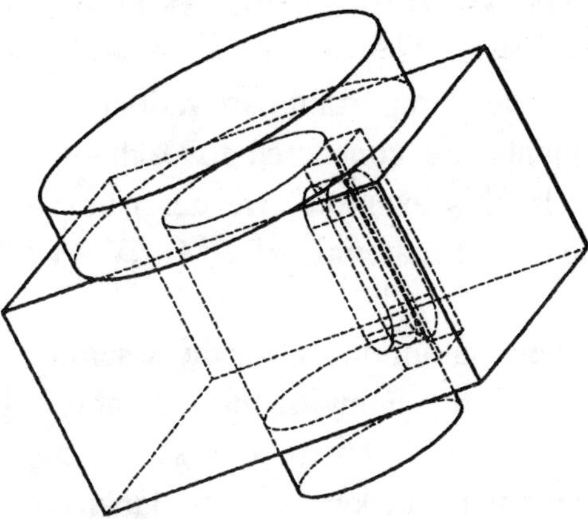

Figure 4-136 The Final Assembly of the Block, Shaft and Key

If you shade this view of the assembly, the key will not be visible because it is hidden in the block. An exploded view is a more suitable representation of the entire assembly, especially when several components are involved.

To create an exploded view, choose **Explode State** from the **ASSEMBLY** menu. Select **Create** to create a new explosion view. Enter a name for the file (the default name is EXP001). Choose **Entity/Edge**, and all of the components are repositioned parallel to the edge you select. In this example, select the axis of the shaft for the repositioning of the components.

Use the left mouse button to select the shaft. The shaft model then moves with the cursor. Drag the shaft to a new position and fix it by pressing the left mouse button again. Next, use the mouse to drag the block away from the shaft. All three components are clearly shown on the screen as indicated Figure 4-137.

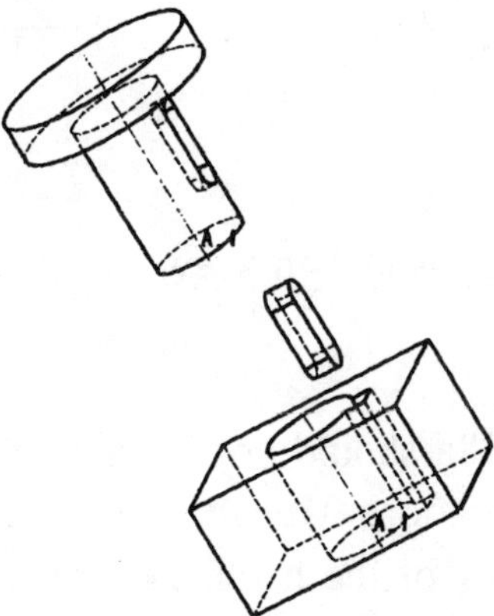

EXPLD STATE: EXP0001

Figure 4-137 An Exploded View of the Assembly

If return to the original view of the assembly, choose **View** from the **Main** menu, then choose **Cosmetic**, **Un-Explode** from the **View** menu.

4.4.4 Example 4-16: Preparation of Assembly Drawings

An assembly drawing is a very important part in documenting an engineering design. It demonstrates how the components are connected to each other, and shows how the entire assembled system functions. In this example, we describe the procedure for preparing an assembly drawing and also the bill of materials, (BOM). We will use the assembly created in Example 4-15.

First create ADTM1, a new datum plane, for the assembly. This datum plane is used to aid in the preparation of a cross section view of the assembly. In Pro/ENGINEER, only the assembly datum plane may be selected to create a cross section. ADTM1 passes through the axis of the shaft and the center plane of the key. It is coplanar with plane DTM1 for the shaft.

In drawing mode, follow the same steps described previously to create the front and top views of the assembly. The front view is a cross-section view, and ADTM1 must be used to create this view. The front and top views are shown in Figure 4-138a.

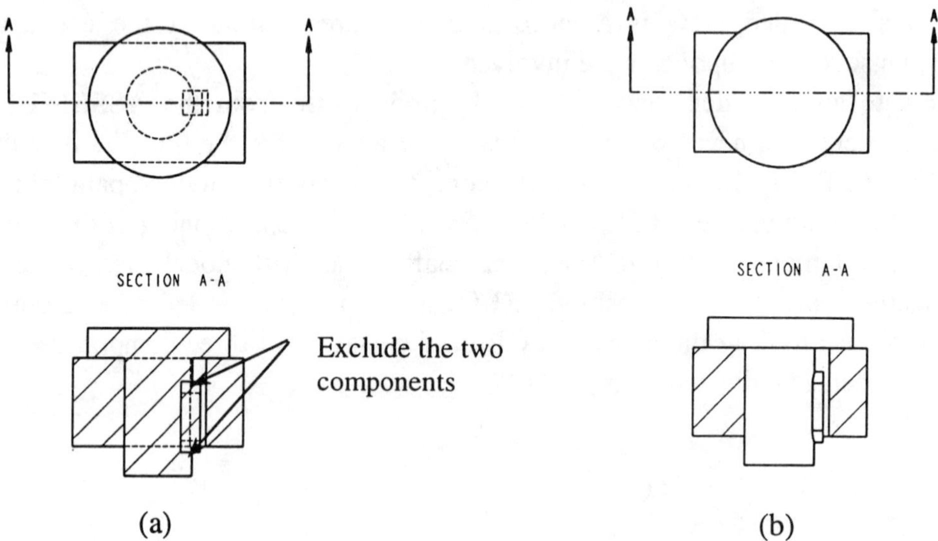

Figure 4-138 Create the Front and Top Views of the Assembly

The shaft and the key in Figure 4-138a should not be shown with a hatch pattern. To delete the hatch lines from these two parts choose **Modify**, **Xhatching**, then select the cross-sectional view and **Done Select**. If the color of the hatch pattern for the part *support* is red, choose **Next Xsec**. Now the color of the hatch pattern for the part *key(shaft)* is highlighted in red. Choose **Excl Comp** to exclude the hatching pattern from the *key(shaft)*. Choose **Next Xsec, Excl Comp** again to exclude the part *shaft(key)*. The drawing shown in Figure 4-138b represents the hatching pattern after the modifications. The spacing and the angle of the lines forming the hatch pattern may be modified by choosing **Spacing** and **Angle** in the **MOD HATCH** menu. Next, choose **Environment**, **No hidden** to delete the hidden lines in the drawing. The modified views of the cross section of the assembly are shown in Figure 4-138b.

BOM (bill of materials) is an essential part of most assembly drawings. Pro/ENGINEER provides a variety of functions to prepare a BOM through **PART LIST**. We demonstrate the technique for creating a BOM for an assembly drawing in he following paragraphs.

Before a part list is created, relevant information that is to be included in the BOM must be assigned to each component in the assembly. In this example, suppose we want to include in the BOM the part No., the name of the component, the quantity, price and specifications for the material. Some of these items, such as the name and quantity, may be extracted by Pro/ENGINEER automatically. For other items, the designer is required to define parameters and provide added information.

In the **Part** mode, retrieve the part *shaft.* Choose **Relations**, **Add Param, String**. Enter PARTNO. Now we have defined a parameter *PARTNO* to save the part number information. Enter X_1000 for this parameter. Choose **Add Param**, **Number**. Enter ***PRICE*** and 100 for this parameter. Repeat above steps to define parameter **MATERIAL**. Assume that the material for the shaft is an aluminum alloy --- 2014-T6.

Retrieve the two parts, the support and the key, and define the same parameters as:

Support:	PARTNO-X_1001;	PRICE-125;	MATERIAL-2014-T6.
Key:	PARTNO-X_1002;	PRICE-20;	Purchase.

In Pro/ENGINEER, the BOM is represented by a part list, which is defined by using a table. The first step in developing the parts list is to draw the table and specify its format. Choose **Table**, **Create, Ascending** (rows progress upward), **Rightward** (Columns progress to right), **By Num Chars** (define the width and height of the cell), **Pick Pnt.** Pick the starting point at the upper left corner of the title block. In the horizontal direction, click 6 for width of the first column. Click 9, 9, 8, 8 for the following four columns and leave a suitable space for the sixth column. Choose **Done** and then specify the height of the row. Click 2, meaning to define two rows and **Choose** Done. A table has been established in the area above the title block as shown in Figure 4-139.

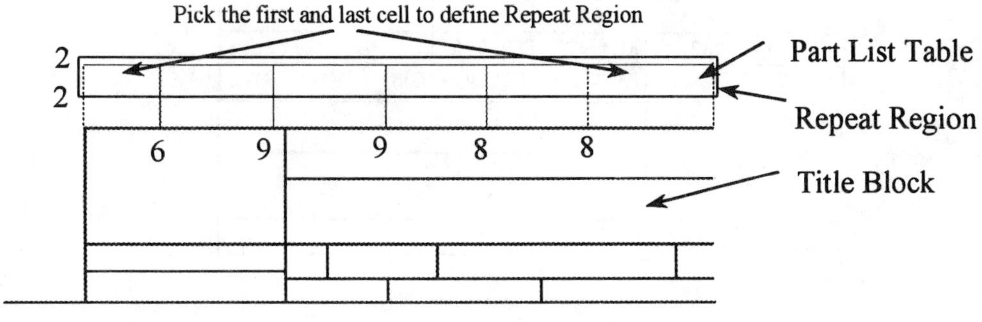

Figure 4-139 The Parts List Table

We will add headings for each column. Choose **Enter Text**, **Keyboard**, and pick the first cell. Type **INDEX**, and this name is added to the first cell. Add text to the other five cells such as: **PT NUMBER, PT NAME, PRICE, QTY** and **MATERIAL**. The part information is added into the table by defining the Repeat Region that indicates to Pro/ENGINEER from which parameter the related information will be extracted. Choose **Repeat Region**, and **Add**. Click on the first and last cells of the top row. The top row now will be highlighted. Choose **Attributes**, select the repeat region, **No Duplicates, Recursive** and then select **Done/Return**. **Enter Text, Report Sym**, and pick the first cell in the repeat region. Choose **rpt..., index**. The text rpt.index will show in the cell. Pick the third cell and select **asm..., mbr...,** and name. This action provides the name of each component in the assembly. Pick the fifth cell and select **rpt..., qty** to show the quantity information for each component as illustrated in Figure 4-140.

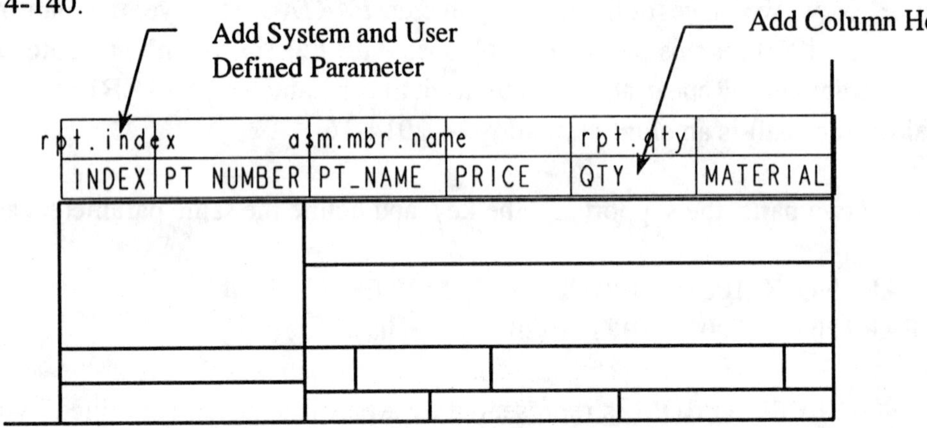

Figure.4-140 Add Column Heading and Define Repeat Region

Next, we are going to apply the user defined parameters to corresponding columns. Pick the second cell and select **asm..., mbr**..., and **User Defined**. Type *PARTNO*. Repeat the above steps to add *PRICE* and *MATERIAL* into the fourth and sixth cell. **Regenerate** if necessary. The part list table is shown in Figure 4-141.

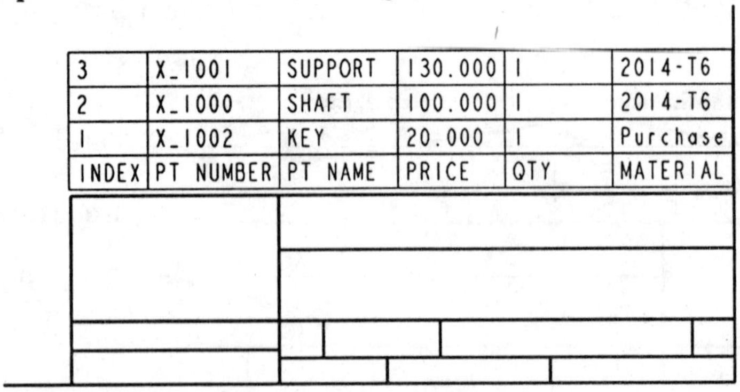

3	X_1001	SUPPORT	130.000	1	2014-T6
2	X_1000	SHAFT	100.000	1	2014-T6
1	X_1002	KEY	20.000	1	Purchase
INDEX	PT NUMBER	PT NAME	PRICE	QTY	MATERIAL

Figure 4-141 Part List Table

After the part list table is complete, the assembly drawing may be modified by using **Modify** command.

To show the BOM balloons in the assembly drawing, choose **Table, BOM Balloon, Set Region**, and select a location inside the table. Choose **Show**, **By View**, and pick the front view. The balloon symbols appear on the assembly drawing. However, they may not be properly placed. Choose **Detail**, **Mod Attach**, and pick the leader of the second balloon. Choose **On Surface**, **Dot** and pick a new position for this balloon. Using the same procedure move the other two balloons to a suitable location as shown in Figure 4-142.

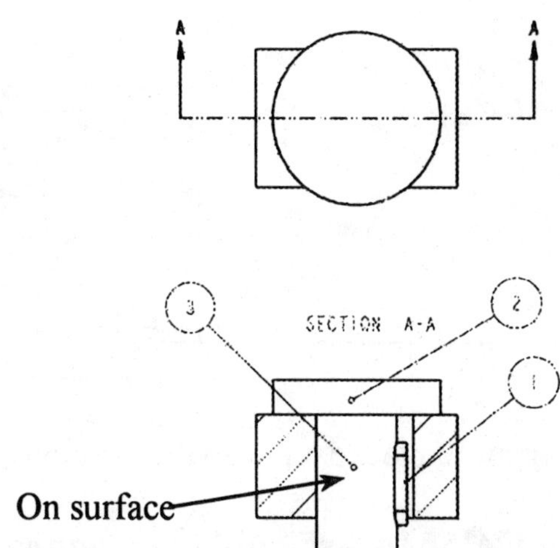

Figure 4-142 Add Balloons to Assembly Drawing

In Pro/ENGINEER, an exploded view may also be shown in the assembly drawing. Choose **Views**, **General**, **Exploded, Scale**, and **Done**. Select a center point for the exploded view. Choose EXP0001 from the **SEL STATE** menu to retrieve the exploded view we have already created in this example. Enter 0.05 for the scale. The exploded view of the shaft, key, and block are shown in Figure 4-143.

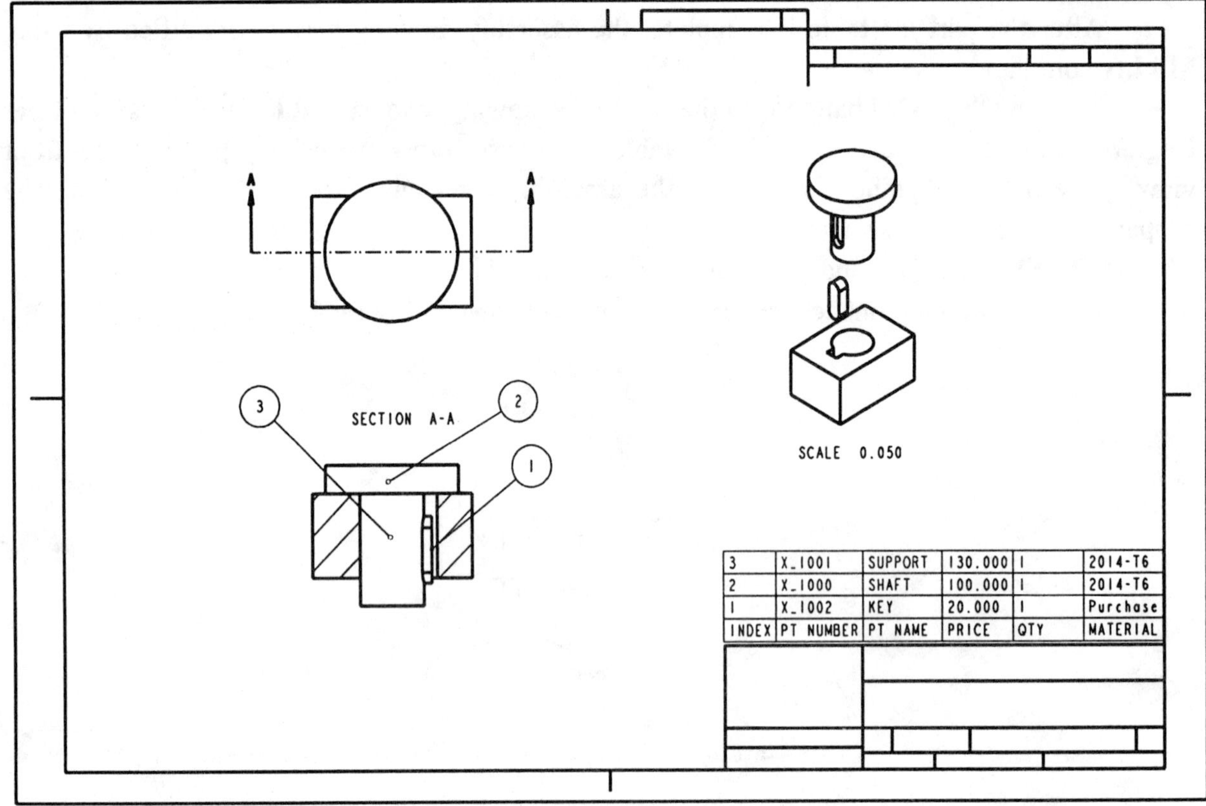

Figure 4-143 Adding an Exploded View to the Assembly Drawing

4.4.5 Example 4-17: Adding Color to Assembly Drawings

In example 4-15, we assembled three components, a block, a shaft and a key. When there are only three components, it is easy to distinguish each component in the assembly. However, consider a more complex assembly, such as an automobile engine, which has a hundred or more components. It is extremely difficult to identify each component in the assembly. The question is --- How may we clarify the 3-D assembly drawing? A natural way is to represent different components with different colors. In this example, we will use the assembly from Example 4-15 to demonstrate the method for adding color to assembled components.

To begin the process, we retrieve the assembly created in Example 4-15 by following the command sequence: **Main**, **Mode**, **Assembly**, **Retrieve**. Type the file name of the assembly drawing created in Example 4-15 in the message window. This assembly is displayed in the drawing window for the assembly.

The Pro/ENGINEER system has only one default color, --- white. Choose **Environment** from **MAIN** menu, and then select **Shading** and **Done** from the **Environment** menu. The assembly displayed on the monitor screen is shown in Figure 4-144. Before adding color to each of the three parts, we must define the colors we intend to employ.

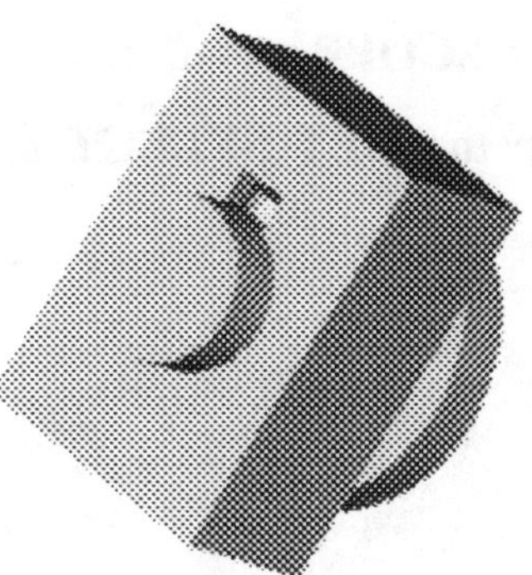

Figure 4-144 Default Shading of the Assembly

To define your new color, you just follow the procedure described previously in Example 4-9. In this example, we define three colors, red, green and yellow. After completing the definition of the colors, we may add color to each component in the assembly. Choose **Set** from **APPEARANCES** menu, and the **USER COLOR** window appears. Choose one color, such as red, from the **USER COLOR** window and a new menu named **SCOPE** under **APPEARANCE** menu appears. Choose **Subassembly** from **SCOPE** menu. Use the mouse to select the part whose color is to be changed to red. The best technique to choose the component is through **Model Tree**, especially when many components are included in the assembly. Choose **Done Sel** from **GET SELECT** menu, and the color of the part selected turns to red. Following the same procedure, assign different colors to the three components to generate the 3D colored illustration presented in Figure 4-145.

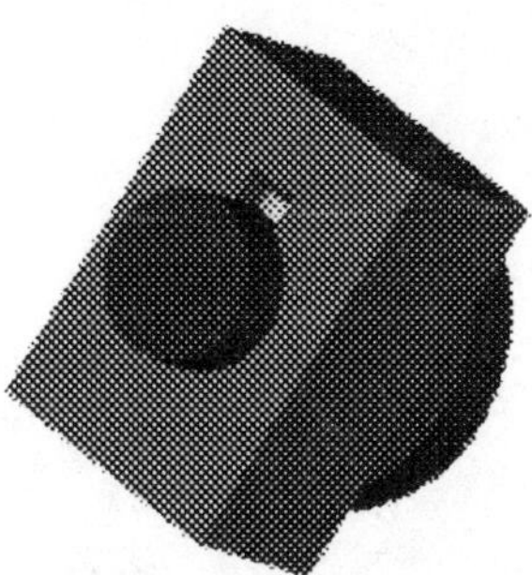

Figure 4-145 A 3D Assembly Drawing with Color[2]

---

[2] Again the illustration in Figure 4-145 is represented in this printing in shades of gray. The use of color in low-volume and low-cost publications is prohibitively expensive.

If the color is not suitable and the representation is to be returned to either white or shades of gray, choose **Unset** from the **APPEARANCES** menu. The **SCOPE** menu appears, and then choose **Subassembly** from the **SCOPE** menu. Select the part whose color you want to revoke, and choose **Done Sel** from the **GET SELECT** menu, and the color of the part returns to its original tone.

4.5 Special Case Studies

In the old days, say 1940s, 1950s, and 1960s, design engineers used pencils and drawing instruments to prepare drawings. They had difficulty in drawing many of the mechanical components that incorporated gears and threads. Using a T-square and triangles to draw orthogonal projections of these features seemed impossible, and was not realistic. Therefore, symbols were adopted to simplify the representation of those features. For example dashed lines and specific symbols have been used for decades to represent gears and threads. This practice has become a convention in engineering design, and it has been implemented in most of the commercial CAD systems.

However, such a symbolic representation of these features causes difficulties when creating solid models for assembly drawings if the assembly includes gears and threaded fasteners. In addition, new manufacturing technologies, such as EDM and rapid prototyping, require detailed information on the geometric characteristics of surfaces on an object to be manufactured. Replacement of the symbolic representation of features by true geometric shapes is required.

In this section, four case studies are described. The first two deal with drawing threads. The third case deals with preparing engineering documentation for a gear. The fourth case describes the techniques for preparing drawings for a component fabricated from sheet metal.

4.5.1 Example 4-18:Create a Solid Model for External Threads Using Helical Sweep

In this example, we demonstrate techniques used to create external threads. Before you create the external threads, you must first create a cylinder to provide the base part into which the threads are cut. The geometry and dimensions of a cylinder prepared for the threads are shown in Figure 4-146. It is important to note that the location of the coordinate system is located on the smaller end of the cylinder. Remembering the orientation of the coordinate system with respect to the part geometry is essential. The datum systems are arranged as follows:

DAT3 x-y plane   DAT1 y-z plane   DAT2 z-x plane

The procedure to create the cylinder geometry is listed below:

(1) Establish a reference system and a coordinate system.

(2) Create the shaft by selecting the **PART** mode, then choose **Create, Solid**, **Protrusion**, **Extrude**, and **Solid**, **Done**. Next choose **One Side** from the **Attributes** menu, and then **Done**. Choose DTM3 as the **Sketching Plane**, which indicates that the x-y plane will be the initial sketching plane. The arrow indicating the direction of the protrusion should point toward the positive z direction; choose **Okay**. Next, select DTM2 as the **Top** reference of the Sketch View.

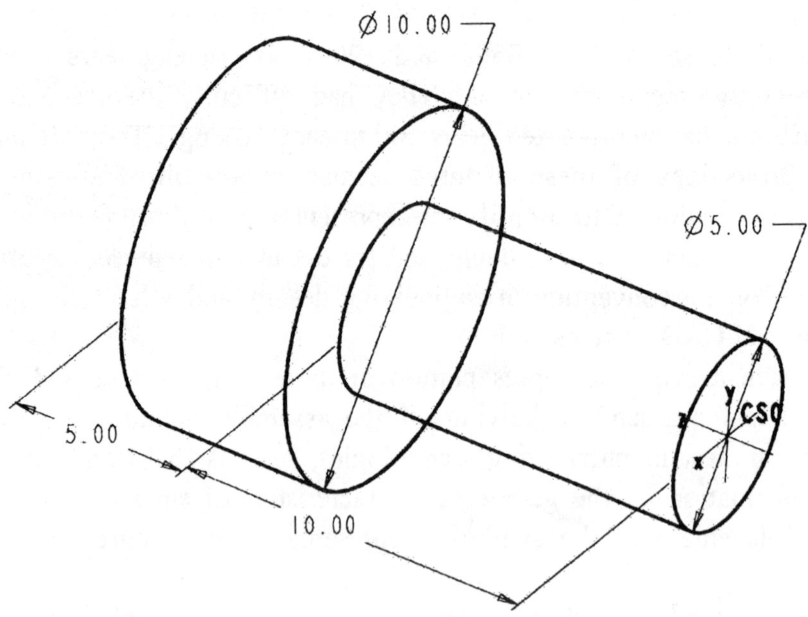

Figure 4-146 Geometry and Dimensions of the Cylinder

(3) Select **Mouse Sketch**, and sketch a Circle (use the MMB). Select the intersection of DTM1 and DTM2, or the CSO, as the center of the circle. **Align** the circle to the CSO, or to DTM1 and DTM2. Modify the diameter of the circle to be 5 and then **Regenerate**. Choose **Done** from the **SKETCHER** menu, and then **Blind**, **Done** from the **SPEC TO** Menu. Type in 10 to respond to the prompt in the message window as the length of the cylinder. Finally, select **OK** from the ***Element Inform*** Window to complete the creation of the shaft portion of the fastener body.

(4) To create the disk at the end of the fastener, begin with the **FEAT** menu and follow the command sequence:

Create, Solid, **Protrusion**, **Extrude**, and **Solid**, **Done**.

Choose **One Side** from the **Attributes** menu, and then **Done**. Choose the other end of the shaft, which is parallel to DTM3 as the **Sketching Plane**, and DTM2, again, as the **Top** reference plane. Pay careful attention to the **Direction** of the extrusion, and be certain that it is pointing away from the shaft.

Select **Mouse Sketch**, **sketch** and draw a circle to represent the disk using the same procedure as employed to draw the circle for the shaft. **Align** the circle to the **CSO**. After aligning and dimensioning the section, **Regenerate** the Model and then **modify** the diameter of the disk to 10. **Regenerate** again and choose **Done**. Choose **Blind** and then define the length of the shaft by entering 5. Select **OK** from the ***Element Inform*** Window, to complete the disk. The drawing on the monitor screen appears as illustrated in Figure 4-146.

After you have created the cylinder, you cut the **THREAD**. There are two types of threads, namely right-handed and left-handed. The following procedure demonstrates techniques used to create both types, and describes how to orient and locate the cutting tool used in forming the threads.

First let's cut a right-handed thread. When cutting a thread using Pro/ENGINEER, you must always know the orientation and location of the cutting tool with respect to the shaft. The kinematics of cutting a thread on a lathe must be followed when you form threads under the Pro/ENGINEER design environment. Figure 4-147 illustrates the orientation and location of the cutting tool related to the cylinder (shaft) when forming right-handed threads. Note the direction of movement of the cutting tool as indicated by the direction of feed.

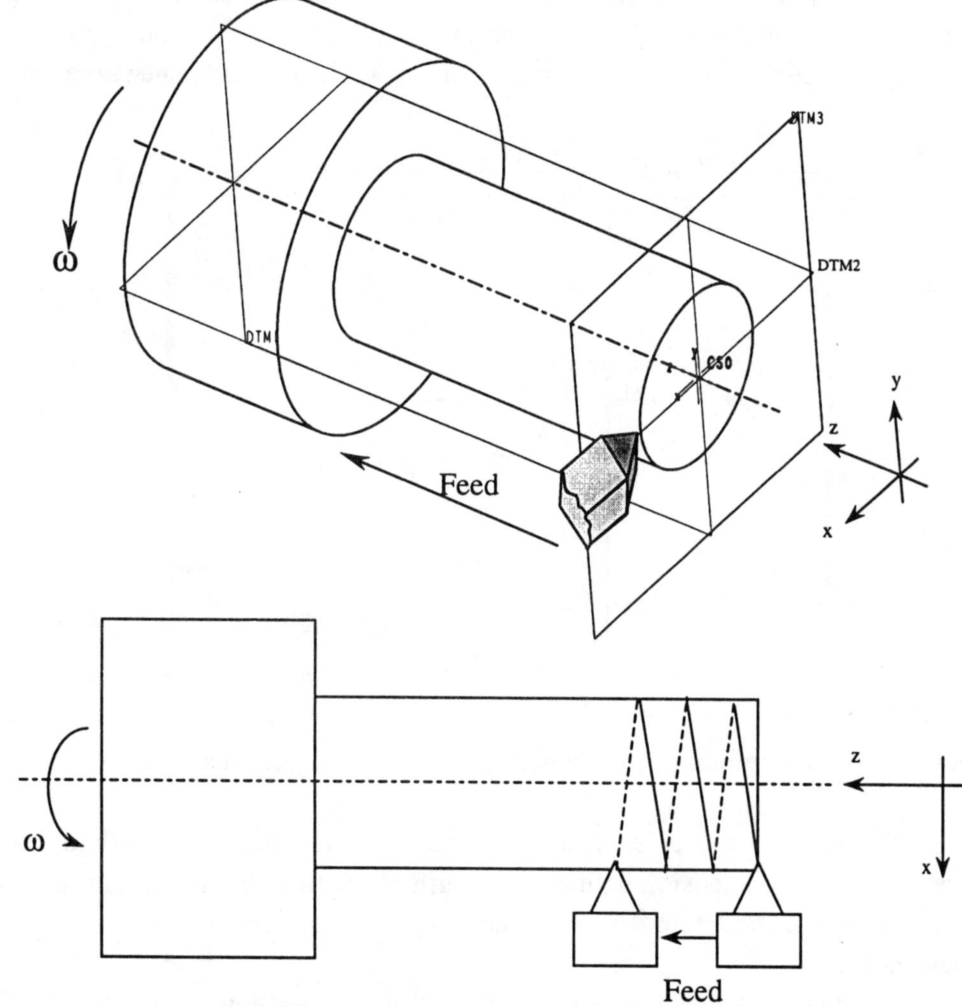

The steps used to create a thread are listed below:

(1) Choose **Solid** from the **FEAT** menu and **Cut** from the **Solid** menu.
(2) Choose **Advanced**, **Solid**, **Done** from the **SOLID OPT** menu.
(3) Choose **Helical Swp**, **Done** from the **ADV FEATOPT** menu.

(4) Choose **Constant**, **Thru Axis**, **Right Handed**, **Done** from the **ATTRIBUTES** menu. By choosing constant, the distance between the coils (the pitch in this case) will be constant. Also note that the sweep trajectory will follow the right hand rule.

(5) Choose DTM2 (the x-y plane) on the shaft as the **Sketching Plane.** (The sketching plane should always be through the axis of rotation of the shaft.)

(6) Select the direction of DTM2, and note that the direction of the arrow attached to DTM2 is in the negative y direction. Then click **Okay**. Choose the outside surface of the disk as the **left** reference plane.

(7) Draw the rotation axis, which coincides with DTM 1. Select **Line**, **Center Line** from the **GEOMETRY** menu, and **Align** the centerline with DTM1.

(8) Draw a solid line to denote the feed direction and cutting side. The first point you pick in drawing this line will be the beginning point of the sweep, as illustrated in Figure 4-148. It also defines the rotation direction of the workpiece as the negative z direction.

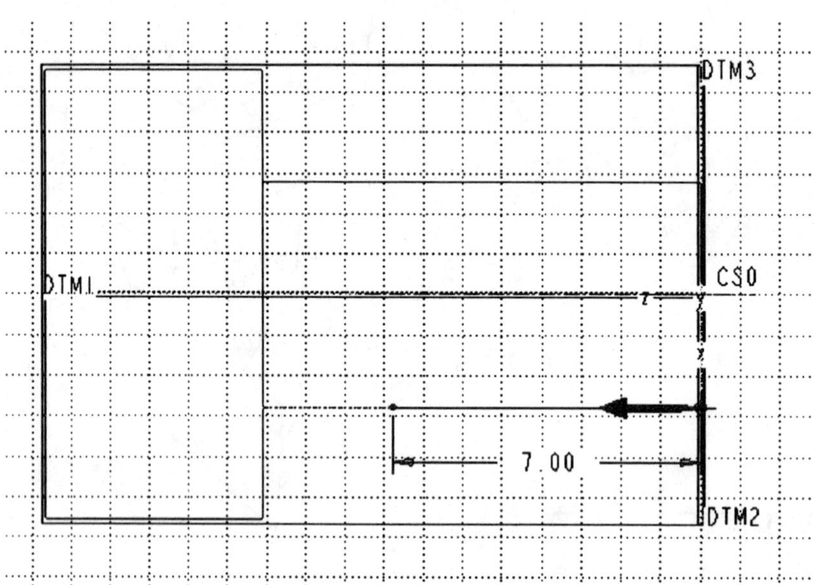

Figure 4-148 Definition of the Centerline, the Feed Direction and the Cutting Side

(9) **Align** this line to the side of the cylinder. **Dimension** the length of the line for sweeping. (This dimension defines the length of the threaded portion of the shaft.)

(10) **Regenerate** the model and **Modify** the length of the solid line to be 7. **Regenerate** again and **Done**.

(11) **Enter** the **Pitch** value as 0.5. The pitch must be at least equal to the width of the cross section of the cut to produce a full height thread. In this example, the pitch is 0.1 larger than the base width of the thread.

(12) Select the **Sketching View** menu, and. **Sketch** the section profile of the tool used in cutting the treads. The cutter is a triangular with equal sides. Note the dimension of 0.10 defined in Figure 4-149, which is the distance between the centerline of the tool and the intersection of the two dotted lines. Also note that the tool form must be a closed loop.

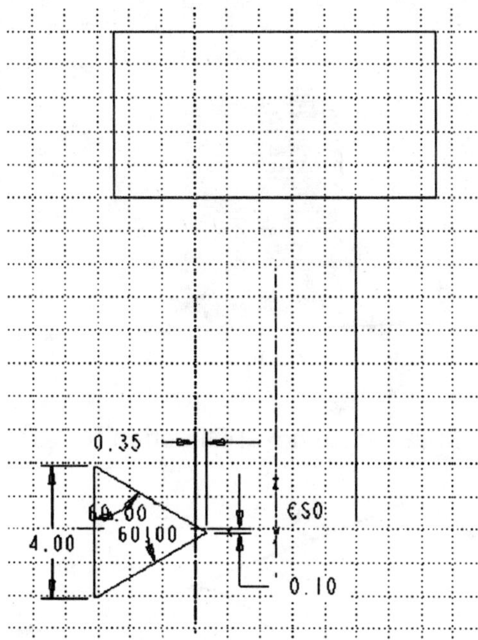

Figure 4-149 Sketch View and Dimensions of the Cutter.

(13) **Dimension** the cross section as shown in Figure 4-149. The starting point for the cut is very critical. If the tip of the cutter is started from inside of the shaft, some uncut material on the end of the shaft remains that caused problems in assembly. On the other hand, if the cutter is located too far from the shaft to contact the end surface of the cylinder, the thread also may not be generated.

(14) Choose the direction of the area to be removed to form the threads. The correct direction is shown in Figure 4-150. Select **Okay**.

(15) Select **Preview** from the ***Element Inform*** Window and then **OK**. The shaft with threads cut following the directions specified above is shown in Figure 4-151. Note the thread geometry at the starting point indicates the cutter tip started from the end of the shaft, and no redundant material remains to interfere with the assembly operation.

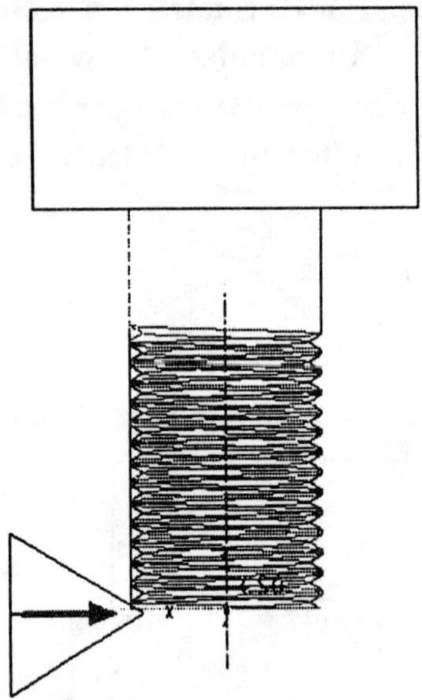

Figure 4-150 Material Removal Direction for Right-handed Thread

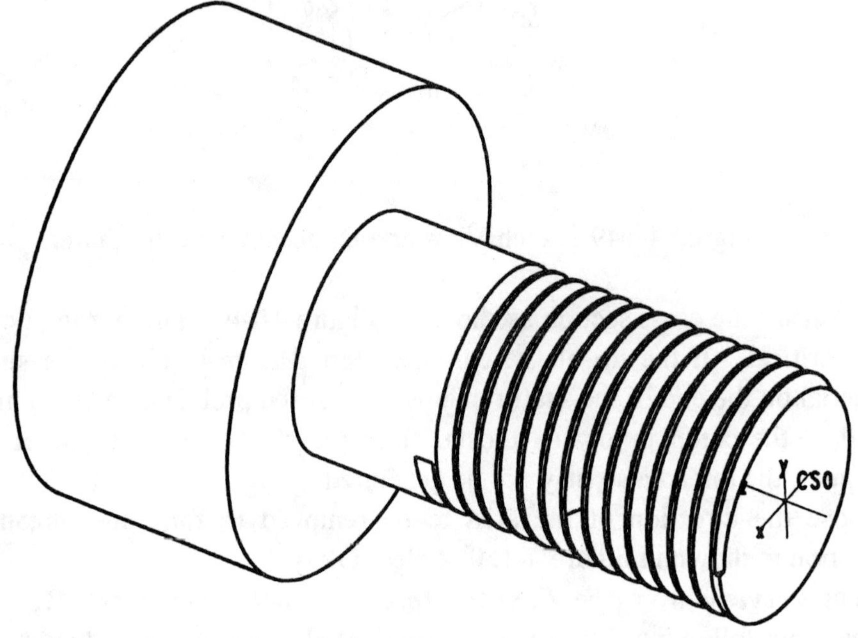

Figure 4-151 A 3D View of a Right-handed Thread Generated in Pro/ENGINEER

The procedure to create left-handed threads is similar to that of creating right-handed threads. However, some differences occur, such as the location of the cutting tool and cutting side. Figure 4-152 illustrates the location of the cutting tool and feed direction with respect to the cylinder when generating left-handed threads.

(1) Choose **Solid** from the **FEAT** menu and **Cut** from the **Solid** menu.
(2) Choose **Advanced**, **Solid**, **Done** from the **SOLID OPT** menu.

(3) Choose **Helical Swp**, **Done** from the **ADV FEATOPT** menu.

(4) Choose **Constant**, **Thru Axis**, **Left Handed**, **Done** from the **ATTRIBUTES**. Note that the sweep trajectory follows the left hand rule.

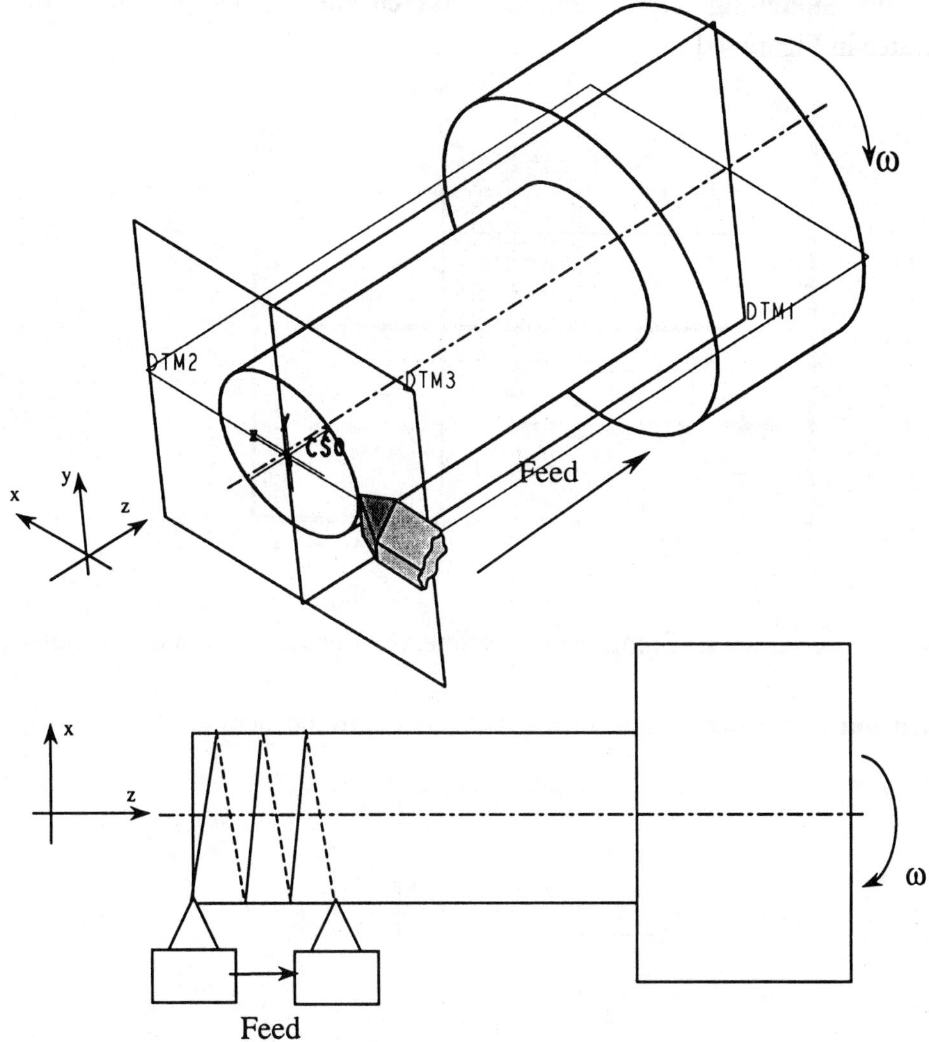

Figure 4-152 The Thread Cutting Process for External Left-handed Threads

(5) Choose DTM2, the x-z plane, on the shaft as the **Sketching Plane**.

(6) Select the direction of DTM2, and note that the direction of the arrow attached to DTM2 is pointing in the negative y direction. Click **Okay**. Then choose the outside surface of the disk as the **Right** reference plane.

(7) Draw the rotation axis coinciding with DTM1 for the sweep, and choose **Line**, **Center Line** from the **GEOMETRY** menu. Align the centerline to DTM1.

(8) Draw a solid line to indicate the feed direction and cutting side. The first point selected for drawing the line is the beginning point of the thread cutting sweep as illustrated in Figure 4-153. It also defines the direction of rotation of the workpiece to be the positive z direction.

(9) **Align** the line to the side of the shaft. **Dimension** the length of the sweep line.

(10) **Regenerate** the model and **Modify** the length of this line to be 7. **Regenerate** again and **Done**.

(11) **Enter** the **Pitch** value as 0.5.

(12) Select the **Sketching View** menu and **Sketch** the section profile of the cutter as illustrated in Figure 4-153.

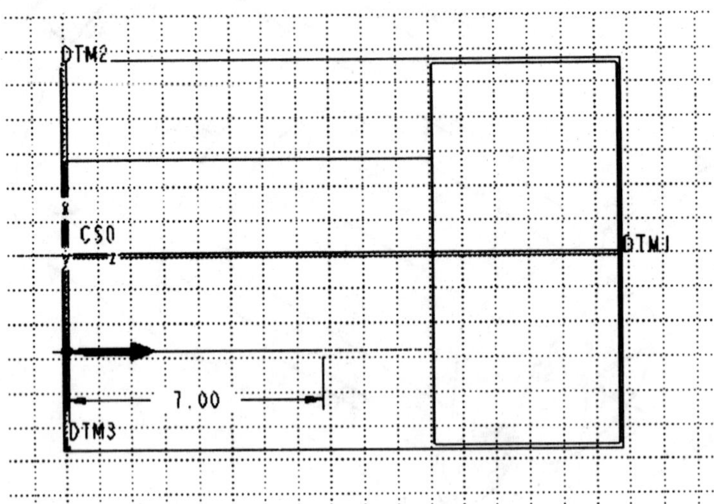

Figure 4-153 A Sketch Defining the Centerline and Feed Direction of the Cutting Tool

(13) **Dimension** the section of the cutting tool relative to the shaft as shown in Fig. 4-154.

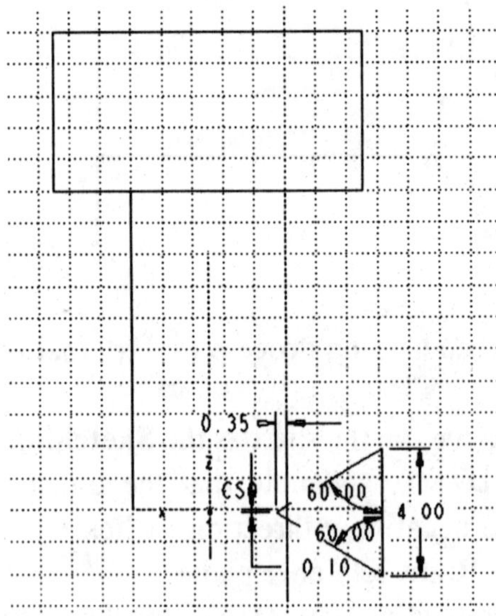

Figure 4-154 Sketch of the Section Profile of the Thread Cutting Tool

(14) Choose the direction in which the area is to be removed. The correct direction is shown in the Figure 4-155. Select **Okay**.

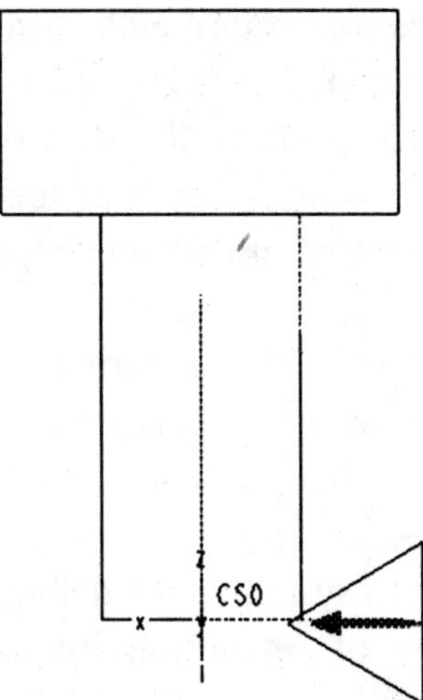

Figure 4-155 Direction of Material Removal for a Left-handed Thread

(15) Select **Preview** from the ***Element Inform*** Window and then **OK**. The shaft with a left-handed thread shown in Figure 4-156 is displayed on the screen of the monitor.

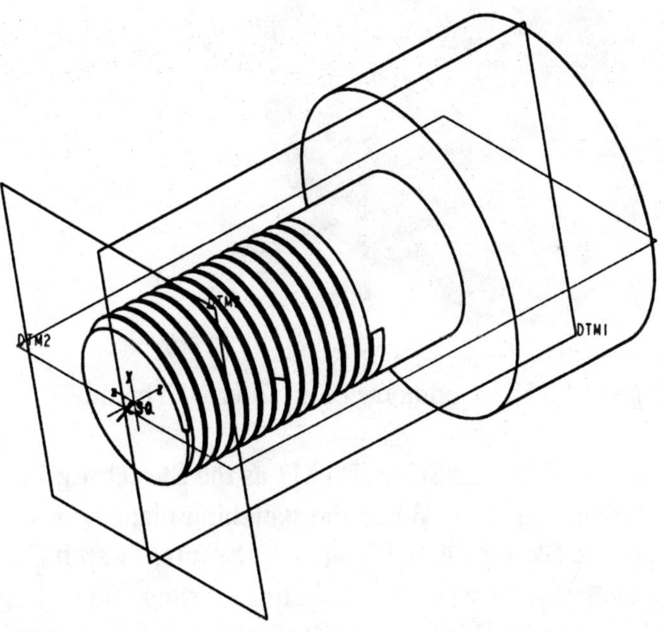

Figure 4-156 Completion of the Left-handed Thread

4.5.2 Example 4-19: Drawing an Internal Thread

The procedure for drawing an internal thread is similar to that employed in creating external threads. If you understand the orientation and location of the cutting tool during the thread cutting operation, which was described in the previous example, it is easy to cut threads regardless of whether they are external or internal. The only signifimayt difference is the orientation of the cutting tool, which is located inside of the cylinder instead of outside. The purpose of this example is to provide an additional description of the helical sweep, and to demonstrate the use of a new function called **spline**. The geometry of an internal thread is illustrated in Figure 4-157. The first step in generating an internal thread is to create the component containing the thread. The sequence presented below is employed to produce the body defined in figure 4-157:

(1) Establish the default datum planes system and a default coordinate system.
(2) Create the shape of the body of revolution by using **Create, Solid**, **Protrusion**, **Revolve**, and **Solid**, **Done**.

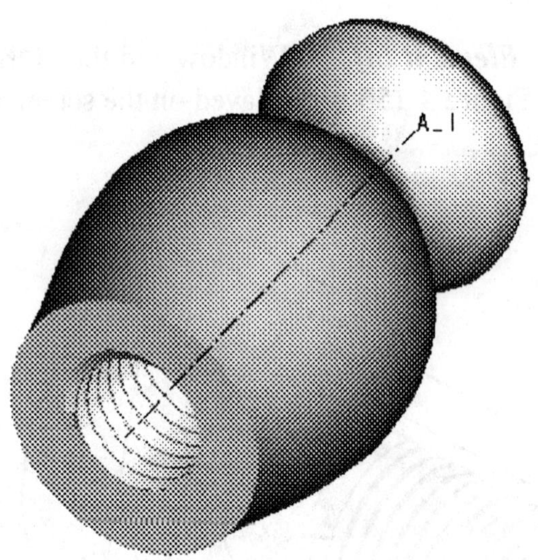

Figure 4-157 Geometry of an Internal Thread

Choose **One Side** and then **Done**. Select DTM3 as the **Sketching Plane**, **Okay**, and choose DTM2 as the **Top** reference plane. When the sketching plane appears, start from the **SKETCHER** menu, and choose **Sketch**, **Adv Geometry**, **Spline**, **Sketch Points**, and **None**. Sketch six points by using the left mouse button, including starting and ending points for the spline feature as illustrated in Figure 4-158. Click MMB to disengage the spline sketch.

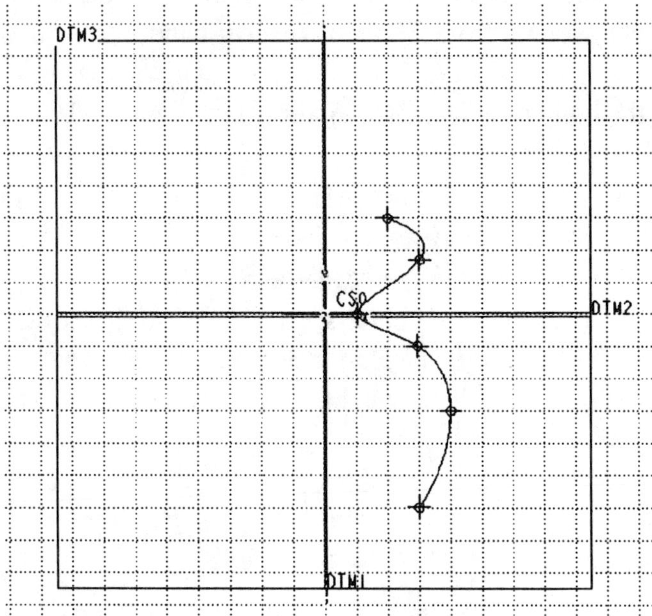

Figure 4-158 Sketching Points for the Spline

Sketch three straight lines by using **Mouse Sketch** to close the 2D view shown in Figure 4-158. Among them, two lines should be parallel to DTM2 and another should be collinear with DTM1 or the y-axis. Also draw a **Center line** collinear with DTM1 as the axis of rotation. Click the y-axis (DTM1) four times to **Align** the centerline as well as the vertical line to DTM1. **Align** the point, which is located on or near the x-axis to DTM2. **Dimension** all six of these points defining positions along both the x and y-directions. The dimensions on the 2D view are shown in Figure 4-159.

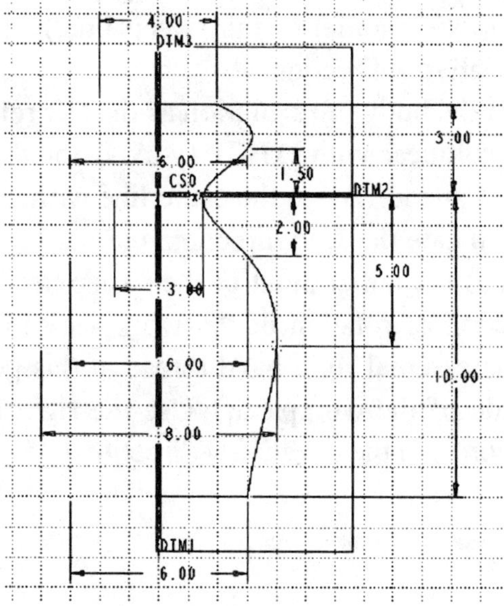

Figure 4-159 Dimensions of the 2D View

Select **Regenerate** and then **Done**, and define the rotation angle as 360°. Select **Done** again and choose **OK** from ***Element Inform*** window.

(3) Create a hole in the body of revolution, as shown in Figure 4-160.

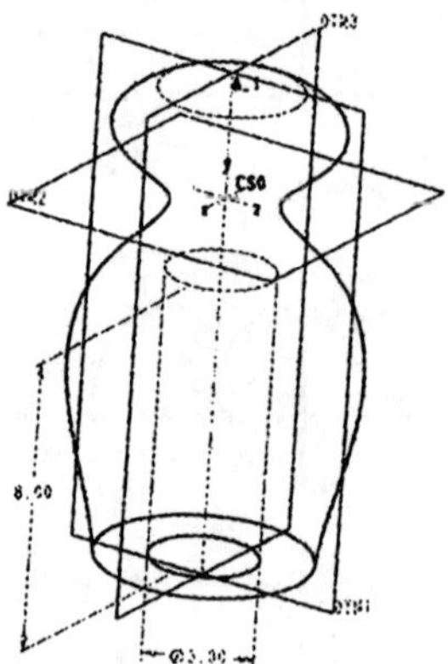

Figure 4-160 3D Drawing of a Body of Revolution

The next task is to **THREAD** the hole inside the cylinder. The procedure is similar to the creation of external threads. Begin by employing the following sequence of commands:

(1) Choose **Create, Solid** from the **FEAT** menu and **Cut** from the **Solid** menu.
(2) Choose **Advanced, Solid, Done** from the **SOLID OPT** menu.
(3) Choose **Helical Swp, Done** from the **ADV FEATOPT** menu.
(4) Choose **Constant, Thru Axis, Right Handed,** and **Done** from the **ATTRIBUTES** menu.
(5) Choose DTM3, the x-y plane, as the **Sketching Plane**. (The sketching plane must always be through an axis parallel to the sweep line.) The direction of the arrow is pointed toward the negative z direction.
(6) Select the small end of the body of revolution as the **left** reference plane.
(7) Draw a rotation axis collinear with DTM1 (y-axis) for the sweep: **Line, Center Line** from the **GEOMETRY** menu. **Align** the centerline with DTM1.
(8) Draw a solid line to indicate the feed direction and cutting side. The first point selected to draw this line defines the initial point of the sweep, as illustrated in Figure 4-161. It also defines the direction of rotation of the workpiece to be the negative y direction. Note, when cutting the external right-hand thread at this position, the rotation direction is also pointing outside of the workpiece. Use the right hand rule to verify this fact. The direction of rotation is in the negative y direction in Figure 4-161.

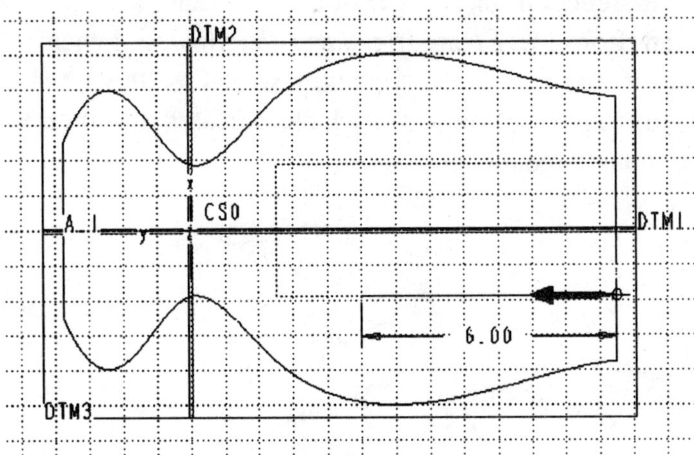

Figure 4-161 A Sketch Showing the Center of Rotation and the Cutting Direction

(9) **Align** the cutting line to the side of the cylinder. **Dimension** the length of the line for sweeping.

(10) **Regenerate** the model and **Modify** the length of the solid line to be 6. **Regenerate** again and **Done**.

(11) **Enter** the **Pitch** value as 0.5. The pitch must be at least equal to the width of the cross section of the cut. In this particular example, it is 0.1 in larger than the base width of this section.

(12) Select the **Sketching View** menu to prepare a sketch of the tread form. **Sketch** the section profile of your cutter, noting that the section defining the thread shape is a closed loop. **Dimension** the section as shown in Figure 4-162, and observe that the starting point of the cut is located external of the cylinder.

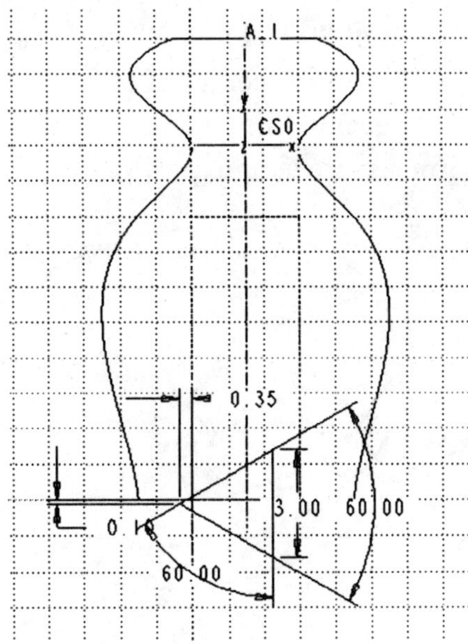

Figure 4-162 Sketch of the Section Profile of the Thread (Cutter)

(13) Choose the feed direction for the cutting tool. Select **Okay**.

(14) Select **Preview** from the ***Element Inform*** Window and then **OK**. A drawing of an internal thread is displayed on the monitor screen as illustrated in Figure 4-163. Note the thread shape at the starting point, which indicates the cutter tip start from outside of the body to eliminate redundant material.

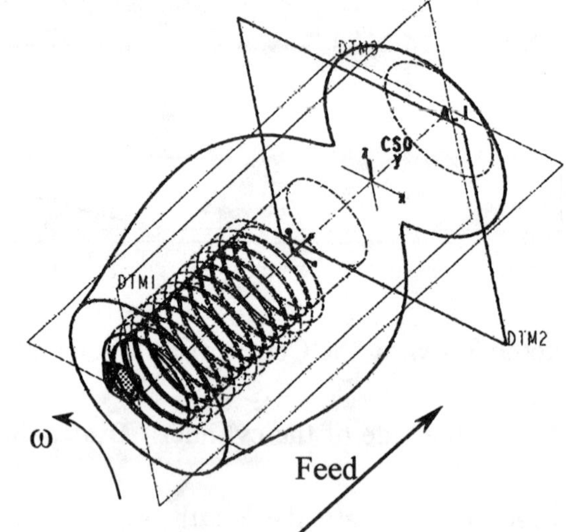

Figure 4-163 A Right-handed Internal Thread

4.5.3 Example 4-20: Drawing of a Spur Gear

In this example, we will draw a spur gear to demonstrate the use of **Pattern** in Pro/ENGINEER. The gear under consideration has 30 teeth as illustrated in Figure 4-164. To begin, create a disk with a diameter of 20 and a depth of 5, as shown in Figure 4-164. Choose the **Two Sides** option to place DTM3 in the center of the disk.

For simplicity, we will not generate the profile of the involute gear teeth because of their complexity. Instead, we will form a more simple profile.

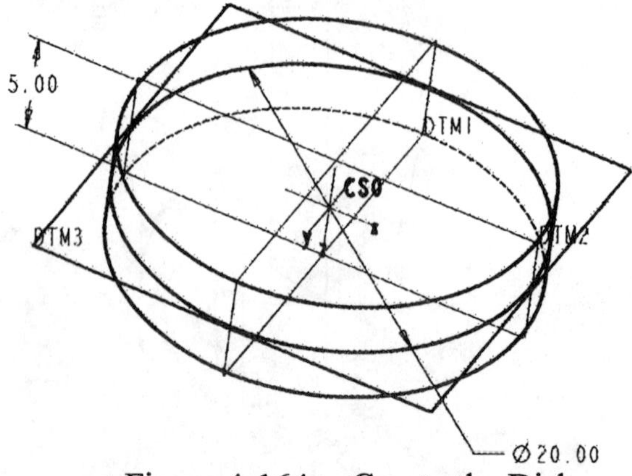

Figure 4-164 Create the Disk

From the **FEAT** menu, choose **Create, Cut, Extrude, Solid, Done, Two Sides, Done**. Pick DTM3 as the sketch plane, and choose **Top**, DTM2 to set its orientation. Now sketch the section to be cut from the circumference of the disk. Choose **Arc, Concentric**, and select some location on the circumference of the disk. Place the starting point above DTM2 and the end point below this axis. Draw two lines at an angle to DTM2. Align the end points of the two lines to the circumference of the disk. Dimension the section as indicated in Figure 4-165a. Choose **Regenerate, Done** and select **Thru All** twice. Figure 4-165b shows the new geometry.

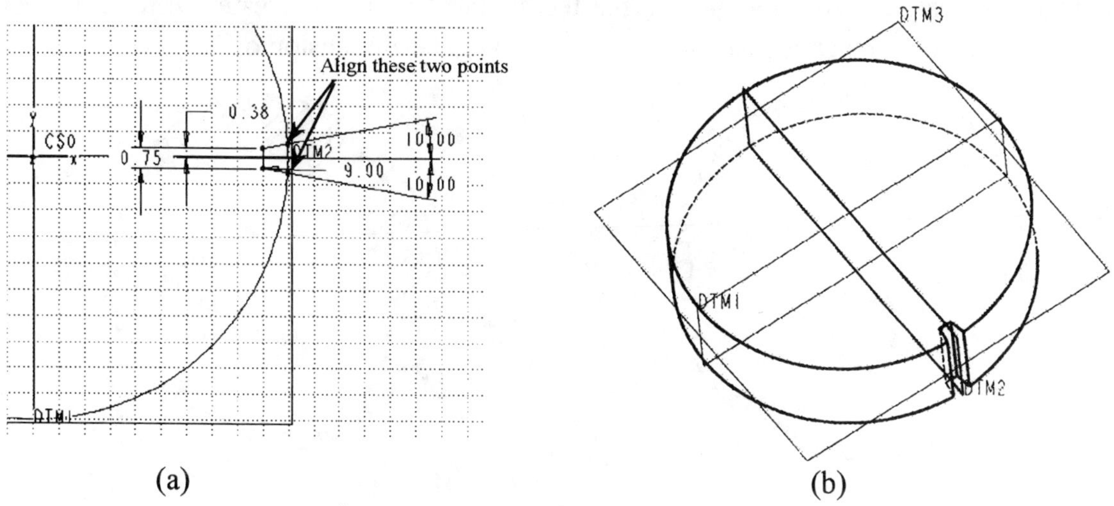

(a) (b)

Figure 4-165 Generating the First Cut

We may copy the slot shown in Figure 4-165b to form the gear teeth. However, an angular dimension is needed to define the location of the adjacent cuts. To determine the angular dimension, we will create an identical cut adjacent to the first one.

Choose **FEAT, Copy, Move, Select, Dependent, and Done**. By choosing **Dependent**, the profile of the additional cuts will depend on the original cut, making it easier to modify the tooth profile. Select the first cut and choose **Rotate**, **Crv/Edg/Axis** and pick the axis of the disk as the axis of rotation. Choose the rotation direction based on right hand rule. Enter the angular dimension of 360/30=12°. (30 teeth are cut into the gear). Choose **Done Move, Done** and the second cut is made as shown in Figure 4-166.

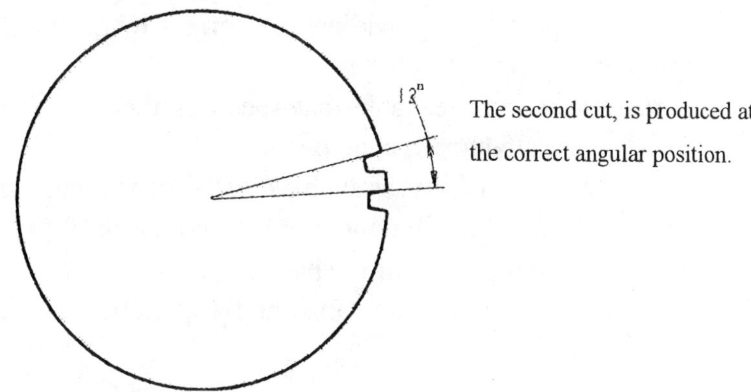

Figure 4-166 A Second Cut is Made at the Correct Angular Dimension

With the angular dimension established, we may now pattern (repeat) the cut. Select **Pattern** from the **FEAT** menu and select the second cut. Choose **General** from the **PAT OPTION** and **Done**. Choose the 12° dimension as the **FIRST** direction. Enter the increment of 12°. When the message prompts selection of another dimension for the **FIRST** direction, choose **Done**, and then enter 29 for the total number of cuts in this direction. Since this is a rotational pattern, there is no requirement to select a **SECOND** direction. Choose **Done** to complete the pattern. The gear is shown in Figure 4-167(a) and (b). You may also try to select the angular dimension of 10° for placing the teeth. Pro/E will not create the teeth properly, because the second cut must be made to establish the angular dimension of 12°.

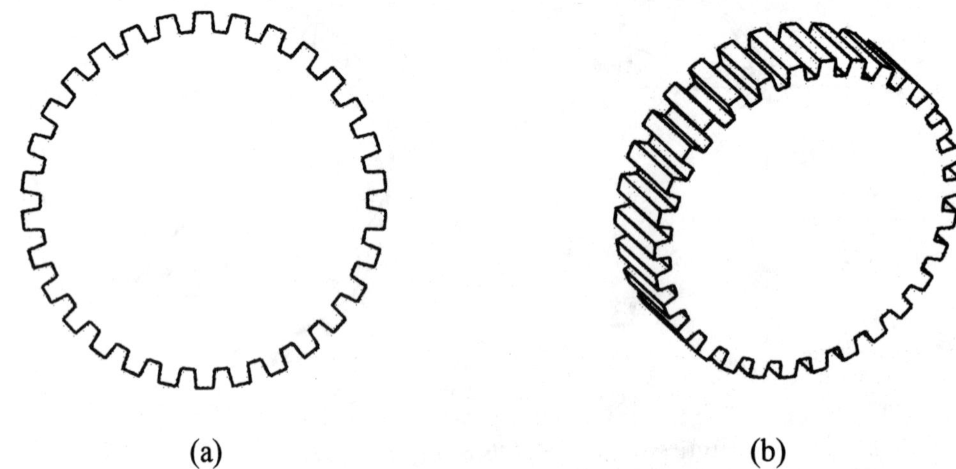

(a) (b)

Figure 4-167 Drawing of a Spur Gear Generated in Pro/ENGINEER

4.5.4 Example 4-21: Creation of a Punched Part Using Pro/SHEET METAL

Pro/SHEET METAL is an optional module for Pro/ENGINEER that greatly facilitates the design of a sheet metal part. With this module, a user is able to:

1. Design sheet metal parts defining the volume and the support of components in an assembly.
2. Create bend tables providing the correct length of sheet for bends of different radii and material thickness.
3. Create a bend order table that specifies the order, bend radius and bend angle used to manufacture a component.
4. Add sheet metal features such as walls, bends, cuts, punches, notches and forms to the part in either an unbent or bent condition.
5. Create the flat pattern of the part.
6. Create a drawing that contains the flat pattern, the sheet metal part, and the bend order table.

In this section, we present an example to demonstrate the procedure for employing Pro/SHEET METAL in the design process.

The shape and dimensions of the sheet metal component to be designed in this example are shown in Figure 4-168.

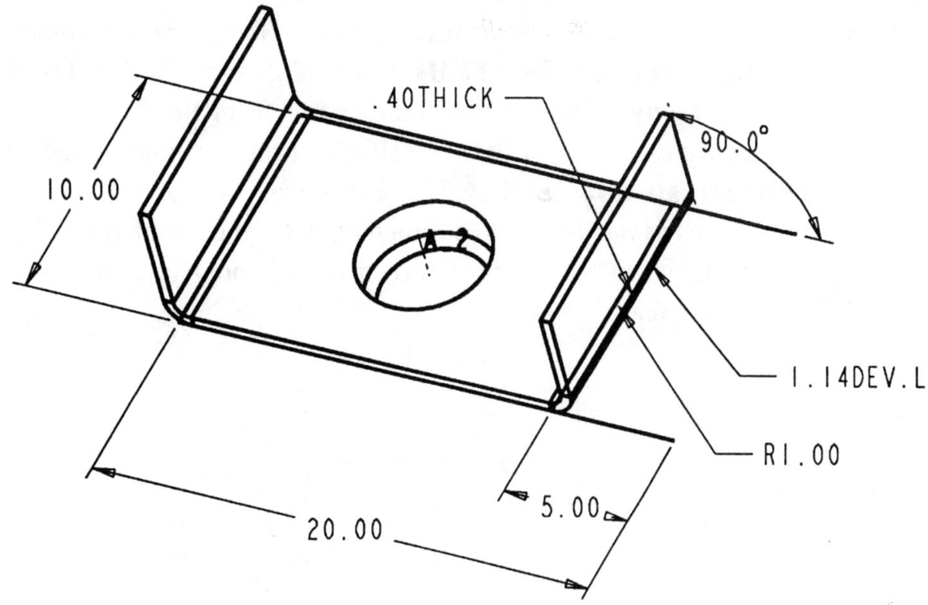

Figure 4-168 Dimensions and Geometry of a Sheet Metal Component

In this example, we start by creating a wall using the function called **Wall** in **Sheet Metal** and then use several other functions in Pro/SHEETMETAL, such as **Bend, Unbend, Bend Back and Extruded Wall** to create other features.

To begin create the necessary features with the following command sequence:

First Step: Create the first wall for the sheet metal component.
Second Step: Use **Bend** to get the "L" shape of this part.
Third Step: Use **Unbend** and **Bend Back** to adjust the shape of the feature.
Fourth Step: Use **Extruded Wall** to create the second wall of the part. This is an alternative technique for forming the "L" shape.
Fifth Step: Use **Form** to generate a punched hole feature.

First, let us create the first wall. From the **MODE** menu, select **Sheet Metal**, then **Create**, and type the file name *sheet* at the prompt in the message window then hit enter. The entire name of the file is sheet.prt.

From the **PART** menu, select the following command sequence to create a reference system: **Feature, Create, Datum, Plane, Default**

To begin forming the sheet metal part, select the following commands starting from the **FEAT** menu: **Create, Sheet Metal, Wall, Flat, Done**.

Leave **Setup New, Plane, Pick** as the default value, then elect DTM3 as the **Sketching Plane**. Select **Okay** when the arrow is pointed away from you. Note, this is a different process than that employed in previous examples.

Choose DTM2 as the horizontal referencing plane by picking **TOP** in **SKET VIEW** menu and clicking on DTM 2. We are ready to sketch the geometry of the sheet metal part.

In this example, let's first draw a rectangle. Choose **Rectangle** from the **Geometry** menu, choose two points to indicate the diagonal of the box. Sketch the rectangle along DTM 1 and DTM2. Choose **Alignment** from **SKETCHER**, and align the two edges of the rectangle to DTM1 and DTM2 respectively. Then choose **Dimension** from **SKETCHER**, and specify the dimensions for the rectangle. You now have specified all the necessary dimensions as well the geometry. Select **Regenerate** from the **SKETCHER** menu.

Modify the dimensions as needed to give dimensions for the rectangle of 20 x 10. Then **Regenerate** again. The display on the monitor screen is presented in Figure 4-169. Choose **Done** from the **SKETCHER** menu.

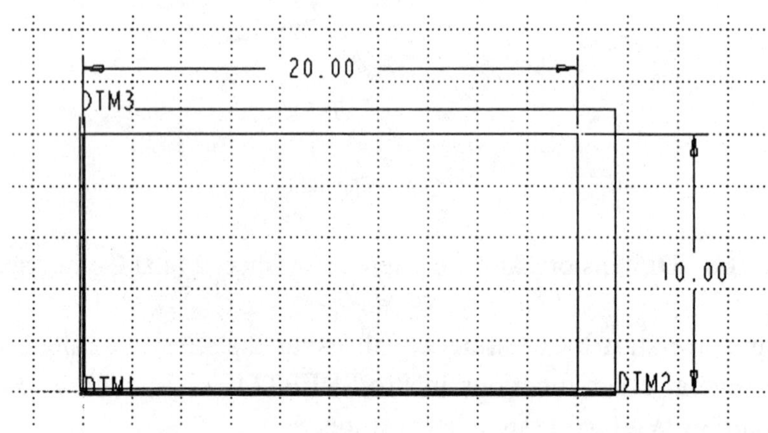

Figure 4-169 Sketch of a Portion of the Sheet Metal Component

Specify the thickness for the sheet as 0.4. Choose **Done** from the ***Element Inform*** Window, and the display on the screen appears as indicated in Figure 4-170.

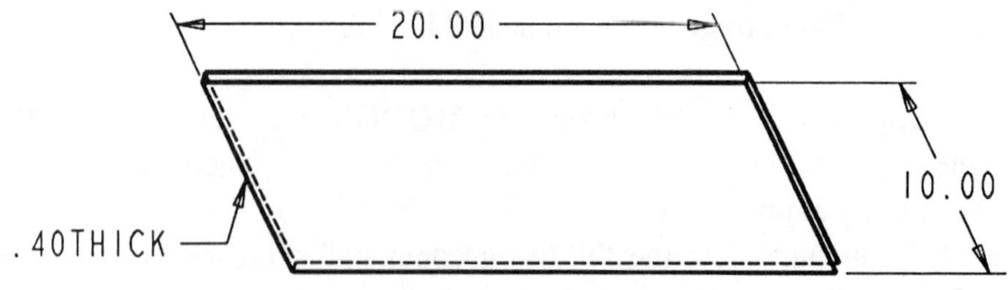

Figure 4-170 3-D view of the Metal Sheet

The part on the screen appears with green and white surfaces. These colors aid in visualizing the part because sheet metal parts are comparatively thin. Because sheet metal parts have a constant thickness, Pro/ENGINEER creates the white surface by offsetting the material thickness from the green color. Remember that the orientation of the white surface is toward the green one. The side surfaces are not added until the part is fully regenerated.

The next step is to create one of the side walls based on the rectangular metal sheet just created. Use **Bend** in this step by selecting the following commands from the **FEAT** menu:

Create, Sheet Metal, Bend, Angle, Regular, Done.

Choose **Part Bend Thl** from the **USE TABLE**, then **Done**. Choose **Outside Rad** from **RADIUS SIDE** menu and then **Done**.

Leave **Setup New, Plane, Pick** as the default value, and select DTM3 or the green side of the part as the **Sketching Plane**. Select **Oka**y when the arrow is pointing into the plane. Choose DTM2 as the horizontal reference plane by picking **TOP** in **SKET VIEW** menu and clicking on DTM2. We are ready to sketch the geometry.

First we must identify the bend line by choosing **Line** from the **Geometry** menu and **Vertical** from the **LINE TYPE** menu. Select one point at the upper edge of the rectangle and the other at the bottom edge to indicate the start and finish positions of the bend line. Click the MMB to finish the line drawing. Choose **Alignment** from **SKETCHER**, and sequentially align the two end points of the bend line to the two side edges. Then choose **Dimension** from **SKETCHER** menu, and specify the distance between the bend line and one of the end edges of the rectangle.

Regenerate, then **Modify**, following the same procedure as previously described. The distance is 5 for the bend line. Choose **REGENERATE** again, then choose **Done** from **Sketcher**. The dimension for the bend line is shown in Figure 4-171:

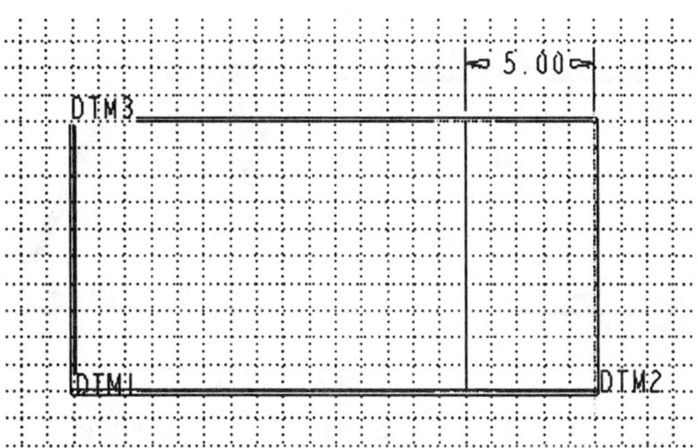

Figure 4-171 Sketch of the Bending Line

Hit **Oka**y from **BEND SIDE** when the arrow direction is pointing to the right. If not select **Flip** then **Okay** from **DIRECTION** menu to reverse the direction of the arrow point so that the fixed area is the larger portion of the metal sheet

Choose **No Relief** from **RELIEF** menu then **Done**.

Choose **90.000°** from **DEF BEND ANGLE** menu, then **Done**.

Choose **Enter Value** from **SEL RADIUS** menu, type in 1 in the message window, and hit **Enter**.

Choose **OK** from ***Element Inform*** Window, to obtain the form shown in Figure 4-172.

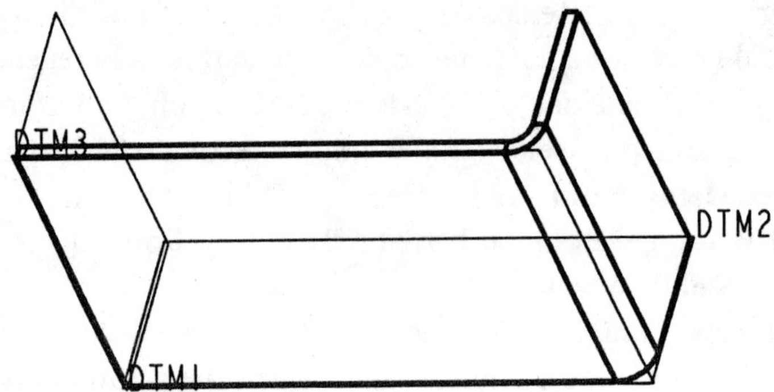

Figure 4-172 Bending One Side of the Sheet Metal Component

The third step is to **Unbend** and **Bend Back** the bending feature we just created. Follow the command sequence below to activate the **Unbend** feature:

Feature, Create, Sheet Metal, Unbend, Regular, Done

Select the portion of the component that stays fixed during unbending, by moving the cursor onto the larger portion of the plate (the left side in Figure 4-172). Choose **Unbend All** from **UNBENDSEL** menu, then **Done**.

Select **OK** from ***Element Inform*** Window, to display the drawing presented in Figure 4-173.

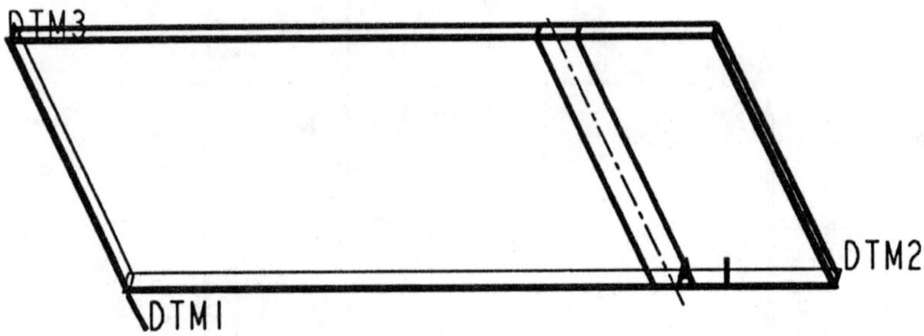

Figure 4-173 Unbending the Side of the Sheet Metal Component

Follow the command sequence listed below to bend back the side:

Feature, Create, Sheet Metal, Bend Back, Done

Select the plane that will stay fixed during the bending back operation by moving the cursor to a location the larger part of the green surface. Choose **BendBack All** from **BENDBACKSEL** menu, then **Done**.

Select **OK** from ***Element Inform*** Window, to display the previous drawing (Figure 4-172).

The Fourth step is to create the other wall (side) at the opposite end of the sheet. Of course, you may use the **Bend** function to create the second wall, but we introduce another function of **Sheet Metal**, called **Extrude Wall** to accomplish this task.

Follow the command sequence below to employ the **Extrude Wall** feature:

Feature, Create, Sheet Metal, Wall, Extruded, Use Radius, Done

Choose **Part Bend Thl** from **USE TABLE,** then **Done.**

Choose **Outside Rad** from **RADIUS SIDE** menu and then **Done**.

Now go to **Pick**, and select one side edge at the opposite end of the bending wall to attach the wall.

Select **Okay** from the **DIRECTION** menu, if the direction of the arrow is pointing inward toward the part. This direction causes the extrusion to attach to the first wall of the metal sheet part. The **SKETCHER** window is illustrated in Figure 4-174a.

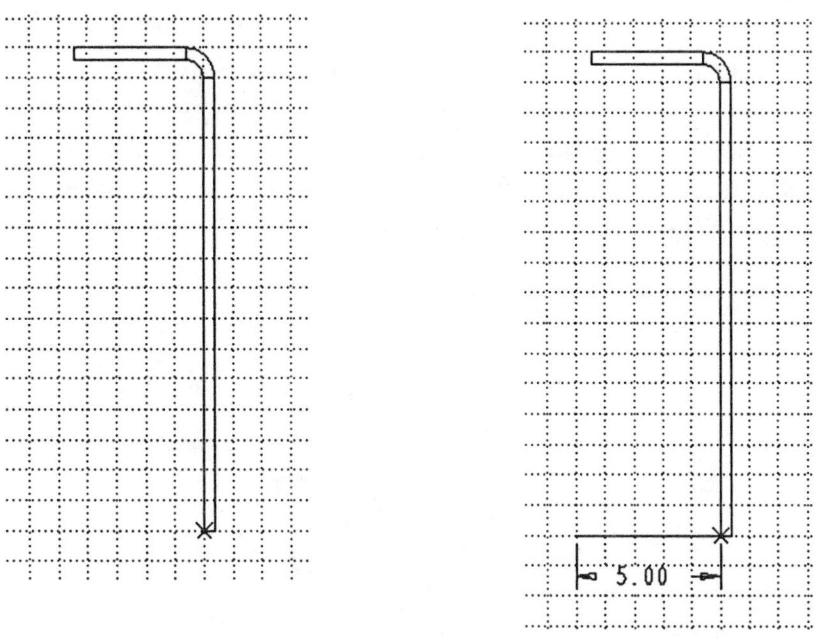

a) The Initial Sketch Plane

b) The Extrusion Line for the Second Side

Figure 4-174 2D Sketch for the Second Side Wall

First we create an extrusion line, by choose **Line** from the **Geometry** menu. Select one point at the upper left corner of the part and then extend the line horizontally to somewhere on the right. Note that you need to click the MBM to complete the line drawing. Choose **Alignment** from **SKETCHER**, and align the start point to the green surface of the part. Then choose **Dimension** from the **SKETCHER** menu, and provide the dimension for the length of this line.

Regenerate, then **Modify**. The width of the wall will be 5, choose **Regenerate** again then **Done** from **SKETCHER**. This line is shown in Figure 4-174b. Choose **No Relief** from **RELIEF** menu then **Done**.

Choose **Enter Value** from the **SEL RADIUS** menu, type 1 at the prompt in the message window, then hit **Enter**.

Choose **OK** from ***Element Inform*** Window. The second side wall is generated as illustrated in Figure 4-175:

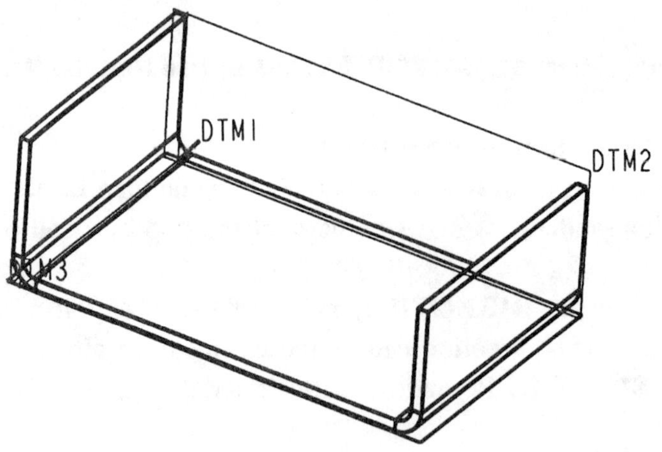

Figure 4-175 The " U " Shaped Sheet Metal Component

For the fifth Step, we use the **Form** function to create a punch form. However, it is necessary to develop a reference form for the punch before we may perform this task. The dimensions and shape of the reference form is shown in Figure 4-176:

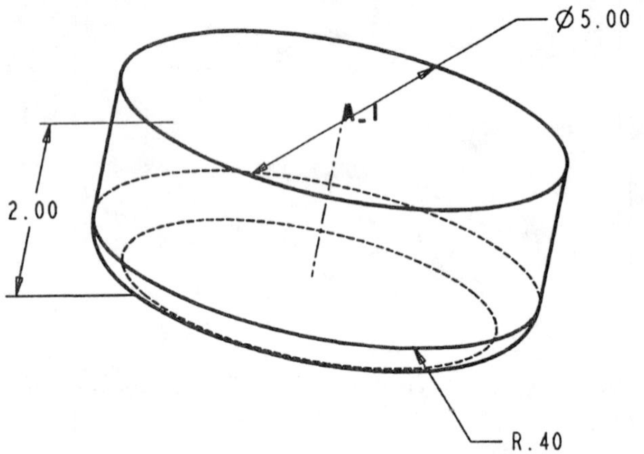

Figure 4-176 The Punch Reference Form

Name the file for the reference form you intend to draw, for example, *punch.prt*. It is important to note its orientation, because it is related to the following form process. To create the form reference part, follow the command sequence listed below:

Feature, Create, Datum, Plane, Default
Create, Datum, Coord Sys, Default, Done

To start drawing the reference form for the punch, select the following commands from the **FEAT** menu:

Create, Solid, Protrusion, Extrude, Solid, Done. Choose the **One Side** option from the **SIDES** menu, and select DTM3 as the **Sketching Plane**. Also choose DTM2 as the horizontal referencing plane by selecting **TOP** in the **SKET VIEW** menu

Choose **Circle** from the **Geometry** menu, and select a point near the origin of the coordinate system as the center of the circle and another point on the sketch plane to define the circle geometry. Choose **Alignment** from **SKETCHER** and align the center of the circle to the origin. Double click the circle to provide a dimension for the diameter. **Regenerate** then **Modify** the diameter to 5 and **Regenerate** again.

Choose the **Blind** option from the **SPEC TO** menu and then **Done**. Enter a depth of 2 for the punch form. Choose **OK** from the ***Element Info*** Window.

To create the **Round** feature for the punch form, employ the following the sequence of commands: **Create, Solid, Round, Done.** Choose **Simple** then **Done**. Select **Constant**, **Edge Chain** from the **RND SET ATTR** menu. Pick the surface of the cylinder, which is opposite to DTM3, and then **Done.**

Enter, New Value, type in the round radius: 0.4, then hit **Enter**. Choose **OK** from the ***Element Info*** Window. The reference form for the punch is illustrated in Figure 4-179.

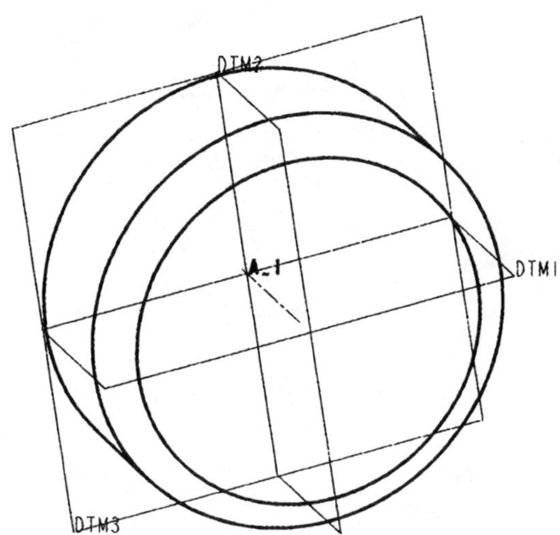

Figure 4-177 Geometry of the Punch Form and Orientation with the Datum Planes

Now we **Change Window** to sheet.prt. The sequence to create the form feature begins with the **FEAT** menu:

Create, Sheet Metal, Form, Punch, Reference, Done

Enter the file name for the reference form *punch.prt*. Now two part windows and the Components Placement window are shown on the screen. The procedure to employ the form feature is similar to the assembly process.

Choose **Mate** from the **Place** menu, and then **Select**, **Pick**. From the *punch.prt* window, select DTM3. From the **DTM ORIENT** menu, select **Yellow**, indicating that the direction of DTM3 is directed outward. Pick the white surface on the sheet metal part as the mating surface at sheet.prt. Remember that the direction of white surface is directed inward. The orientation of these two surfaces is such as to guarantee that the form feature may be generated. Otherwise, you will see an error message when you try to execute the form function.

Choose **Align Offset** from Place menu, pick DTM2 at *sheet.prt* and choose **Red** from the **DTM ORIENT** menu. Also pick DTM2 at *punch.prt* and choose **Red** from the **DTM ORIENT** menu. Then type 5 to specify the offset.

Choose **Align Offset** from Place menu, pick DTM1 at *sheet.prt* and choose **Red** from the **DTM ORIENT** menu. Also pick DTM1 at *punch.prt* and choose **Red** from the **DTM ORIENT** menu. Enter 7.5 to specify the offset, and then **Done**.

Choose **Okay** from the **DIRECTION** menu, which shows the arrow pointing to the outside of the sheet part. The finished sheet metal component and its shaded model are shown in Figure 4-178 and Figure 4-179, respectively.

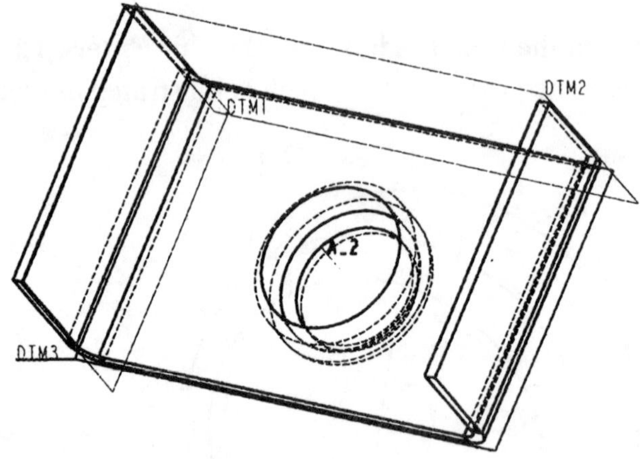

Figure 4-178 Completed Geometry of the Sheet Metal Model

Figure 4-179 Solid Model of the Sheet Metal Part

4.6 Utilization of Pro/LIBRARY

4.6.1 Example 4-22: Design by Using Pro/ENGINEER Libraries

When an engineer is in the process of designing parts and/or products, a fundamental principle he/she has to follow is "Standardization". "Standardization" means the use of components, which exist in the market. As we all know, companies do not make bolts and nuts on their own. Instead, they purchase bolts and nuts from those companies that are specialized in making standardized components, such as bolts and nuts.

Under the Pro/ENGINEER design environment, standard parts can be retrieved from the BASIC LIBRARY. The BASIC LIBRARY is a collection of commonly used parts, such as bolts, nuts, rivets, screws, and so on. In this example, we present a case study to demonstrate how to find and use the parts or features that are available to the Pro/ENGINEER users.

In this example, two plates are to be fastened together by four pairs of bolt and nut at the four corners as illustrated in Figure 4-180. The dimensions of the two plates are 2 x 2 x 0.5. Four 1/4-20 hex bolts are used to connect the two plates. The length of the bolt is 1.25", based on the thickness of the two plates.

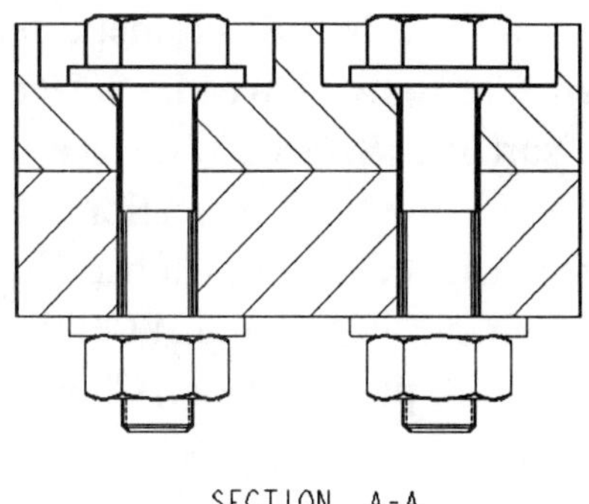

Figure 4-180 An Assembly of Two Plates and Four Sets of Bolts, Washers and Nuts

We will follow the following steps to finish this assembly. First assemble the two plates together. Then drill the four counter bore holes. Finally, we insert the fasteners including the washers, bolts and nuts to complete the assembly.

1. Create the Plate

Figure 4-181 shows the geometry of the two plates that are identical to each other.

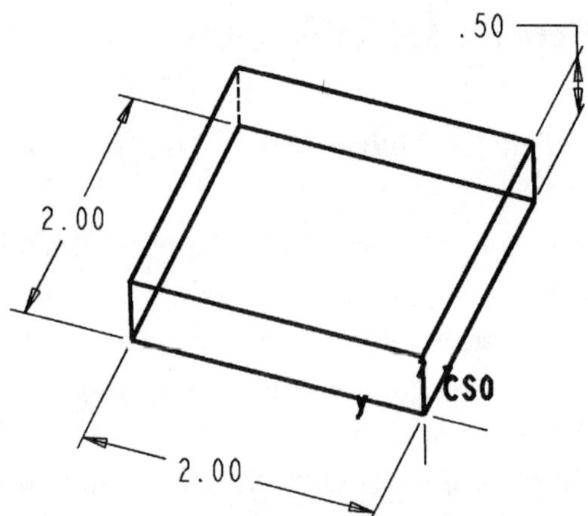

Figure 4-181 Two Identical Plates

2. Retrieve the Bolt from Pro/Library

 Part, Search/Retr, Pro/Library, /eng_prt_lib, /sqr_hex_bolt, rhb.prt (This is the part family for the hex bolt), **SelByParams, Select All, Done Sel**, .25, **SelByParams, Select All, Done Sel**, 1.250000 (L). The bolt is shown in Figure 4-182. However, only the length of the thread is defined by the library and other information about the thread has to be defined by the user.

 In Pro/ENGINEER, cosmetic feature can be used to create thread. Choose **Feature, Create, Cosmetic, Thread.** Select the cylindrical surface within the thread length as the thread surface. Select the surface where the thread begins as the start surface. Select the direction for the feature creation. The arrow should point to the thread creation direction. Under **SPEC TO** menu, choose **UpTo Surface** and select the chamfer surface. Enter the major diameter of the thread 0.225. Go to **Mod Params** to custom the parameters of the thread such as pitch and etc. Choose **Preview, Ok** to finish.

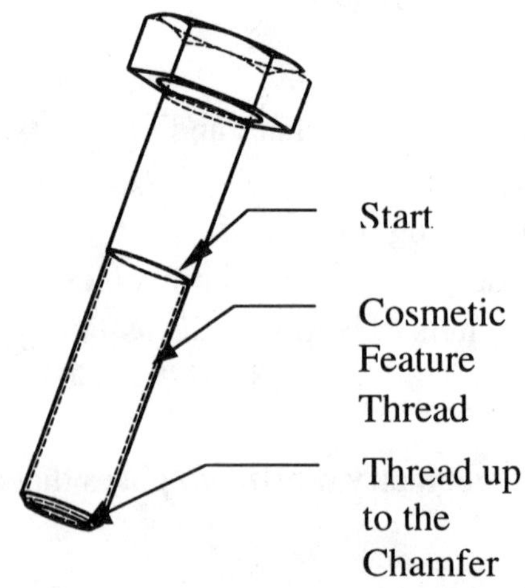

Figure 4-182 Hex Bolt

Save the bolt as a new part. Choose **Dbms, Save As, Enter** and give the name: *mybolt.*

3. Retrieve the Washer from Library

   Part, Search/Retr, Pro/Library, /eng_prt_lib, /plain_wash, wa.prt, SelByParams, Select All, Done Sel, .25N (N stands for narrow). The washer is shown in Figure 4-183. Save the file as *mywasher*.

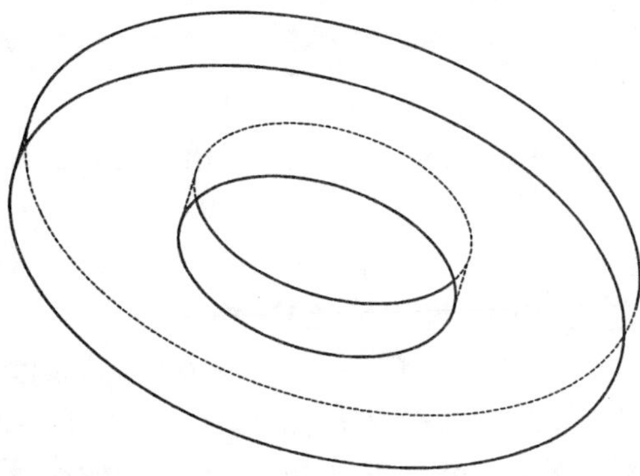

Figure 4-183 Washer

4. Retrieve the Nut From Library

   Part, Search/Retr, Pro/Library, /eng_prt_lib, /sqr_hex_nut, rhn.prt (This is the part family for hex nut), **RHN009, SelByParams, Select All, Done Sel**, .25. The nut is shown on the screen (Figure 4-184). The internal thread has to be defined.

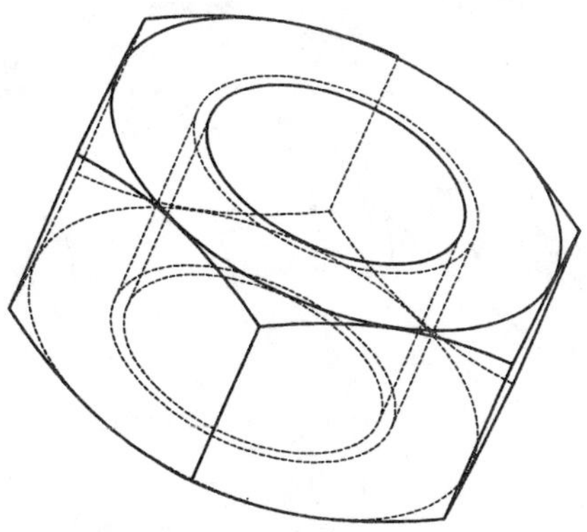

Figure 4-184 Hex Nut

Again, from **Feature, Create, Cosmetic, Thread**, Select the inner hole surface of the nut as the thread surface. Set the depth option of the thread as in step 2. The major diameter of the thread is 0.2750. Save the file as *mynut*.

5. Assembly the Two Plates

   Attach the two plates together, as shown in Figure 4-185.

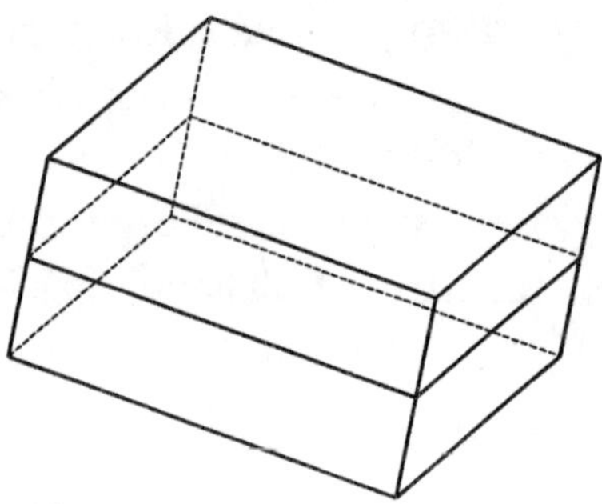

Figure 4-185 Assembly the Two Plates

6. Create the Counter Bore Hole

Besides standard parts, Pro/ENGINEER also offers a lot of useful features in its libraries. We can use these features in part or assembly design.

A datum point should always be created first to locate the feature that is going to be created. Under **ASSEMBLY** menu, select **Create, Feature, Datum, Point, On Surface**. Select the top surface of the top plate. Select two edges for reference. The distances to the two edges are 0.5 and 0.5 (Figure 4-186). We will use **pattern** to create the other three datum points.

Under **ASSY FEAT**, choose **Pattern** and select the datum point as the feature to be patterned. Select one of the dimension value 0.5 as the first pattern direction. The dimension increment is 1. Choose **Done** under EXIT menu and type 2 for the number of the total instances in the first direction. Repeat the above procedures for the second pattern direction. The dimension increment is also 1.

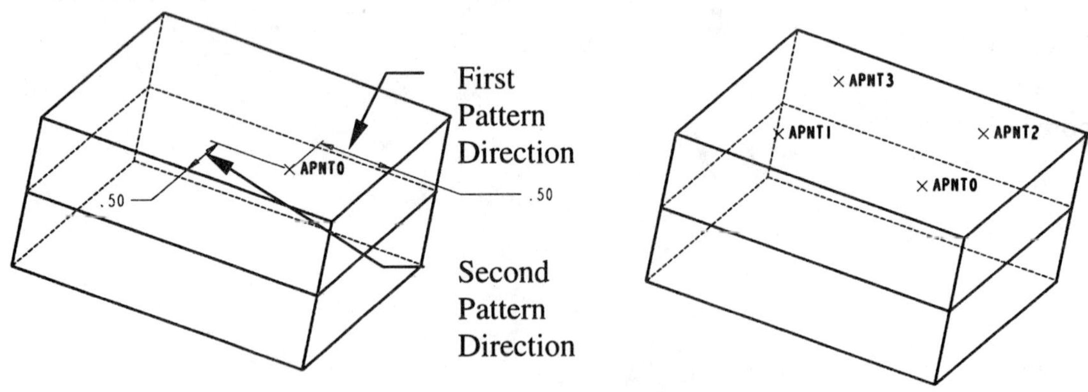

(a) Create the First Datum Point (b) Pattern the First Datum Point

Figure 4-186 Create and Pattern the Datum Point

To create the counter bore hole, under **ASSY FEAT** menu, choose **Create, User Defined, Search/Retr, Pro/Library, /featurelib, /design_udf, /design_udf_st, /hexhead_scr_hole, ushnp.gph**, Enter Y to retrieve reference part (The hole can not be shown without a reference part). Choose **SelByParams, Select All, Done Sel,** 1/4, **UDF**

Driven, Done, SameDims, Done, Read Only, Done. Select the datum point APTN0 and select the top plane of the top plate as the ref. Datum plane. Choose **Manual Sel** and select both of the plates, **Done**. The arrow should point away from the assembly (Select twice).

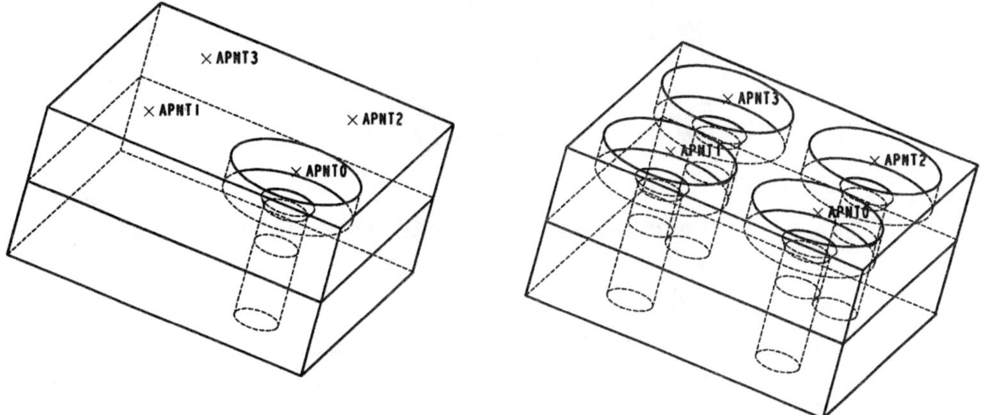

(a) Create the Counter Bore Hole (b) Pattern the Counter Bore Hole

Figure 4-187 Create and Pattern the Counter Bore Hole

Again, we can use pattern to create the other three counter bore holes.

From **ASSY FEAT** menu again, choose **Group** (The counter bore feature is actually a group). **Pattern**, and select the counter bore hole (Group USHNP) from the model tree window(If it is not displayed, choose **Tree** menu bar in Model tree window, and check off **feature** under **show** menu). We can simply choose **Ref Pattern** because this pattern will follow the previous pattern of the datum point. The assembly after finish step 6 is shown in figure 4-187.

7. Insert the Washer

Component, Assemble, Retrieve, mywasher . Assemble the washer into the counter bore hole at APTN0 as shown in Figure 4-188.

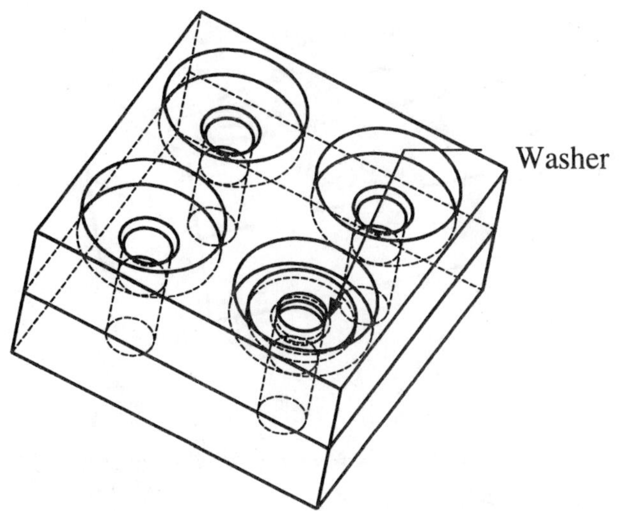

Figure 4-188 Insert the Washer

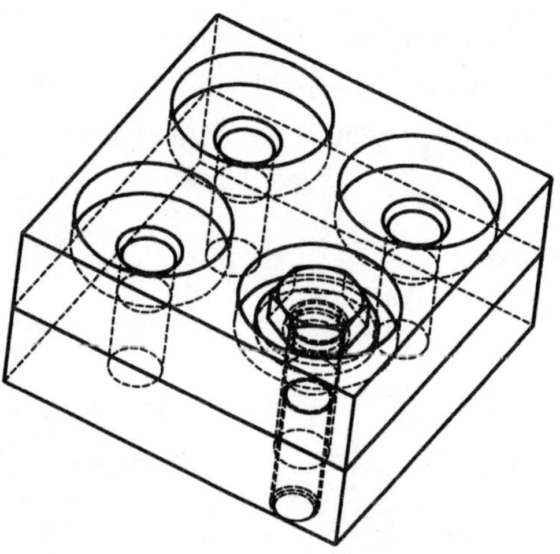

Figure 4-189 Insert the Hex Bolt

8. Insert the Bolt

   Component, Assemble, Retrieve, mynut and assemble it as shown in the Figure 4-189.

9. Insert the Washer

   Following the same step in 7, insert another washer at the other end of the bolt.

10. Insert the Nut

    Component, Assemble, Retrieve, mynut and assemble it as shown in Figure 4-190.

11. Group the Bolt, Washers and nut

    Four pairs of bolt and nut will be put into the assemble. One easy way to finish this is first group them together and then patter the group.

    To group the bolt, nut and two washers, select **COMPONENT, Adv Utils, Group, Create, Local Group**. Give a name for this group: g1. Select the last four features in the Model tree window, **Done**.

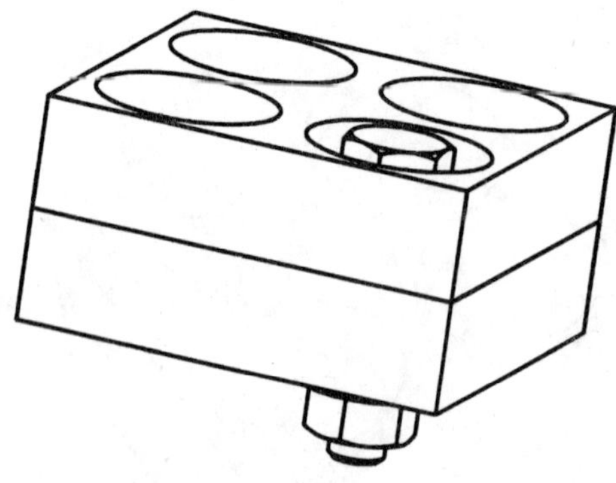

Figure 4-190 Insert a Washer and a Nut

12. Pattern the Group Feature

Now we can patter the group. **COMPONENT, Adv Utils, Group, Pattern**, Select the group we just created. Select **Ref Pattern** to follow the patter of the four counters bore holes.

Now the assembly is finished. It is shown in Figure 4-191. A new datum plane ATM5 is also created that goes through the axes of two counter bore holes. ADTM5 will be used in drawing mode. From this example, it can be demonstrated that by using libraries, the design process can be speed up.

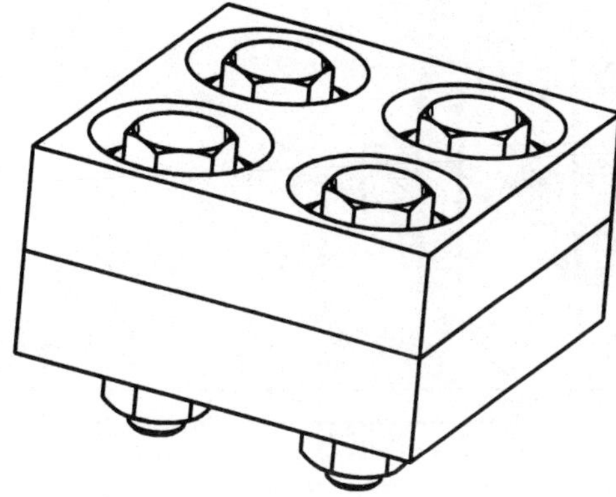

Figure 4-191 Pattern the Bolt, Nut and Washer

13. Create the Drawing

First create a top view of the assembly and then create a cross-section view along the datum plane ATM5 we created in the assemble mode.

The display of threads should follow certain standard. Choose **Draw Setup**, and **Modify**. A window will display and let you modify the setup file. Change the option: "hlr_for_threads" to "yes" and "thread_standard" to "std_ansi_imp_assy" . This will display the threads according to ANSI standard. You can also set the "thread_standard" to "std_iso_imp_assy" to display the threads according to ISO standard. Choose **Environment** and select **No hidden**.

14. Modify the Hatch Line

In the cross section view in Figure 4-192, the bolt, nut and the washers should not be sectioned and there should be no hatch line for these parts. Choose **Modify**, **Xhatching** and select the hatch line in the cross-section. The space, angle, line style can be modified by choosing the corresponding menu. To remove the hatch line for the nut, choose **Pick Xsec** and select one nut, Choose **Excl Comp** and then the hatch line will be removed. Choose **Next Xsec, Excl Comp** to remove the hatch lines of all the bolts, nuts and washers in this view.

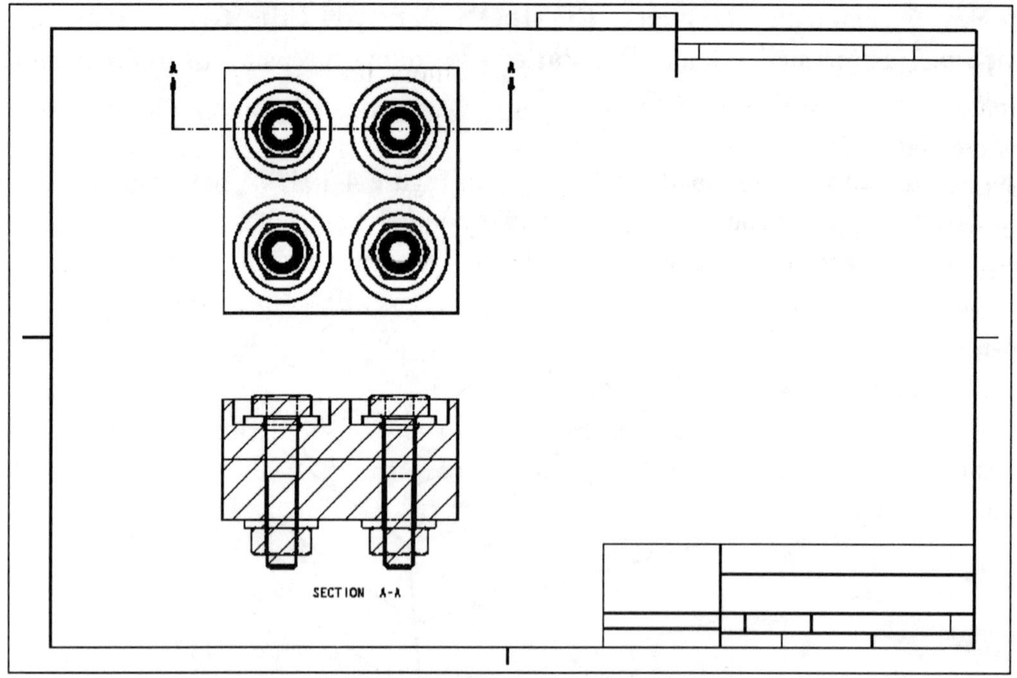

Figure 4-192 Add Views to the Drawing

Although the hatch line of the fasteners have been removed, the drawings of the fasteners in the cross-section may be still not right. Choose **Views, Disp Mode, View Disp**, and select the cross-section view. Choose **No Hidden** and the drawing of the cross section is shown in Figure 4-193.

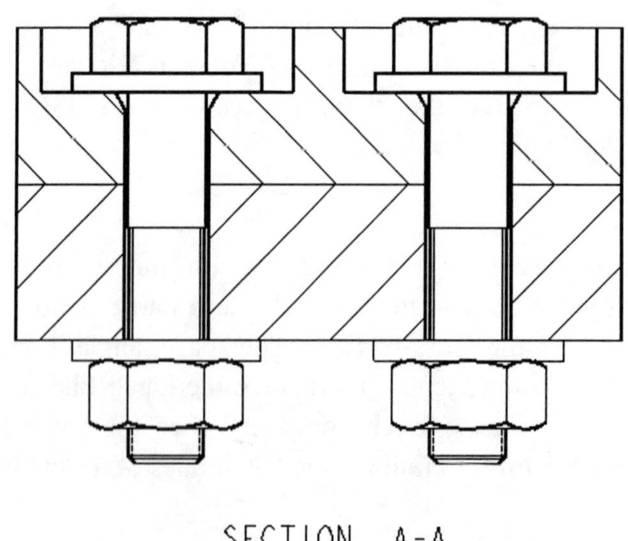

Figure 4-193 Drawing of the Bolt, Nut and Washer

4.6.2 Example 4-23: Replacement of Part Drawings after Assembly

In Pro/ASSEMBLY, you can create assembly features, such as datum planes, axes, coordinate systems, and solid features as in the **Part** mode. It is important to note that those entities created in Pro/ASSEMBLY differ from those created in the **Part** mode because they belong to an assembly and not to a specific part. Pay attention to subtractive assembly-level features intersecting a part in an assembly, the defaults set by the system are to generate instances of the intersected components, which are invisible in the part model unless you change their settings. In order to depict this concept, an example is used for the purpose of demonstration.

1. Create two plates (components)

The geometry and dimensions of the two plates are illustrated in Figure 4-194. They are the same as those shown in Example 4-13. The dimension of the bottom plate is 3 x 3 x 1/2 inches and the upper one is 2 x 2 x 0.5 inches.

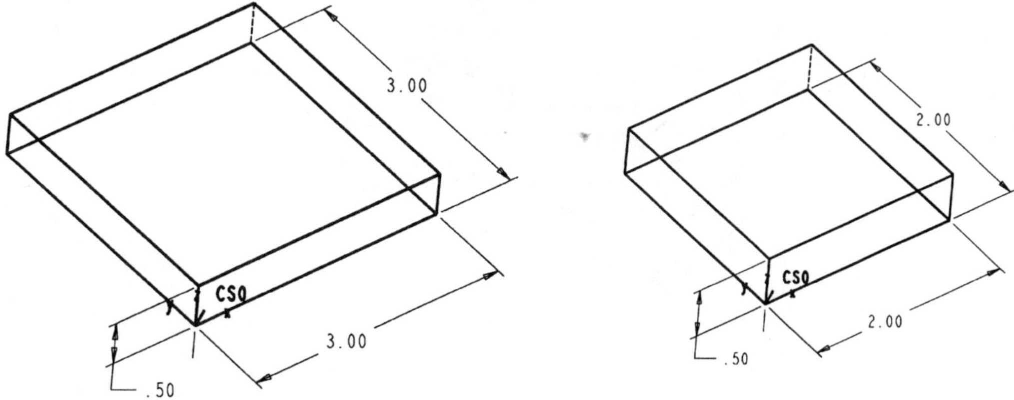

(a) The Bottom Plate (b) The Upper Plate

Figure 4-194 Geometry and Dimensions of the Two Plates

2. Assembly of the two plates

Those two plates are assembled as shown in Figure 4-195.

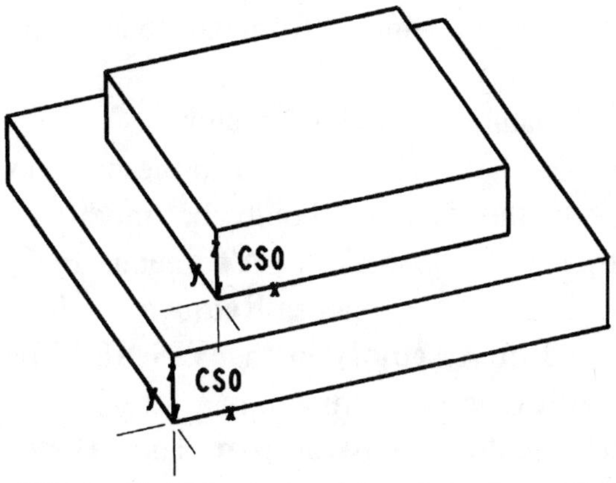

Figure 4-195 Assembly of the Two Plates

3. Create a subtractive assembly feature

To create a hole in the center of the assembly, under **ASSY FEAT** menu (select **Feature** in **ASSEMBLY** menu), choose **Create, Solid, Hole, Straight, Done, Linear, Done. Pick** the upper surface as the placement plane, then **Pick** up two perpendicular edges of the upper plate to define the location of the hole. Enter the distance from reference as 1 twice. Then **One Side, Done. Thru All, Done**. Enter the diameter of the hole: 0.5. Then **INSTRCT OPER** menu will appear. Select **Add Model, Auto Sel, Confirm**, which means that the assembly-level hole will intersect with both of the two plates. In order to demonstrate the difference of the assembly to a part-level feature, at this moment, we can just use the default settings of the system for other items in this menu. Choose **Done** and then **OK** from ***Element Inform*** window to accept the definition you have done. The screen will show the hole in the assembly as below (Figure 4-196).

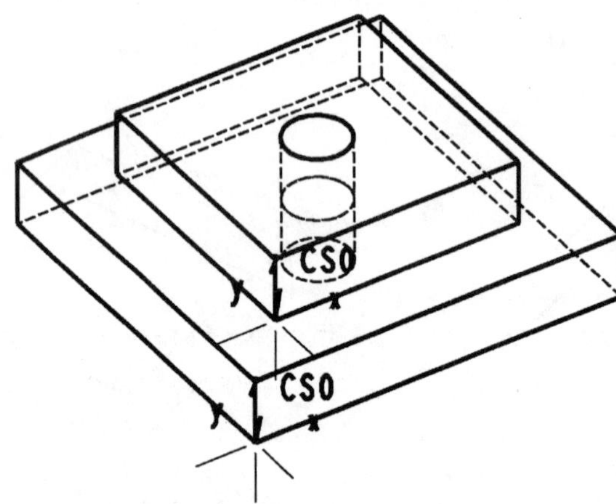

Figure 4-196 Position and Visibility of a Hole in the Assembly

However, if we go back to check the part mode, which can be accessed by choosing **Mode**, **Part, Search/Retr** under **MAIN** menu. Double click the file name representing the upper plate, another window appears, and the part you have selected will be displayed in this window, but the hole doesn't show up in the part model.

Choose **ChangeWindow** from **MAIN** menu, and then click somewhere within the assembly window to switch the working window from part to assembly.

4. Transfer assembly-level feature to part-level feature

In order to make the hole visible and attached to the part models, which comprise the assembly, some definitions of the feature should be modified. Choose **Feature** from **ASSEMBLY** menu, then **Intersect** in **ASSY FEAT** menu. **Pick** the hole to redefine it. **INSTRCT OPER** menu will appear again. Choose **Remove Model, All, Confirm**, then select **Vis Level**. Choose **Part** instead of **Assembly** in the **VIS SETTINGS** menu, and **Done** to accept the switch from the default visibility setting to part level.

Choose **Add Model, Auto Sel, Confirm** and **Done**. Now if you choose **ChangeWindow** to switch from assembly window to the part window, select **Regenerate** from

PART menu, now the hole feature will show up in the part model, and in the part model tree it is called Assembly Cut.

5. Create a Part-level feature and Update the associated Assembly

A part-level feature is created in Part mode, it is located in the upper plate and its dimensions are illustrated in Figure 4-197.

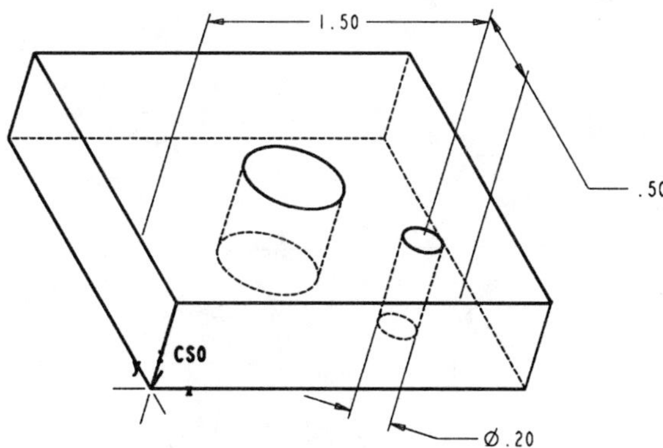

Figure 4-197 Dimensions of a Small Hole in the Upper Plate

After creating the hole in the **Part** mode, **ChangeWindow** to the assembly window, the small hole doesn't show up at this moment. Choose **Regenerate** from the **ASSEMBLY** menu, and **Automatic**, the modified components of the assembly will be automatically detected and updated. Now the small hole will appear in the assembly window as illustrated in Figure 4-198.

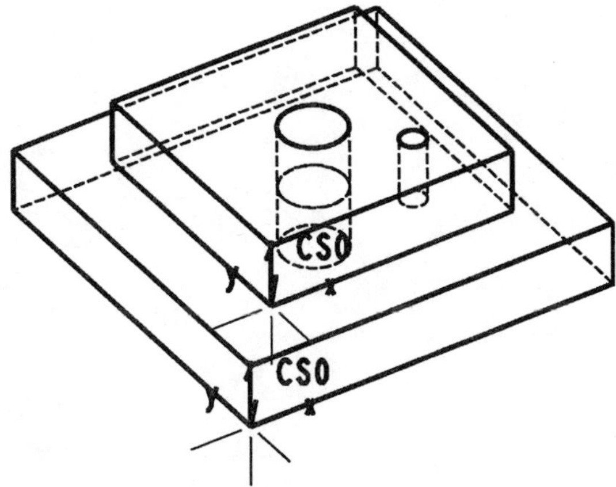

Figure 4-198 Updated Assembly Geometry with a Small Part-level Hole

6. Manipulate the assembly-level feature in Part mode.

An assembly feature visible in the part level is a part feature visible in the **Part** mode. When you suppress an assembly cut in Part mode, it does not affect the parent assembly feature. However, when you delete an assembly cut in Part mode, which can be approached following the same procedure as in the delete part-level features, the system removes the intersection from the parent assembly feature. Figure 4-199 illustrates the parent assembly geometry after removing the assembly-level in Part mode.

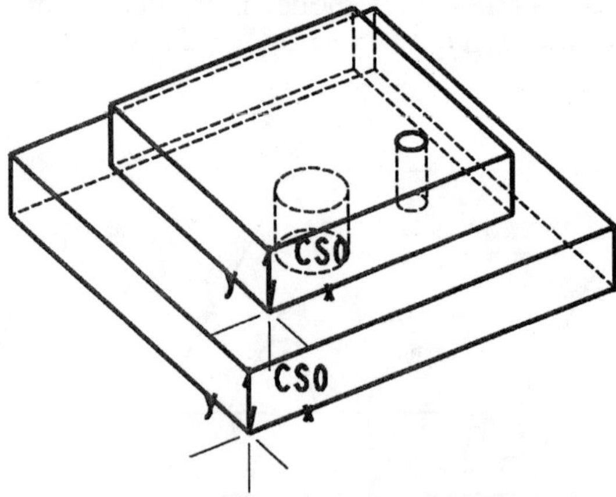

Figure 4-199 Parent Assembly Geometry after Removing the Assembly-level in Part Mode

4.7 Concept of Parametric Design

After having learned the examples through Section 4.6, readers may have felt that they are able to model almost anything. This could be true. As the learning experience accumulates, their ability to work independently increases accordingly. They should be able to expand their knowledge beyond the scope covered by the book. In this section, we introduce readers a methodology called Parametric Design. The central part of the methodology is to build quantitative relations between key design parameters in the design process. These relations capture the design intents and characterize the product functionality. When modifications or redesigns to the original designs are called on, efforts to make adjustments or changes can be minimized

4.7.1 Example 4-24: Parent – Children Relation in a Part Design

The following is an example utilizing Pro/Engineer's "Parent-Child" relationships in part modeling. We will create the crank end shown in Figure 4-200 with a pattern of dowel pin holes on its face. We will then modify one dowel pin hole and see how the rest of the pattern is updated.

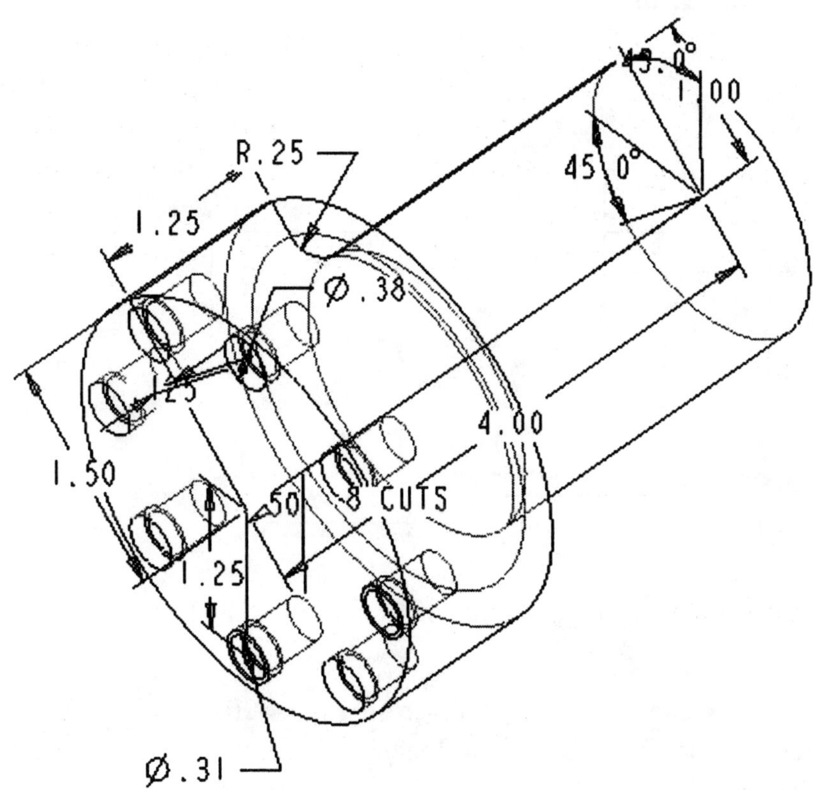

Figure 4-200 Complete Model of Crank End

Select **PART-Create** and enter a filename, say *crank.prt*. Select from the **FEAT** menu: **Create-Datum-Plane-Default,** then, **Create-Datum-CoordSys-Default-Done**. We will create the general shape of the crank end using the Revolve command. Select from the **FEAT** menu: **Create-Solid-Protrusion**, then from SOLID OPTS: **Revolve-Solid-Done.** From **ATTRIBUTES** select **One Side and Done**. Then, **SETUP SK PLN-Setup New-Plane-Pick** and use the cursor to pick DTM1. Make sure the arrow from the DTM1 is in the +x direction, (use **Flip** or **Okay** from the **DIRECTION** menu). From **SKET VIEW, pick Top-Plane-Pick** and click on DTM2. Now we are ready to sketch the revolved section. From SKETCHER, pick **Sketch-Line-Geometry-2 Points**, and draw the revolved section as shown in Figure 4-201 followed by the middle mouse button to end the chain.

To define the axis of revolution we must draw a centerline, **select Sketch-Line-Centerline**, and draw a centerline along DTM2, which should be at the bottom of the sketch. Align the centerline, DTM2, and the bottom of the part section by double clicking on the centerline. Align the left side of the part to DTM3 by double clicking on left side of the part. Now create the fillet at the step-down part of the crank end. Select **Sketch-Arc-Fillet-Pick** and follow the directions in the message window to draw the fillet. That is, pick the two edges of the surface that intersect at the fillet location, the vertical step-up and the long thin part of the shaft. Now, dimension the part by selecting **Dimension-Pick** and using the mouse button sequence: left-left-middle to create a dimension between two lines.

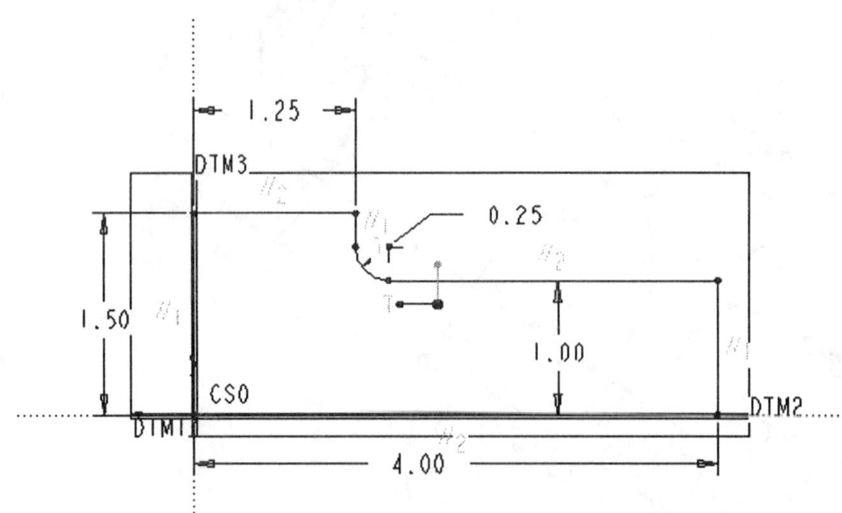

Figure 4-201 Dimensioned Cross-Section of Crank End to be Revolved

The resulting dimensions should resemble those in Fig. 4-201. Select Regenerate, Modify the Dimensions as shown in Fig. 4-200, then Regenerate-Done. In the **REV TO** menu select **360-Done**. Then, **Preview-Okay**. The part should look like Fig. 4-200.

Now we will create the dowel pin holes on the face of the crank end. Start by modeling the first dowel pin hole. Select **Feature-Create-Solid-Cut**. From the **SOLID OPTS** menu, select **Extrude-Solid-Done**. Then select from **ATTRIBUTES**: **One Side-Done**, and from **SETUP SK PLN**: **Setup New-Plane-Pick** click on DTM3, the plane on the surface of the crank end and make sure the arrow points in the -z direction. Now, we do something different. For the pattern functions to occur properly, an angular reference plane must be created through the part.

Select from the SETUP SK PLN menu: Top-Make Datum-Through-AxisEdgeCrv-then pick the long cylindrical part of the model as the object the new datum plane will pass through. Then, to fully constrain the new plane we must also give an angular reference with respect to one of the existing planes. Select from DATUM PLANE: Angle-Plane-and pick DTM2. Select Done and Enter Value of 45 degrees. We now have a new datum plane to reference the angular spacing for the circular pattern of holes we will create. From SKETCHER select: Sketch-Circle-Geometry-Ctr/Point. Draw the circle at an angle as shown in Figure 4-203a.

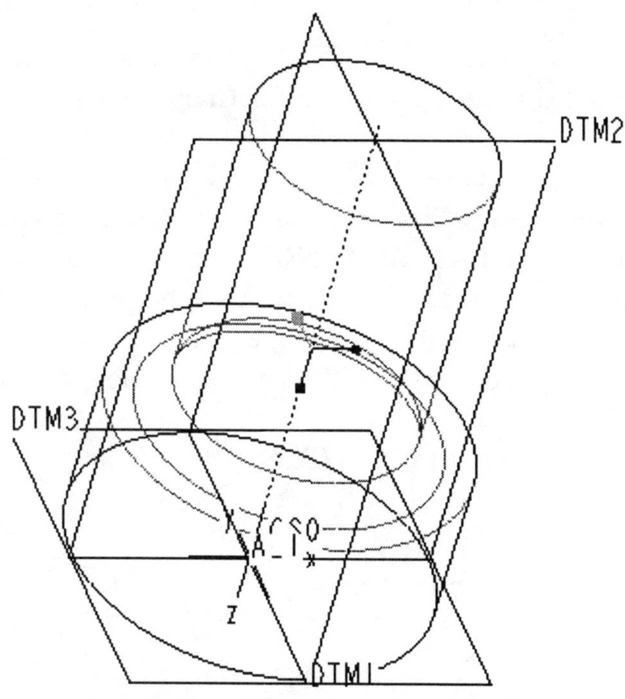

Figure 4-202 Revolved Crank End

Align the center of the circle to DTM4 (the new datum): Alignment-double-click the center of the circle. Dimension the circle's diameter and position with respect to the center of the crank. You will need to pick the center of the circle then the intersection of the datum planes at the center of the crank. After you choose a location for the dimension using the middle mouse button, choose Slanted to dimension the hole's radial distance from the center of the crank. Select **Regenerate, Modify Dimensions**, and **Regenerate** again (the hole is 5/16 diameter at a radial distance of 1.25). Select Done-make sure the cutting arrow points to the inside of the hole-Okay. From **SPEC TO**, select **blind-Done,** and Enter depth of 0.5. Select **Preview-Okay**. The part should resemble Figure 4-203b.

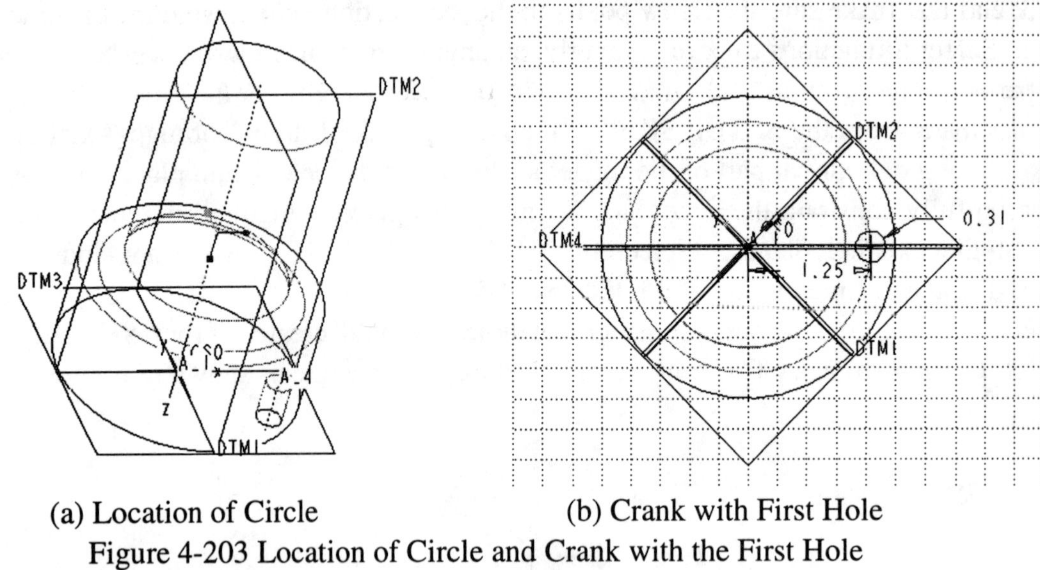

(a) Location of Circle (b) Crank with First Hole

Figure 4-203 Location of Circle and Crank with the First Hole

Now we will pattern the dowel pin hole around the perimeter of the crank end. Select from **FEATURE**: **Pattern-Select-Pick** (pick the hole feature), then from **PAT OPTIONS** select: **Identical-Done**. From **PAT DIM INCR** select: Value and pick on the angular reference for the hole that we created with the new datum plane earlier; it should be 45 degrees. Enter the default dimension increment of 45 degrees then **Done**. Enter the total number of instances (including original) of 8 and **Done**. The model should now resemble Figure 4-204.

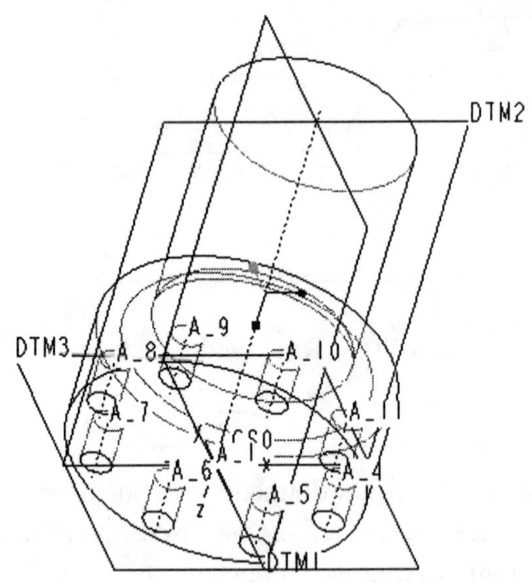

Figure 4-204 Completed Pattern of Dowel Pin Holes

Now, we will explore the Parent-Child functions of Pro/Engineer. We will modify all the dowel pin holes by simply changing the first one, or the "Pattern Leader". We will add a larger diameter opening to all of the dowel pin holes. Select **Create-Solid-Cut-Extrude-Solid-Done-One Side-Done**. Pick the face of the crank as the sketching plane (DTM3) and

DTM2 as the Top reference plane. Now, create a concentric hole around the original dowel pin hole. Select **Sketch-Circle-Concentric-pick** on the perimeter of the dowel hole named A_(#), where # is lowest numbered circle, i.e., the first circle. This is the first hole we made and defines the rest of the pattern.

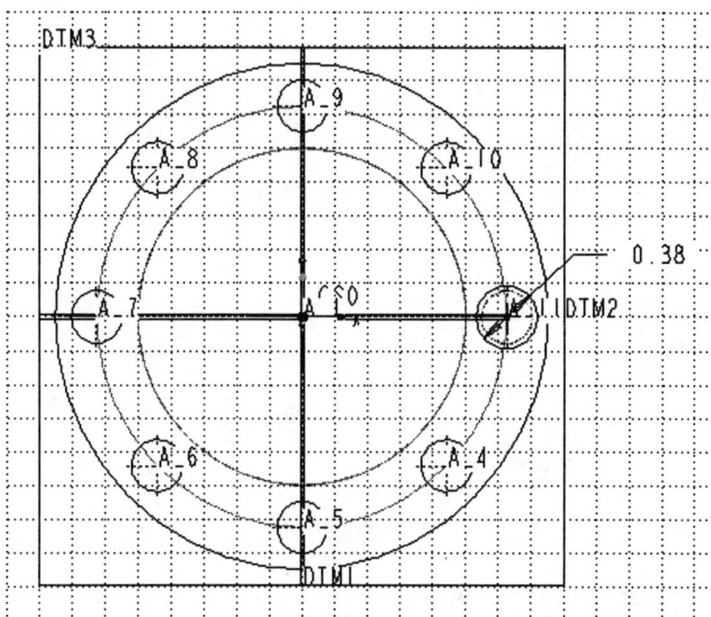

Figure 4-205 Concentric Circle around the Parent Dowel Pin Hole

Use the left mouse button to form a concentric circle slightly larger than the original hole as shown in Figure 4-205. Then give it a diameter dimension (3/8) and Regenerate the model. Select done-blind-(make sure the arrow points toward the center of the hole)-Enter Depth of 0.125. Then select Preview, Okay.

Now, to update the rest of the holes in the pattern, select Pattern-(pick the concentric circle that was just created)-Ref Pattern-Done. The other holes should be updated and the resulting part should look like Figure 4-200.

4.7.2 Example 4-25: Parent – Children Relation between Two Assembled Parts

Now we will explore the "parent-child" relations of Pro/Engineer for part assemblies. The crank end from the previous example will be assembled with and a dowel pin that will be created now. Select **PART-Create**, and enter a name for the dowel pin, say *dpin.prt*. Follow the same routine to establish a default coordinate system and planes as in the previous example: **FEATURE-Create-Datum-CoordSys-Create-Datum-Plane-Default-Done**. Now, model the dowel pin. Select: **FEATURE-Create-Solid-Protrusion-Revolve-Solid-Done**. From **ATTRIBUTES** select **One Side-Done**. Select DTM1 as the sketch plane, (make sure the arrow points in the +x direction). Then, select DTM2 as the Top reference plane. Now, from **SKETCHER** we will draw the surface to revolve. Select Sketch-Line, and draw the section as shown in Figure 4-206.

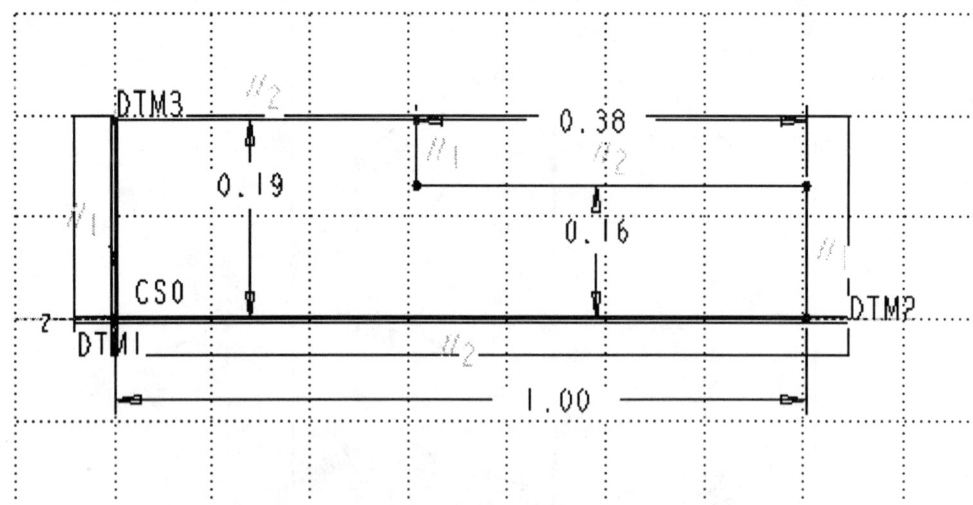

Figure 4-206 Cross-Section of Un-revolved Dowel Pin

Also, draw a centerline along the bottom edge of the section to be revolved. Choose **Sketch-Line-Centerline**, and use the left mouse button to define the ends of the centerline. Finally, align the centerline and bottom edge of the part to DTM2, the bottom, and horizontal plane. Now, Dimension and Regenerate the part. Figure 4-206 shows the locations for all the relevant dimensions. Select **Dimension-Normal-Pick**, and select the defining edges with the left mouse button and the location for text with the middle button. After successful Regeneration, Modify the dimensions to the correct values. The radius of the large part of the pin is 3/16", the small radius is 5/32", the total length is 1", and the length of the small radius section is 0.375". Select **Done** and **360-Done** from the REV TO menu after dimensioning, then **Preview-Okay** from the **Revolve sub** menu. The part should like Figure 4-207.

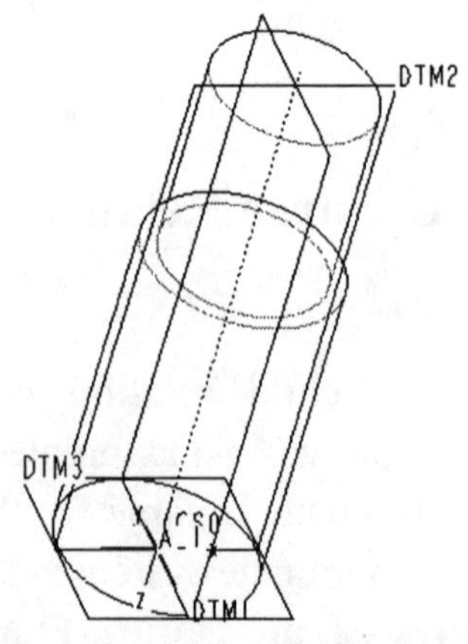

Figure 4-207 Complete, Revolved Dowel Pin

Now, we can assemble the dowel pin to the crank-end from the previous example. Select **Done** from **FEATURE** and **Mode** from **MAIN** to get back to the first menu. Select **Assembly-Create**, and enter a file name for the assembly, say *crankend.prt.* Now, we will add the first component to the assembly. Select **Component-Assemble**, and enter the name of the crank end part. The crank end should appear in the Assembly window. Next, we will add the pin to the assembly. Select Assemble again and enter the name of the dowel pin part. At this point, it should be noted that the title at the top of the window with the pin has a series of asterisks around it, i.e., *******dowelpin.prt*******; this means that it is the active part window and commands are valid for this window. The dowel pin needs to be inserted into the first hole in the crank end. To accomplish this, the pin must be constrained to that hole. Select **Add Constrnt** from the **COMP PLAC** menu and Mate from the **PLACE** menu. The small end of the dowel pin will mate with the bottom of the first hole in the crank. The first hole is the one with the lowest number identifying it, i.e., A_1 and must be used to ensure the proper patterning of the assembly to the other holes. The lowest number may be hard to find unless the part is spun around. So, first select the end of the pin, then select the bottom face of the hole. You may need to use **Query select** to get the correct faces.

The command window should prompt that the component is still not fully constrained, so we must eliminate more degrees of freedom from the pin. **Select Insert-Pick** from the **PLACE** menu and pick the smaller diameter of the pin and the smaller diameter of the first hole in the crank end. The command window should prompt that the component can now be placed, i.e., it is fully constrained. Select **Show Placement** from the **COMP PLAC** menu to make sure the pin is properly inserted in the first hole, then select **Done**. The pin and crank assembly should resemble Figure 4-208.

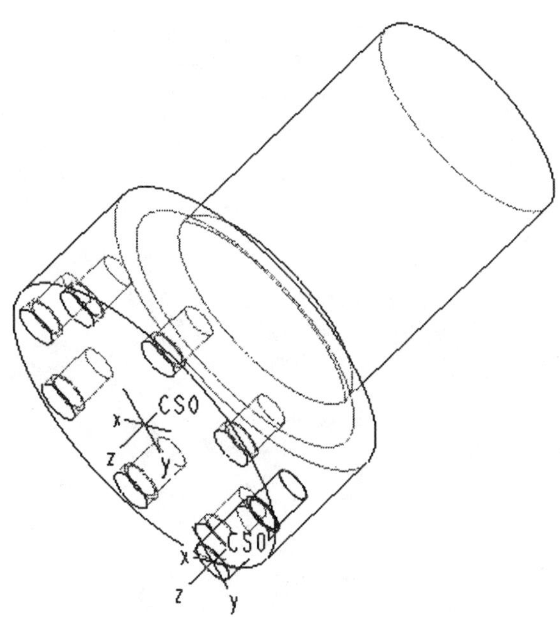

Figure 4-208 Assembly of First Dowel Pin and Crank End

Now, to utilize the parent-child relations of Pro/ENGINEER, the constraints of the single dowel pin and hole assembly will be referenced to give the other 7 holes a dowel pin. Select **Pattern** from the **COMPONENT** menu, then pick one of the holes in the crank. The small radius-section of the holes is the original pattern and defines any subsequent modifications to it. Select **Ref Pattern** and **Done**. All of the holes will be updated with a pin using the constraints of the parent hole. The assembly should resemble Figure 4-209.

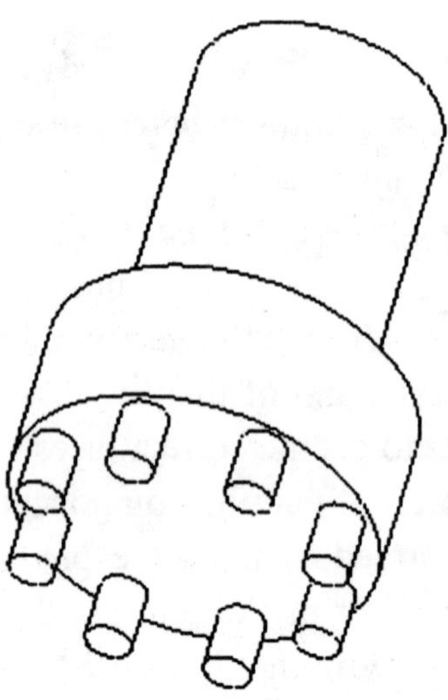

Figure 4-209 Complete Crank End Assembly

REFERENCES

1. F. L. Amirouche, Computer-aided Design and Manufacturing, Prentice Hall, Englewood Cliffs, New Jersey, 1993.
2. R. E. Barnhill, IEEE Computer-Graphics and Applications, 3(7), 9-16, 1983.
3. D. D. Bedworth, M. R. Henderson, and P. M. Wolfe, Computer-Integrated Design and Manufacturing, McGraw-Hill, New York, NY, 1991.
4. H. R. Buhl, Creative Engineering Design, Iowa State University Press, Ames, Iowa, 1960.
5. B. L. Davids, A. J. Robotham and Yardwood A., Computer-aided Drawing and Design, London, 1991.
6. J. Dieter, Engineering Design, McGraw-Hill, New York, 1983.
7. J. H. Earle, Graphics for Engineers, AutoCAD Release 13, Addision-Wesley, Reading, Massachusetts, 1996.
8. J. Encarnacao, E. G. Schlechtendahl, Computer-aided Design: Fundamentals and System Architectures, Springer-Verlag, New York, 1983.
9. G. Farin, Curves and Surfaces for Computer-aided Geometric Design, New York, 1988.
10. S. Fingers, J. R. Dixon, A review of research in mechanical engineering design, Part I: Descriptive, prescriptive and computer-based models of design processes, Research in Engineering Design, 1(1), 51-68, 1989.
11. S. Fingers, J. R. Dixon, A review of research in mechanical engineering design, Part II: Representations, analysis, and design for the life cycle, Research in Engineering Design, 1(2), 121-38, 1989.
12. J. D. Foley, A. Van Dam, Feiner S. and J. Hughes, Computer Graphics, Principles and Practice, 2nd edition, 1990.
13. J. D. Foley, A. D. VanDam, Fundamentals of Interactive Computer Graphics, Addision-Wesley, San Francisco, 1982.
14. J. G. Griffiths, A bibliography of hidden-line and hidden-surface algorithms, Computer-aided Design, 10(3), 203-6. 1978.
15. M. P. Groover and E. W. Zimmers, Computer-aided Design and Manufacturing, Englewood Cliffs, NJ, 1984.
16. C.S. Krishnamoorthy, Finite Element Analysis, Theory and Programming, 2nd Ed., 1995.
17. D. Hearn, M. P. Baker, Computer Graphics, Englewood Cliffs, NJ, 1986
18. P. Ingham P. CAD System in Mechanical and Production Engineering, London, 1989.
19. M. Mantyla, An introduction to Solid Modeling, Rockville IN:Computer Science Press., 1988
20. C. McMahon and J. Browne, CADCAM: from Principles to Practice, Addison Wesley, Workingham, England, 1993.
21. W. M. Newman, R. F. Sproull, Principles of Interactive Computer Graphics, New York, McGraw-Hill, 1979.
22. R. L. Norton, Machine Design: An Integrated Approach, Prentice Hall, Upper Saddle River, New Jersey, 1996.

23. A. A. G. Requicha and H. B. Voelcker, IEEE Computer Graphics and Applications, 2(2), 9-24, 1982.
24. N. P. Suh, The Principles of Design, Oxford University Press, New York, NY, 1990.
25. D. L. Taylor, Computer-aided Design, Reading MA: Addision-Wesley, 1982.

EXERCISES

1. Use Pro/ENGINEER to design a shaft. The diameter of the shaft is 50 mm and the length of the shaft is 100 mm. You need to present, at least, two different approaches to design the shaft. For each approach, describe the manufacturing process associated with it.

2. Use Pro/ENGINEER to design an object. The object consists of
(1) three features;
(2) four features;
(3) five features; and
(4) tabulate these features and suggest manufacturing processes to make these features.

3. Use Pro/ENGINEER to design four types of keys used on a key board. You may create a part model for one type of key. Then modify the designed model and generate the other three designs.

4. Form a team of three or four students and go to a garage or a junk yard. Find an assembly that has, at least, three components. Therefore, share the responsibility of making individual part models and finally assemble them together. Add color!

5. We use scotch tapes in office to stick papers together. Use Pro/ENGINEER to design the housing for scotch tape. Prepare the engineering drawings of the design. Specify the material, present the specifications for the design, describe the manufacturing process(s) to make the housing, and estimate the cost.

6. Use Pro/ENGINEER to design the cover part of a weighing scale used in bath room. Specify the material, present the specifications for the design, describe the manufacturing process(s) to make the cover part. What would the possible failure modes be if the design were not appropriately done?

7. Form a team of three to four students and design a mechanical structure to be used for testing strain gages. Present the design specifications and engineering drawings, including both part drawings and subassembly or assembly drawings? Add color!

8. Owners of a small machining company have decided to market a jack to lift the class of trucks known as sports utility vehicles (SUV). One corner of the SUV is to be lifted

sufficiently high to permit the change of a tire when required. A market survey team has suggested a design incorporating a lifting screw and a thrust bearing. Your team, which consists of design engineers is to complete a detailed design including the assembly drawing, the detail drawings and a three-dimensional shaded view.

To assist your design, certain information on the design of threads is provided. The common element among screw fasteners is their thread. In general terms, the thread is a helix that causes the screw to advance into the work piece or nut when it is rotated. Threads may be external (screw) or internal (nut or threaded hole). Thread forms originally differed in each major manufacturing country but after World War II they were standardized in Great Britain, Canada, and the United States to the Unified- National Standard (UNS) series. The international standards for threads using metric dimensions are also available. Both UNS and ISO threads are in general use in the United States. The ISO threads are not interchangeable with UNS threads; however, both share key geometric features such as the 60° degree included angle, and a thread size defined by the outside (major) diameter of an external thread. The thread pitch is the distance between adjacent threads. The crests and roots are defined as flats. The specifications permit rounding of these flats to reduce the stress concentration, and to accommodate tool wear. The pitch diameter and the root diameter are defined in terms of the thread pitch with slightly different ratios used for UNS and ISO threads.

CHAPTER 5

ENGINEERING ANALYSIS

5.1 Introduction

When engineers are in the process of designing parts and products, they are asked to keep in mind the product specifications. In order to meet the specification requirements, analyzing the product performance based on what has been designed is critical. Very often they try to develop mathematical models to predict whether the product performance meets the required specifications.

The finite element method is a powerful analytical tool for solving engineering problems, such as the deformation and stress of solids, the transfer of heat, the flow of fluids or electrical problems. The major advantage of the finite element method is its versatility. The method can be used to solve most of structural and mechanical problems, from linear to non-linear problems, and for both static and dynamic analyses, that can be described by differential equations. At present engineers are able to use the finite element method to evaluate a designed system's response to loading, cracking, and buckling, to evaluate the failure modes, and to examine operational conditions with models that are closer to reality.

Like other methods of calculation, the finite element method has its inherent error resources, besides human errors. As an approximation method in calculation, accuracy of the results obtained from using the finite element method depends on the appropriateness in selecting the type of mesh, the number of elements, the boundary conditions, etc. The issue of accuracy will be discussed in the case study presented in this chapter.

Under the Pro/ENGINEER design environment, users are provided a direct access to mesh generation, a fundamental step in performing finite element analysis. Also there is a module called Pro/MECHANICA, which is designed to perform finite element analysis under the Pro/ENGINEER design environment. Such an integration of engineering design and engineering analysis offers designers with a productive and powerful tool to meet the design challenge in the meantime to reduce the time from the product design to market. In order to provide readers with a comprehensive picture on how to perform engineering analysis using Pro/ENGINEER, we present two approaches. The first approach is the integration of Pro/ENGINEER and some FEA software, such as ANSYS available on market. The second approach is to use Pro/MECHANICA directly, which runs under the Pro/ENGINEER design environment. Figure 5-1 illustrates the basic structure of these two approaches.

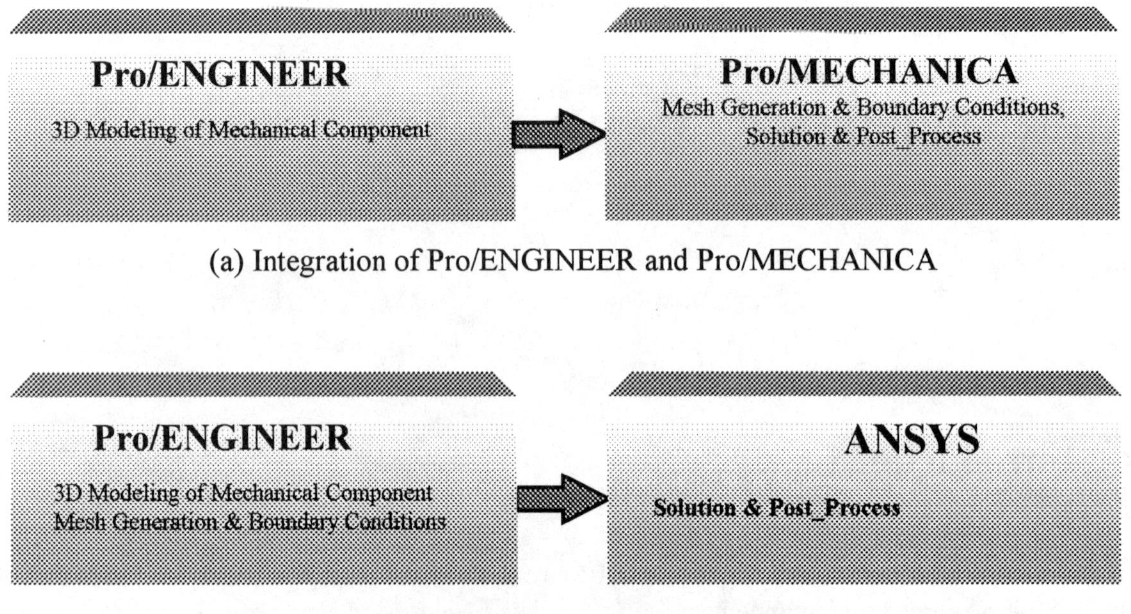

(a) Integration of Pro/ENGINEER and Pro/MECHANICA

(b) Integration of Pro/ENGINEER and ANSYS

Figure 5-1 Two Approaches of Finite Element Analysis

5.2 Structure of Pro/MECHANICA

To meet the need of Pro/ENGINEER users, Pro/MECHANICA, as a multi-discipline computer-aided engineering tool, enables the users to perform engineering analysis of the design in terms of structural, thermal and dynamic requirements. Pro/MECHANICA is a family of three design analysis products, as illustrated in Fig. 5-2. The three products in the Pro/MECHANICA module are:

Pro/MECHANICA STRUCTURE,
Pro/MECHANICA THERMAL, and
Pro/MECHANICA MOTION.

- PRO/MECHANICA STRUCTURE – a structural analysis package that provides structural modeling and optimization capabilities for both parts and assembled. This product features static, modal, buckling, contact, pre-stress, and vibration analyses.

Users can also use the product to determine how sensitive the model is to shape and property changes.

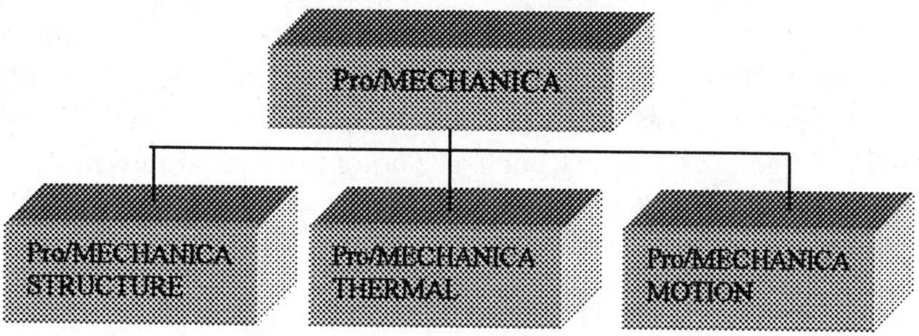

Figure 5-2 Three Major Products of Pro/MECHANICA

- Pro/MECHANICA THERMAL – a thermal analysis package that provides steady state and transient thermal modeling as well as optimization capabilities for both parts and assemblies. Users can also use the product to determine how sensitive the model is to shape and property changes.
- Pro/MECHANICA MOTION – a motion analysis package that provides mechanism modeling and mechanism design optimization capabilities. This product features 3D static, kinetostatic, dynamic, and inverse dynamic analyses as well as interference checking.

There are two versions of Pro/MECHANICA. One version is completely integrated with Pro/ENGINEER and the other version is partially integrated. The material presented in this chapter focuses on the first version of Pro/MECHANICA that is fully integrated with Pro/ENGINEER. In terms of finite element analysis, there are three major tasks:

- Defining the type of analysis, such as stress analysis, strain analysis, etc.
- Running the program to perform the analysis.
- Reviewing the results obtained from the analysis to determine how the part designed responded to, or behaved.

The basic steps for Pro/MECHANICA STRUCTURE are illustrated in Fig. 5-3. The basic steps for Pro/MECHANICA THERMAL is similar to Pro/MECHANICA STRUCTURE, but the load and constraints have to be thermal cases.

5.3 Review of the Key Steps in Finite Element Analysis

Important steps of performing finite element analysis include: discretization, constraints, assembly, manipulation and visualization.

The first step in the finite element analysis is discretization. In this step, a continuous structure is divided into discrete elements, which are connected by nodes. Materials properties, such as mass, Yong's modules, shear modules and temperature distribution, are assigned to each element. Boundary conditions, such as loads and displacements, will be assigned to the nodes as constraints. Each of these elements is governed by equations that dedicate the condition in question, such as stress, displacement or energy conversation. By continuity, all of the elements are linked, as a result the adjacent nodes define the internal boundary conditions of the nodal points. External boundary conditions need be defined. After defining the load conditions and constrains, the finite element analysis is accomplished by the FEM software. Results obtained from FEA are displayed through post-process, such as a color print of an internal stress field.

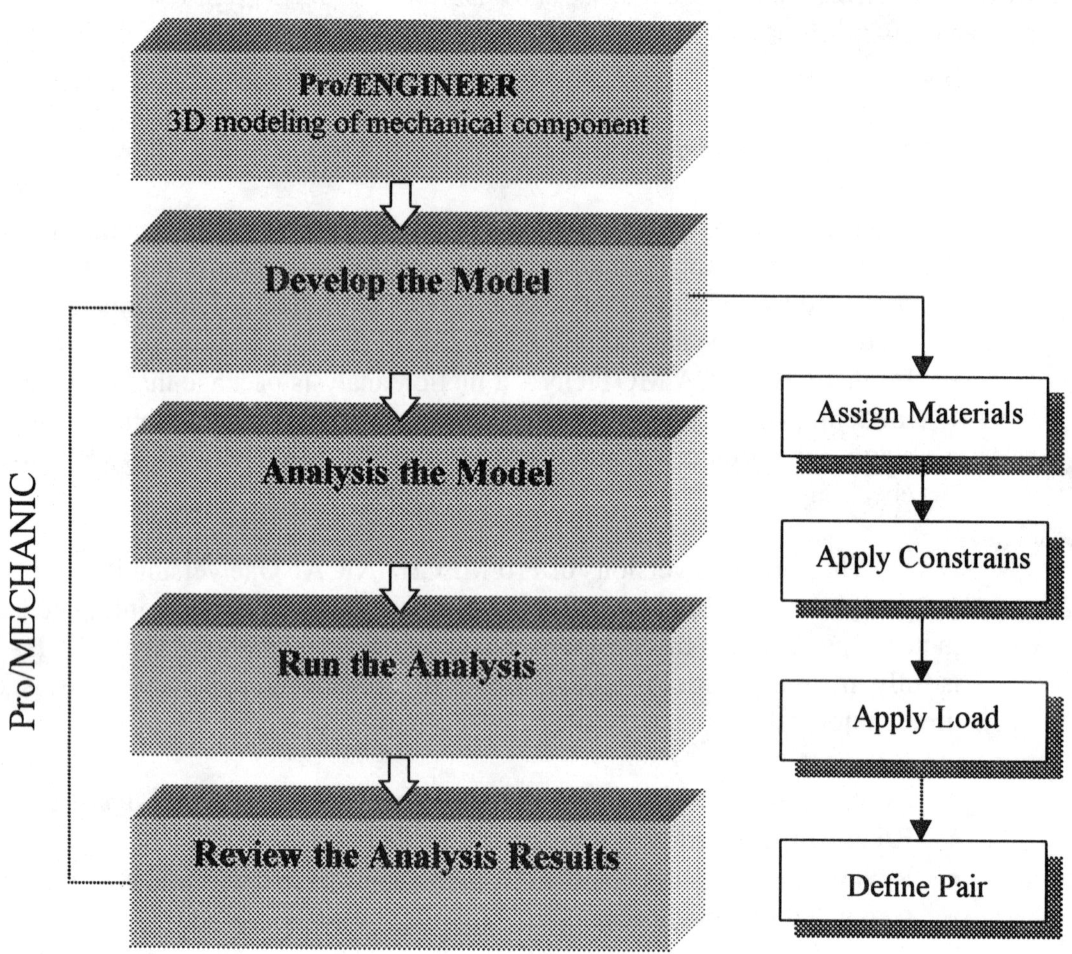

Figure 5-3 Finite Element Analysis Process for Full Integration of Pro/ENGINEER and Pro/MECHANICA

5.3.1 Pro/MESH and Post-Process

Pro/MESH provides the ability to build a finite element model directly from a Pro/ENGINEER part or assembly. When the model is completed with mesh generation, loads and constraints defined, the model can be output in a number of data file formats, such as

ANSYS, NASTRAN, PATRAN for engineering analysis. users can also write the model's mesh data to a file and store it in the Pro/ENGINEER database for later retrieval.
There are three types of mesh generated using Pro/MESH. The common shape for 2-D elements is triangular or quadrangular, and for 3-D element, the common shape is tetrahedral, as illustrated in Figs. 5-4a, 5-4b, and 5-4c where a cantilever beam structure is used for the purpose of demonstration. It is important to note that, by modifying the element size for global control and local control (around the hole), we can control the accuracy of the solution and the computing, simultaneously. To display the results obtained from FEA, Pro/FEM-POST provides post-processing capabilities that enable the user to display the thermal, structural, or modal analysis results on screen in a variety of graphical formats, and to print the results as well.

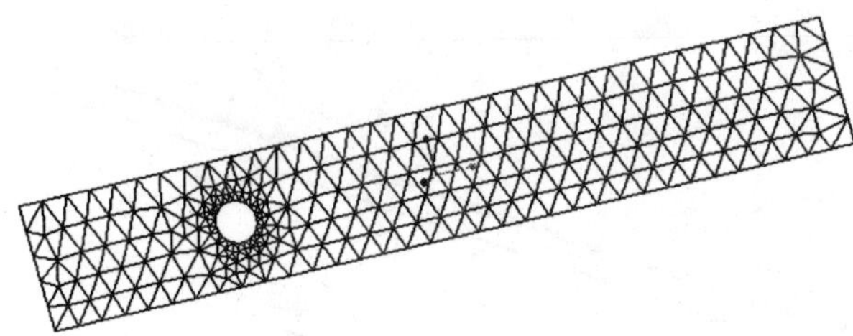

(a) Triangular Element Shell Mesh

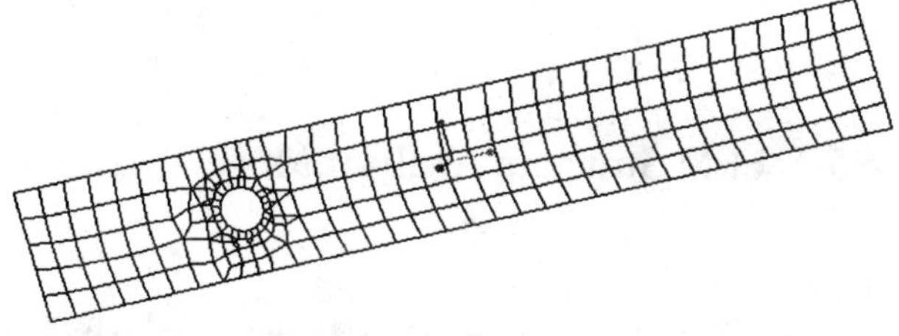

(b) Quadrangular Element Shell Mesh

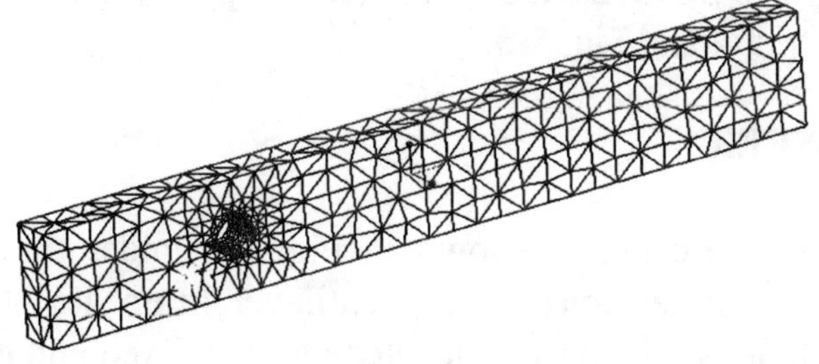

(c) Tetrahedral Element Solid Mesh

Figure 5-4 Three Types of Solid Mesh

5.3.2 Formulation of a FEA Model

Mesh generation is an essential step to perform FEA. However, the loads acting on the part and constraints reflecting the physical limitations are as important as the mesh generation. As illustrated in Fig. 5-5, the beam has a transverse hole located near the fixed end. At the other end, a point force (100 N) normal to the top face is applied. Under the Pro/ENGINEER design environment, users are allowed to incorporate these loads and constraints to formulate a model directly from the part design to meet the fundamental requirements to perform FEA.

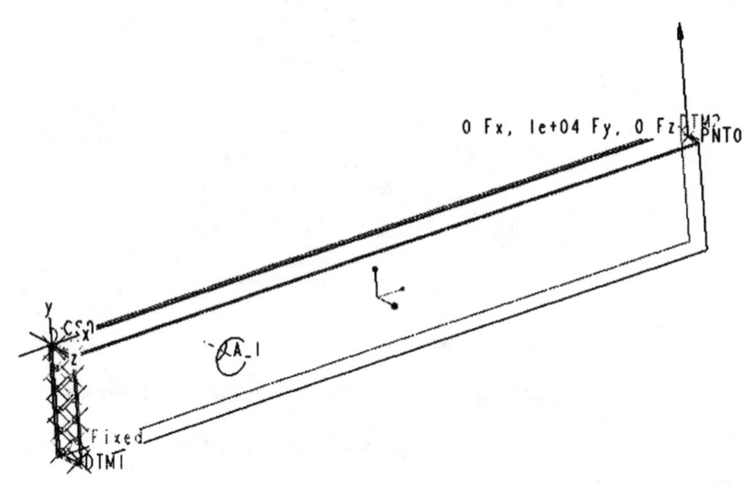

Figure 5-5 FEM Structure Analysis Model - Beam Structure

5.4 Stress Analysis of the Beam Structure Using Pro/MECHANICA

In this section, Pro/MECHANICA is used to perform stress analysis of the beam structure displayed in Fig. 5-4 and Fig. 5-5.

First Step: Pro/ENGINEER

- Generate the geometric model shown in Fig. 5-6 under the Pro/ENGINEER design environment. The dimensions of the cantilever beam are 80 x 12 x 4 inches. The diameter of the hole is 4 inch. It is located close to fixed end of the beam, the distance is 20 inches.
- Choose **MECHANICA** from the **PART** menu
- Choose **Structure** from the **MECHANICA** menu

Second Step: Formulate the Model (Pro/MECHANICA)

- Choose **Model** from the **MEC STRUCT** menu

- Assign material properties to the model.
    1. Choose **Materials** from **STRC MODEL**
    2. Choose **Assign** from the **MATERIALS** menu.
    3. Choose **Part** from the **ASSIGN** menu.
    4. Pick up the whole part as the object, then **Done Sel**.

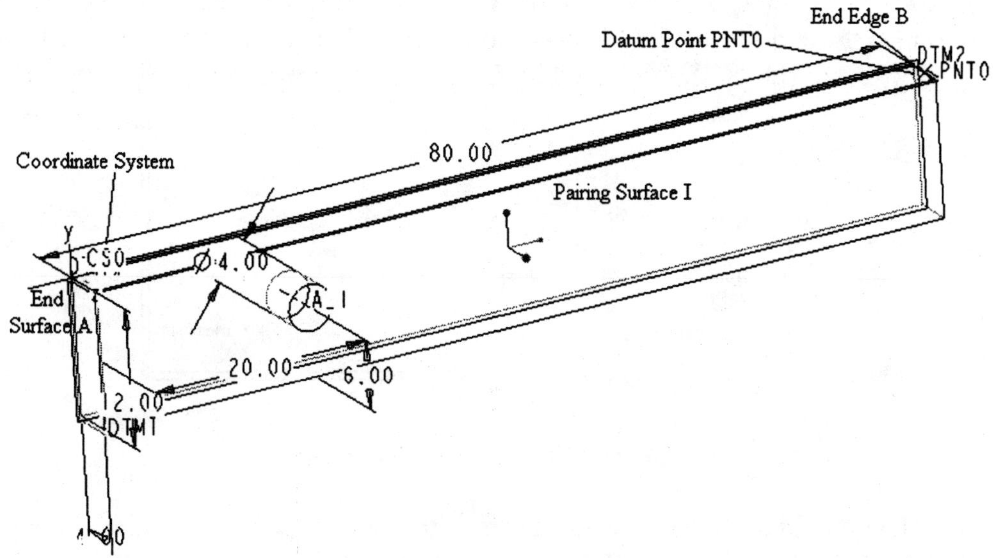

Figure 5-6 Pro/ENGINEER Part - Beam Structure

On the computer screen, Pro/MECHANICA displays the data form as shown in Fig. 5-7. It is the material library. Select one of those materials from the Library.

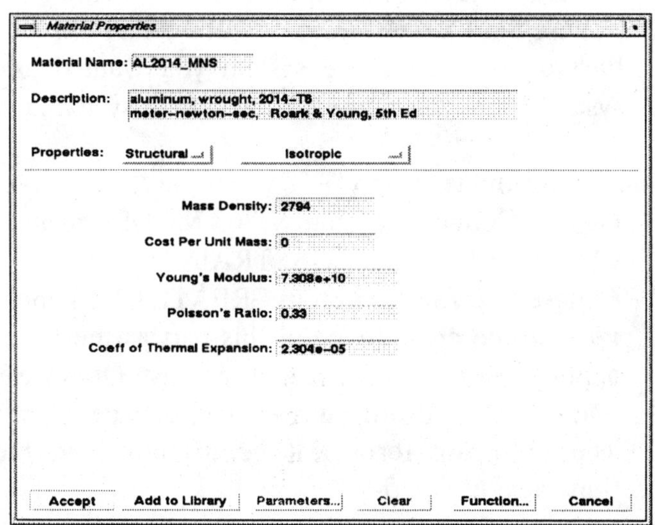

Figure 5-7 Material Library List

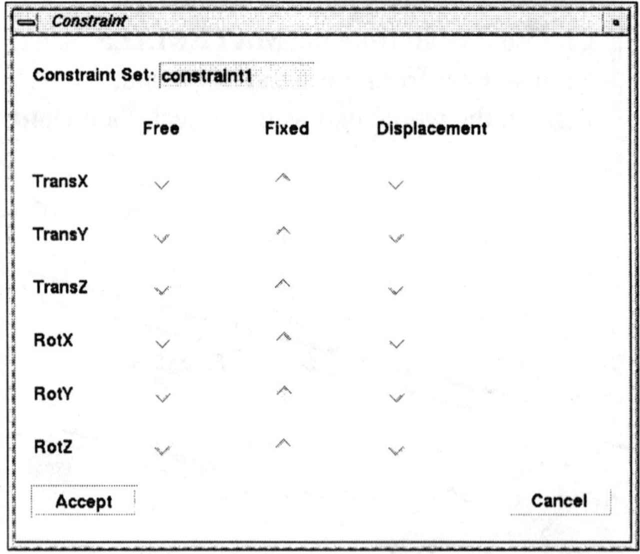

Figure 5-8 Material Properties Setting Form

The material library provides users with options. You can either choose **Accept** or **New Set.** If choosing **New Set,** you can change the properties of the material. You may also add certain new properties to the library so that you can use it next time without resetting.

- Manipulate Coordinate System (Optional)
  1. Choose **Coord System** from STRC MODEL
  2. Choose **Set Current** from COORD SYSTEM
  3. Pick up the coordinate system you wanted as the current coordinate system, here you can select the default system, then **Done Sel**

- Apply constrains on the model.
  1. Choose **Constraints** from STRC MODEL menu.
  2. Choose **Create** from CONSTRAINTS.
  3. Choose **Face/Surface** from CREATE CNST menu.
  4. Pick up the end surface of this part as the location where you want to apply constrains on, here you may use **Query Sel** in order to make sure that the selected surface is the end surface of the beam. Then **Done Sel** . The following form will be displayed on the computer screen, as illustrated in Fig. 5-9.

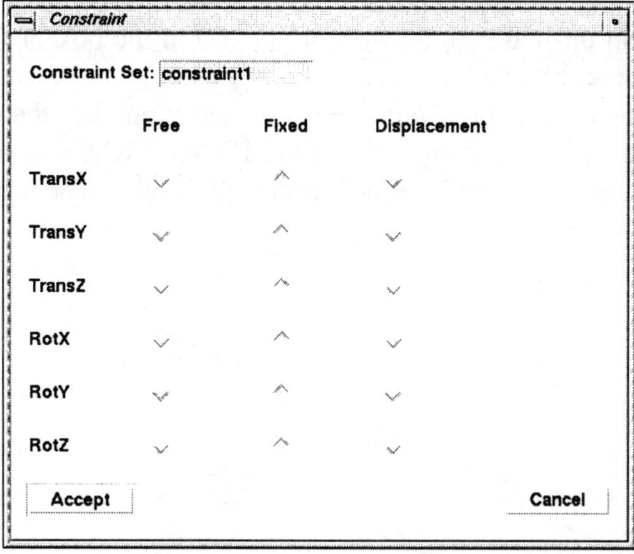

Figure 5-9 Constraint Setting Form

For each degree of freedom, you can change its definition. Here we use default values, that is, Fixed on the whole degree of freedom, choose **Accept**.

- Apply load on the model.
  1. Choose **Loads** from STRC MODEL menu.
  2. Choose **Create** from LOADS .
  3. Choose **Point** from CREATE LDS menu.
  4. Pick up datum point *PNT0* as the location where you want to apply load on, then **Done Sel** . The following form will be displayed, as illustrated in Fig. 5-10.

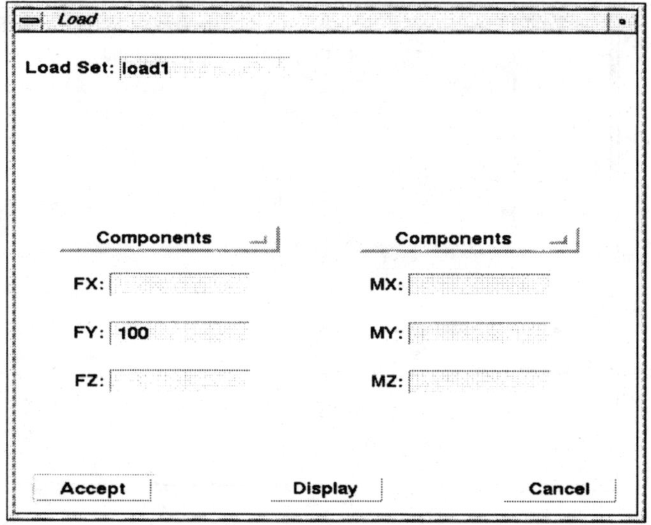

Figure 5-10 Load Setting Form

Fill in this form. In this case study, we just put 100 in the box of FY, and leave other boxes empty. Choose **Accept**.

Under the fully integrated approach, there is no need to modify the size and number of mesh elements as traditional FEA packages do. Pro/MECHANICA will automatically adjust the mesh size and number to fit the defined accuracy level. The established model is illustrated in Fig. 5-11.

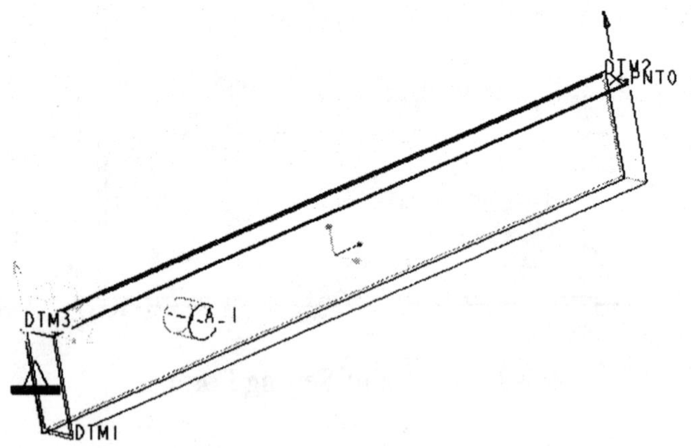

Figure 5-11 Pro/MECHANICA Analysis Model - Beam Structure

Third Step: Analyze the Model (Pro/MECHANIC)

- Choose **Analysis** from MEC STRUCT menu, one window will pop up:

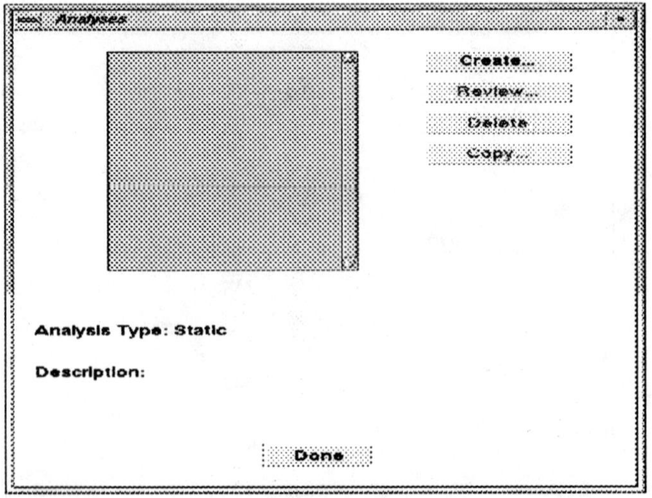

Figure 5-12 Analyses Creation Form

- Choose Create from Analyses window, then another window will show up:

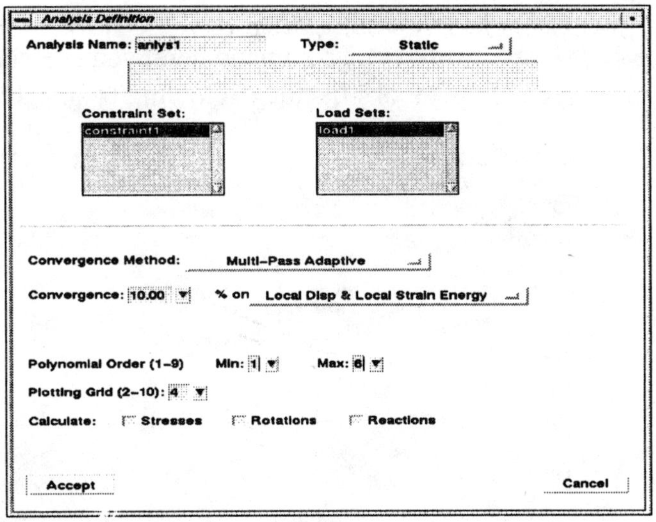

Analysis Definition
Analysis Name: anlys1
Type:
Static
Constraint Set:
constraint1
Load Sets:
load1
Convergence Method:
Multi-Pass Adaptive
Convergence: 10.00
% on
Local Disp & Local Strain Energy
Polynomial Order (1-9)
Min: 1
Max: 6
Plotting Grid (2-10): 4
Calculate:
Stresses
Rotations
Reactions
Accept
Cancel

Figure 5-13 Analysis Definition Form

You need to decide which type of analysis you will use, here we use the default item: Static. Also you need make selection in windows: Constraint Set and Load Set. Leave other items as default numbers. Then **Accept** and return to Analyses Window. Choose **Done** at Analyses window.

Fourth Step: Run an Analysis (Pro/MECHANIC)

- Choose **Analysis** from MEC STRUCT menu, Run window will pop up:

Run
anlys1 (Standard/Static)
Start...
Restart...
Batch...
Stop...
Summary...
Log...
Settings...
Description:
Done

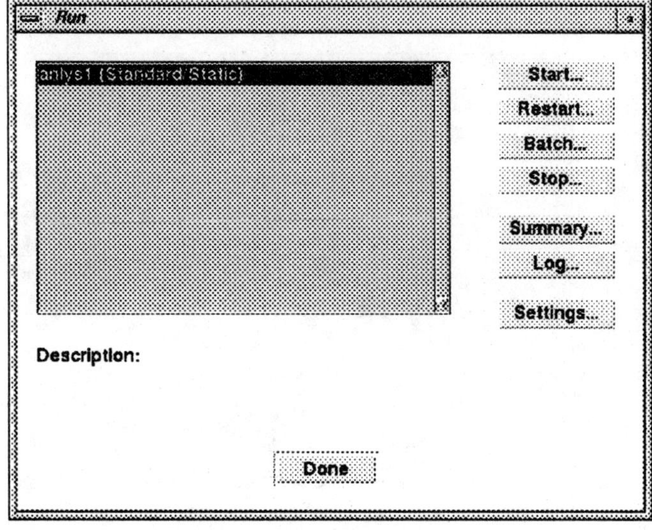

Figure 5-14 Run an Analysis

- Choose **Start...** from this window, following the instruction of the popping up message, an analysis will begin to run. You can also select **Summary...** to monitor the running process. The Summary window is illustrated as follows. After this step is finished (Use **Summary...** command here, you can check the status of solution.). The summary window looks like the following (Fig. 5-15):

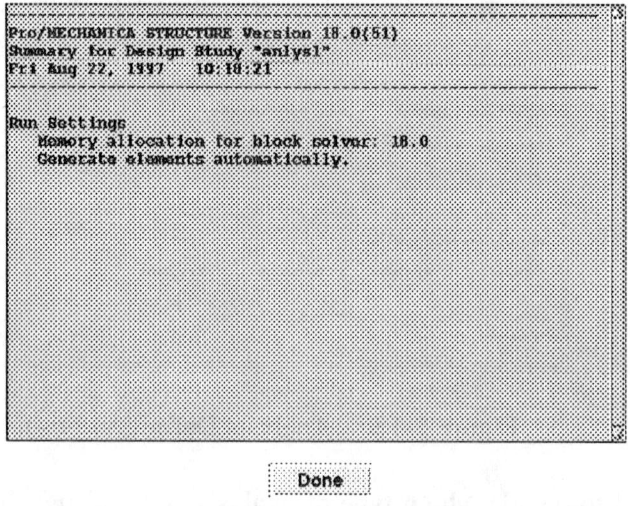

Figure 5-15 Summary Window of Running Status

- Choose **Done,** go to next step.

Fifth Step: Review the Results (Pro/MECHANIC)

- Choose **Results** from MEC STRUCT menu, Pro/MECHANIC switches to displaying model. Meanwhile, Result Window will show up.

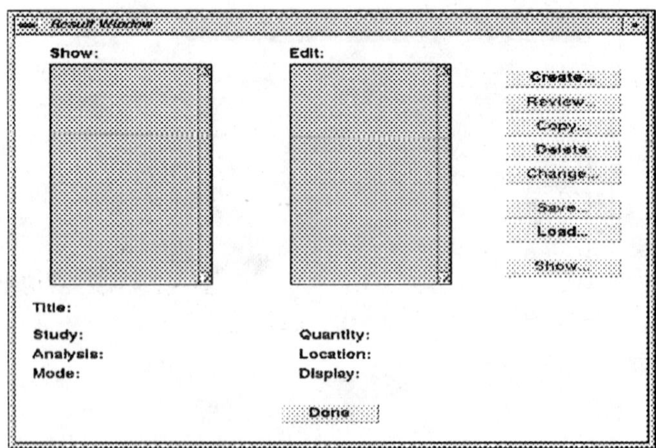

Figure 5-16 Result Window

- Choose **Create** from this window, you will create a new reviewing window, you need to define which study and what kind of results (define the content of the result) you want to demonstrate, as illustrated as the following two forms. Here we select Stress and XX under Quantity item. Turn on Deformed. Then **Accept** and go back to Result Window

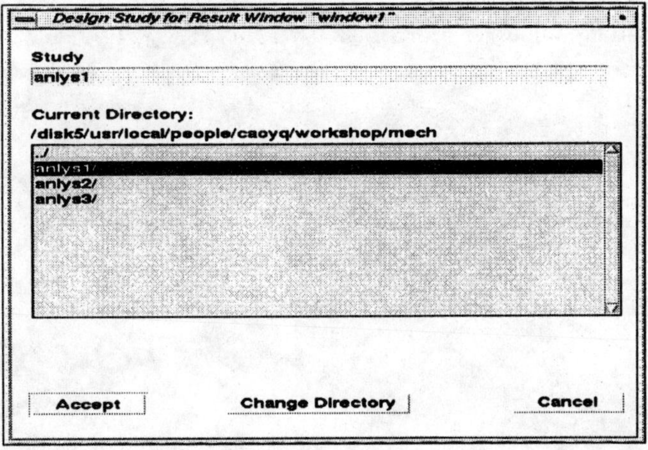

Figure 6-17 Design Study for Result Window

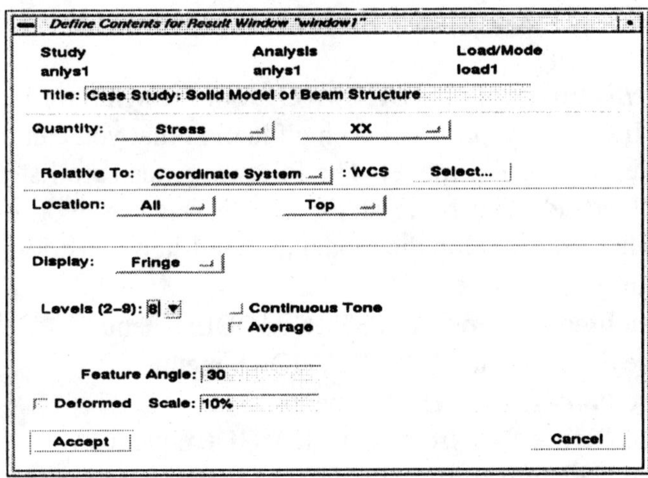

Fig 5-18 Define Contents for Result Window

- Make sure that the same window is highlighted for Show and Edit windows in the Result Window. Select **Show....** Now the result you have designed will display, illustrated in Fig. 5-19. Following the same rule as at Pro/ENGINEER, you can rotate, move or zoom the picture you just obtained.

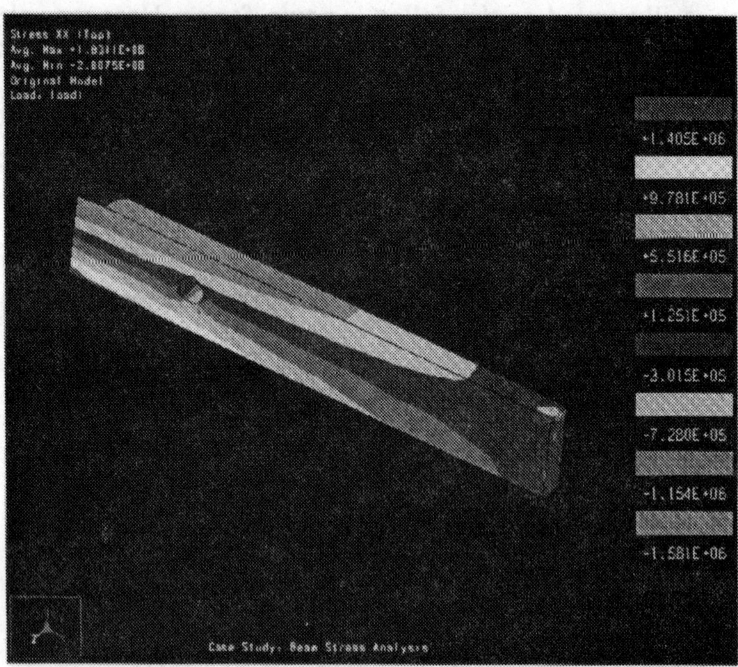

Fig 5-19 X-direction Stress Distribution Plot For Solid Mesh

- Select **Done/Return** from the **SHOW CTL** menu will return to the Result Window. Also, Pro/MECHANIC provides such a function that you can treat your model as shell model, in stead of a solid model. The following steps are listed to show how to work with a shell model. The basic steps are the same as working with a solid model. However, when you develop the model (Second step), an additional step is needed.
- Working with Idealizations
  1. Choose **Idealizations** from STRC MODEL menu.
  2. Choose **Shells** from IDEALIZATIONS menu.
  3. Choose **Pair**s from SHELLS menu.
  4. Choose **Define Pair** from SHELL MODEL menu.
  5. Choose **Constan**t from THICKNESS menu.

You can verify the Use Pairs setting that enables you to switch between the solid and shell modeling. You control this setting through the Use Pairs toggle under the SETTINGS menu, which can be accessed through the MECHANICA menu.

In the Default State, the Use Pairs toggle is off and inactive, which flags the model as a solid model. When you define pairs for your model in Pro/MECHANICA, the software automatically activates and turns on the toggle. Provided you keep the toggle on and have fully paired your model, Pro/MECHANICA analyzes the model as a shell model.

The following window has been separated into two parts, the left half displays the FEM analysis result for the solid model, and is the same as the former window, but the right half displays the analyzed result for the shell model.

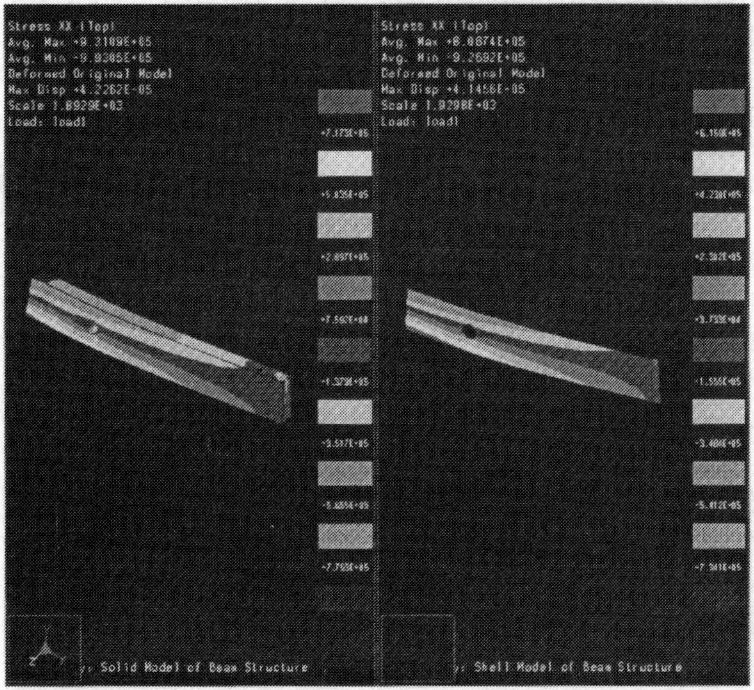

Fig5-20 X-direction Stress Distribution Plot for Solid Model and Shell Model

5.5 Stress Analysis Using Pro/ENGINEER and ANSYS

5.5.1. Basic Steps

1. Pro/ENGINEER
   a. Design and create a part or assembly for FEA.
   b. Prepare the model for FEA
   - Simplify the part geometry; suppress or modify features unnecessary to the FEA.
   - Add coordinate systems for specifying vectors, such as forces and displacements.
   - Add datum points, datum curves or cosmetic curves. The position loads can only be applied on the datum points, datum curves or cosmetic used for creating regions used in Pro/MESH.
   - Define material properties and save the information to a data file.
2. Pro/MESH
   a. Define a FEM model.
   - From file, assign materials data to the model.
   - If needed, create regions for partial loading, constraints or local mesh.
   - Apply load and constraints on the model.
   - Use mesh control to modify the size and number of mesh elements.

On one side, reducing the element size will improve the accuracy of analysis, on the other side, it will increase the computational effort.

- Define pairing surfaces.

b. Mesh the model.

Determine the mesh type. There are there types of mesh: Shell mesh, Solid mesh and mixed mesh. In this example, we select shell mesh.

c. Store the model.

d. Select the data file format, such as ANSYS, and then output the mesh model to one file, the extension name of ANSYS format will be. ans. The pre-process has been done. The next step is to use ANSYS to solve the problem. Generally speaking, you can get the help from the user manuals.

The whole procedure is also illustrated in Fig. 5-21.

5.5.2 Experiment Process for integration of Pro/E and FEA Software

Use the same example as in Pro/MECHANICA depicted in figure 5.6, following is the procedure to implement FEA using integration of Pro/ENGINEER and some FEA software.

First Step: Pro/ENGINEER

- Create the part for FEA, the shape and dimension are illustrated in Fig. 5-6. Then save the part as fem.prt in your directory. Or you can create a subdirectory first (for example, named hw9) and **Change Dir** to *hw9*, all the files you create later will be saved in this directory.
- Add Coordinate Systems.
  1. Choose **Feature** from PART menu, **Create**.
  2. Choose **Datum** from FEAT CLASS menu.
  3. Choose **Coord Sys** from DATUM menu.
  4. From OPTIONS menu, select **Pnt+2Axes**, then **Done**. **Pick** one vertex and two connected edges from the part. Follow the message on the MESSAGE WINDOWS, determinate the coordination.
- Add Datum Points.
  1. Choose **Feature** from PART menu, **Create**.
  2. Choose **Datum** from FEAT CLASS menu.
  3. Choose **Point** from DATUM menu.
  4. Choose **On Curve** from DATUM POINT menu.
  5. From PNT DIM MODE menu, select **Actual Len**, then pick up the End Edge B, enter the number (half of the edge length, that is 2).

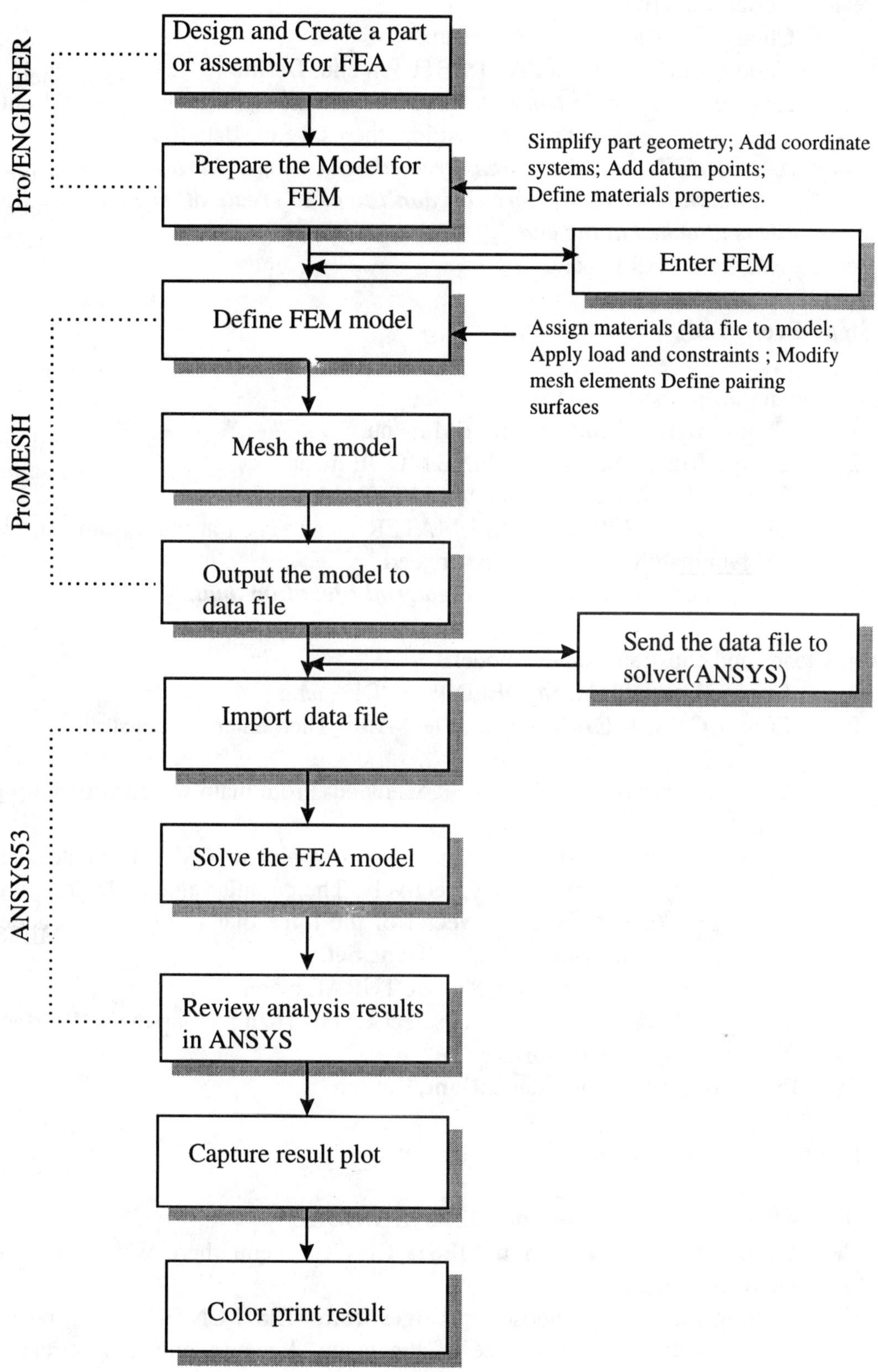

Figure 5-21 Basic Steps of Finite Element Analysis

- Create materials data file
  1. Choose **Set Up** from PART menu.
  2. Choose **Material** from PART SETUP menu, **Define**.
  3. Enter the file name: *fem.mat*. A vi editor window will show up and enter the material properties at proper position, then save the data file.

  Note: A material's file has been created and saved in the database, named alum.mat. So actually here you don't need to create other new data file. The file is attached at the end.
- Choose **FEM** from PART menu

Second Step: Pro/MESH

- Assign materials properties to the model.
  1. Choose **Define Model** from FEM menu.
  2. Choose **Materials** from DEFINE MODE menu.
  3. Choose **Whole Part** from FEM MATERIAL menu
  4. Choose **From File** from USE MATER menu. Then at the fem menu, pick the file fem.mat that you have just created.
     Or you can type in the default material file: alum .mat. .

- Apply loads and constrains on the model.
  1. Choose **Loads/BC** from DEFINE MODE menu.
  2. From CONSTR CASE menu, select **Add**. Then enter the constraint case name: *force*.
  3. Choose **Force** from STRUCTURAL menu. From main window, pick up PNT0, **Done Sel**.
  4. Select coordinate system from main window, pick up CS0. Then enter the X, Y, and Z components of force respectively. The default value is 0. Here, we enter X=0, Y=*100*, Z=0. Now the vector of the force that you have just added will show on the main window, select **Done Sel.**
  5. Choose **Displacement** from STRUCTURAL menu.
  6. From DISPLACEMENT menu, select **Fixed.** Then pick up the end surface A of the beam as the fixed surface.
  7. From Loads/BC menu, select **Done/Return**.

- Modify the size and number of mesh elements.

  1. Choose **Mesh Control** from DEFINE MODEL.
  2. Choose **Global Max** from MESH CNTRL menu, here we enter maximum element size: 2.5.
  3. If using shell mesh, choose **On Edges** from MESH CNTRL menu; pick up the circle on the outside surface of the beam. Because only half circle can be selected per pick, pick twice at opposite position of the circle. Enter number of nodes on each edge: 8. From MESH CNTRL menu, select **Done/Return**.

4. Or if using solid mesh, choose **Local Max** from MESH CNTRL menu, pick up the inside surface of the hole. Enter the maximum element size: 1.5. From MESH CNTRL menu, select **Done/Return**.

- Define pairing surfaces (for Shell Mesh only).
  1. Choose **Shell Model** from DEFINE MODEL menu.
  2. Choose **Define Pair** from SHELL MODEL menu.
  3. Pick up the Pairing Surface I and its opposite surface (the backside Surface) as the shell model pair.
  4. From SHELL MODEL menu, select **Done/Return**.

- Mesh the model.
  1. Choose **Make Model** from FEM menu.
  2. From FEM MESH menu, select **Shell Mesh** or **Solid Mesh**.
  3. Choose **Mesh Pairs** from SHELL MESH menu (for Shell Mesh only).
  4. There are two types of shell models. **Triangular** or **Quadrangular** from SHELL TYPE menu. Or **Tetrahedral** from SOLID TYPE menu.

- Store the model.
  Choose **Store Model** from the FEM menu, then <return> the model will be automatically saved as fem.fmd.

- Output models to certain format according to the solver you selected.
  1. Choose **Output Model** from FEM menu.
  2. Click on the name of the solver you are going to use, such as ANSYS from OUTPUT MESH menu, you will see a mark show ahead of that name.
  3. Choose **Accept**, then enter the file name for output: fem (the extension will be automatically attached, for ANSYS, will be .ans), it will be saved in the file you specified.
  4. Select **Done/Return** from the **FEM** menu. Now **Exit** from Pro/ENGINEER.

Material Table of Aluminum

Alumna Material File (Meter-Newton-Second)

YOUNG_MODULUS	= 7.308e+10
POISSON_RATIO	= 0.33
SHEAR_MODULUS	
MASS_DENSITY	=2794
THERMAL_EXPANSION_COEFFICIENT	= 2.304e-05
THERM_EXPANSION_REF_TEMPERATURE	=
STRUCTURAL_DAMPING_COEFFICIENT	=
STRESS_LIMIT_FOR_TENSION	=
STRESS_LIMIT_FOR_COMPRESSION	=
STRESS_LIMIT_FOR_SHEAR	=
THERMAL_CONDUCTIVITY	=
EMISSIVITY	=
SPECIFIC_HEAT	=
HARDNESS	=
CONDITION	=
INITIAL_BEND_Y_FACTOR	=
BEND_TABLE	=

REFERENCES

1. ANSYS Engineering Analysis System User's Manual, (Rev. 4.3) Vols. 1 and 2, Swanson Analysis Systems, Houston, Pa., 1987.
2. P. E. Allaire, Basics of the Finite Element Method, Solid Mechanics, Heat Transfer, and Fluid Mechanics, University of Virginia , 1985.
3. I. Bruce, A Sohrab, Techniques of Finite Elements, New York, NY, 1980.
4. IDEAS, "Getting Started" Finite Element Modeling and Analysis, SDRC, Milford, OH, 1990.
5. C. T. F. Ross, Finite Element Methods in Engineering Science, 1990.
6. J. R. Whiteman, The Mathematics of Finite Elements and Applications Highlights 1993, 1994.

EXERCISES

1. Use Pro/Mesh to generate finite elements for an object you have selected. Using three types of mesh is required to demonstrate the flexibility provided by the Pro/ENGINEER design environment for engineering analyses.

2. The geometry and dimension of a thin wall plate are shown below. At the central position, there is a hole for positioning a pin to fix the position of the plate. Both ends of the plate are free to deform. Two distributed loading, 100 psi, act in the horizontal direction at both ends. Two concentrated loads, 500 lbs act in the vertical direction at the both ends.

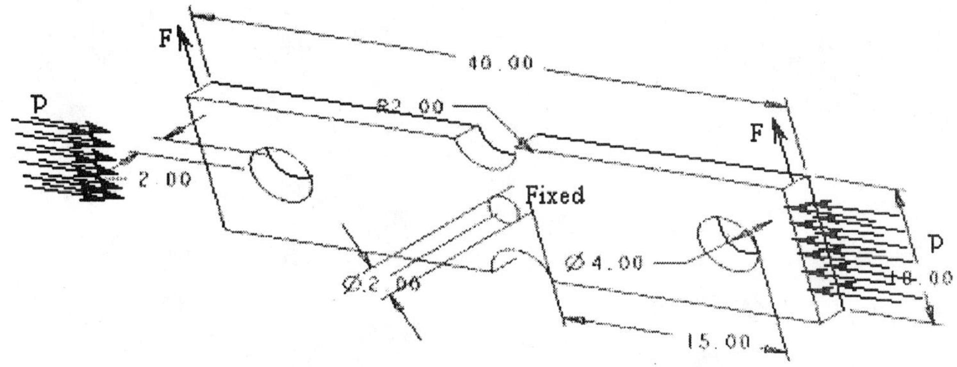

You are asked to use a constraint case including all these constraints.

(1) Analyze the results obtained from a combined use of Pro/ENGINEER and ANSYS codes, or another FEM package. Your work should be as detailed as possible. The minimum requirement includes the stress distribution along the longitudinal direction and the deflection in the vertical direction. Different mesh methods, such as shell mesh and solid mesh, should be used to compare their results and gain a better understanding of the accuracy and computing time. You

element solution available in ANSYS) in your discussion. This will allow you to distinguish the errors associated with Pro/ENGINEER from the errors associated with ANSYS.

(2) Since the geometry and the load case both are symmetrical, you should think about a set of constraints on the central plane to simplify the solution procedure. This is critical when you use FEA. A common approach would be to analyze a half of the whole structure.

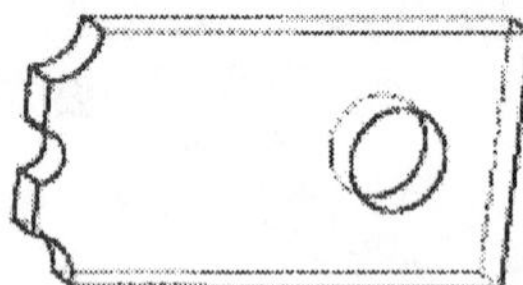

(3) Use a color printer to print your result to illustrate the stress distributions and the beam deflection.

(4) Analyze the thin wall plate problem using Pro/MECHANICA. Your work should be as detailed as possible. The minimum requirement includes the stress distribution along the longitudinal direction and the deflection in the vertical direction.

(5) Compare the results you have obtained from the two different approaches.

CHAPTER 6

ENGINEERING APPLICATIONS

6.1 Introduction

When engineers design parts in their office, they work on a computer. All the information is organized and stored in an electronic format. When the product design is completed, the process to manufacture the designed parts begins. There is a variety of manufacturing processes used by industry to implement the realization of designed parts. Typical manufacturing processes are casting, forging, molding, and machining. In this chapter, we focus our attention on machining to demonstrate the integration of computer-aided design and computer-aided manufacturing under the Pro/ENGINEER design environment.

Traditionally, when the design of a mechanical part is completed, blue print(s) of the part design is given to production engineers who work on the shop floor. They examine the size and geometry of the designed part, paying careful attention to the specifications listed on the blue print(s), such as tolerances, finish quality requirements, etc. They make a plan in which they specify the type and size of the raw material to be used, the steps to manufacture the part, and the proper machine tool for each operation. The procedure to make such a plan is called Process Planning. The objective of process planning is to make sure that the designed part can be manufactured with sufficient efficiency and at an acceptable cost.

In terms of machining a mechanical part, there are five basic steps to follow in process planning:

(1) examine the dimensions, tolerances, and geometry of the part to be machined;
(2) obtain a block of raw material of suitable size and geometry;
(3) select a fixture to hold the part during machining and place the workpiece on it;
(4) select a tool or tools to machine the workpiece. and
(5) determine machining parameters and generate tool paths.

As an integrated system for product design and manufacturing, Pro/ENGINEER offers a module called Pro/MFG. Pro/MFG is designed to map the manufacturing intent of a designer as captured by the NC (numerically controlled) tool paths onto the geometry selected from the designed parts. With Pro/MFG, a user can generate NC code for milling, turning and EDM (electrical discharge machining). In this chapter, an example will be presented to demonstrate the basic steps for using Pro/MFG-MILL to machine a mechanical part.

6.2 Pro/MFG-MILL for NC Machining

6.2.1 Pro/MFG Concept

The flowchart shown in Figure 6-1 illustrates the information flow for performing a manufacturing process in Pro/MFG. In Pro/MFG, the workpiece serves as a start condition. It represents any form of raw stock: bar stock, casting, and so on that will be machined during a manufacturing operation. It can be created by copying the design model and/or by modifying the dimension or deleting/suppressing features to represent the real workpiece.

In Pro/MFG, different operations or sequences can be defined and applied to the workpiece to transform it into a finished part, or the design model. The design model is used as the basis for all manufacturing operations. Features, surfaces, and edges are selected on the design model as references for the tool path generation. Referencing the geometry of the design model sets up an associative link between the design model and the workpiece. By assembling the workpiece and design model, Pro/MFG allows users to define the extents of machining by creating manufacturing sequences and simulating the material removal process. Because of this link, all associated manufacturing operations are updated to reflect changes in the design model. The workpiece, at various stages of machining, also defines the design and configuration of the fixture.

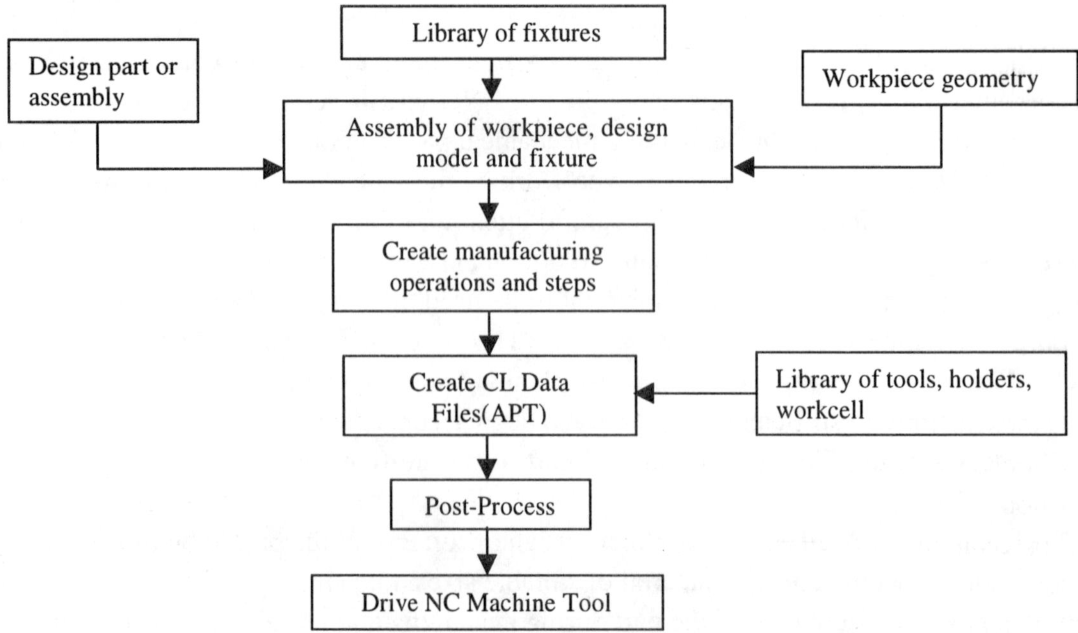

Figure 6-1 Basic Manufacturing Process in Pro/MFG

6.2.2 Case Study of Using Pro/MFG

Case study: Slider Basket

The design model to be machined, the slider basket, is shown in Figure 6-2. We created it in Example 4-1. We will follow the basic steps in Figure 6-1 to perform the machining operation in a virtual environment. For the purpose of demonstration, the NC sequences to be discussed may differ from actual machining sequences performed on the shop floor. In this example, we will demonstrate a face milling process to machine the bottom of the slider, an end milling process to mill the corner slot, and a drilling process to make the hole in the middle of the slider. The unit in Example 4-1 is "mm". In this example, we change the unit to "inch" and keep the size of the part the same through **Set Up**.

Creation of Workpiece Based on the Design Model

As a first step in developing a manufacturing process, we will create a workpiece associated with the design model. The workpiece is a Pro/ENGINEER part and can be created or retrieved in the **Part** mode. When using Pro/MFG, however, it is preferable to create it based on the design model. The steps to create a workpiece are listed below. In this way, whenever the design model is modified, corresponding modifications will be made to the workpiece, accordingly.

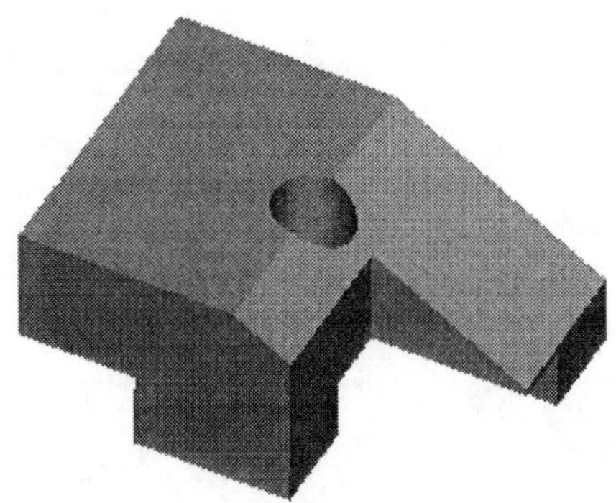

Figure 6-2 Design Model of Slider Basket for Machining Operation

Choose **Manufacture** from the **MODE** menu, **Create**, enter the name: slider_mfg.

Choose **Part**, enter design part name: *slider.*

Choose **Mfg Model** from the **MANUFACTURE** menu. **Create**, **Workpiece**, and enter a workpiece name: *slider_wp.* In general, the workpiece has a small offset from the design model. To create the workpiece, first create datum planes that are offset from the top and bottom planes of the part. Choose **Datum, Plane, Offset,** select the top plane and enter an offset value of 0.04. Choose **Done**. Datum plane DTM1 is created. Refer to Figure 6-3a for an illustration. To create another datum plane, choose **Modify** from the **MANUFACTURE** menu, **Mod Part,** click on the DTM1 datum plane (The one that belongs to the workpiece). Choose **Feature**, **Create, Datum**, and create a second datum offset from the bottom plane of the slider basket by 0.04.

To create solid features, choose **Create, Protrusion, Solid, One Side**. Select the DTM1 datum plane as a sketching plane. Sketch a rectangle that offsets the edge of the design part by 0.04. Choose the depth option **Up To Surface** and select DTM2 as the surface to extrude up to. The workpiece geometry is created. As shown in Figure 6-3b, the design part is embedded in the workpiece. Another way to express this in Pro/ENGINEER is to assemble the workpiece with the design part as shown in Figure 6-3(b). As mentioned before, this will allow the user to easily define the extents of machining by creating manufacturing sequences and dynamically simulating the material removal process.

Assemble Manufacture Model with Fixture

As shown in Figs. 6-4a and 6-4b, the workpiece is clamped by a vise. The bottom surface of the slider basket is positioned in such a way that it faces upward for machining the surface and the slot, and drilling the hole. The vise has been created already and is saved as vise.part (If you don't have the fixture in this example, you can create your own or escape this section). The vise is fixed on a base which is also created and saved as base.prt. Two blocks (3 x ½ x 1/2) are put underneath the workpiece to avoid interference between the tool and vise during drilling.

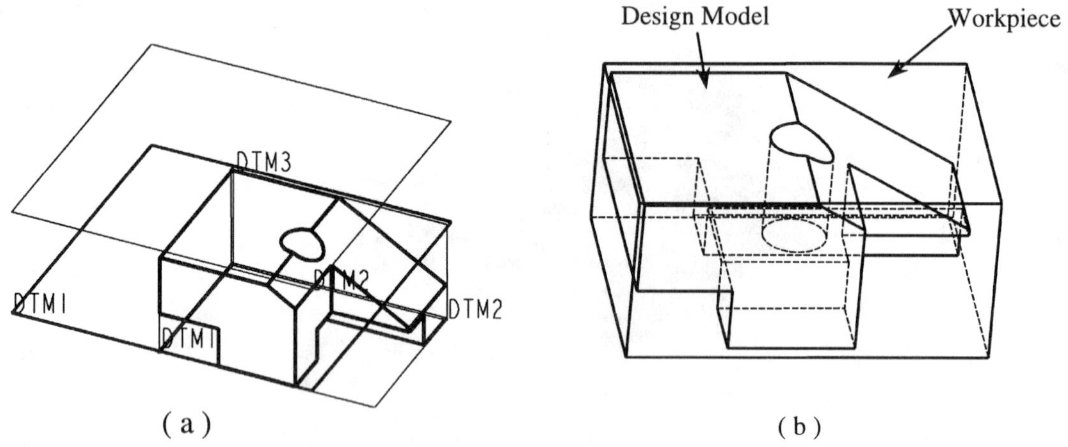

Figure 6-3 Create the Workpiece Geometry

Choose **Mfg Setup**, **Fixture**, **Name** and enter a name for the fixture: *fixture*. Assemble the workpiece with the two blocks. Choose **Component**, **Assemble**, and enter the name: *block*. Assemble the part *block* with the workpiece according to Figure 6-5a. The top surface of the blocks is mated with the workpiece. The other two sides of the block are **Aligned** with the workpiece. Assemble the other block with the workpiece.

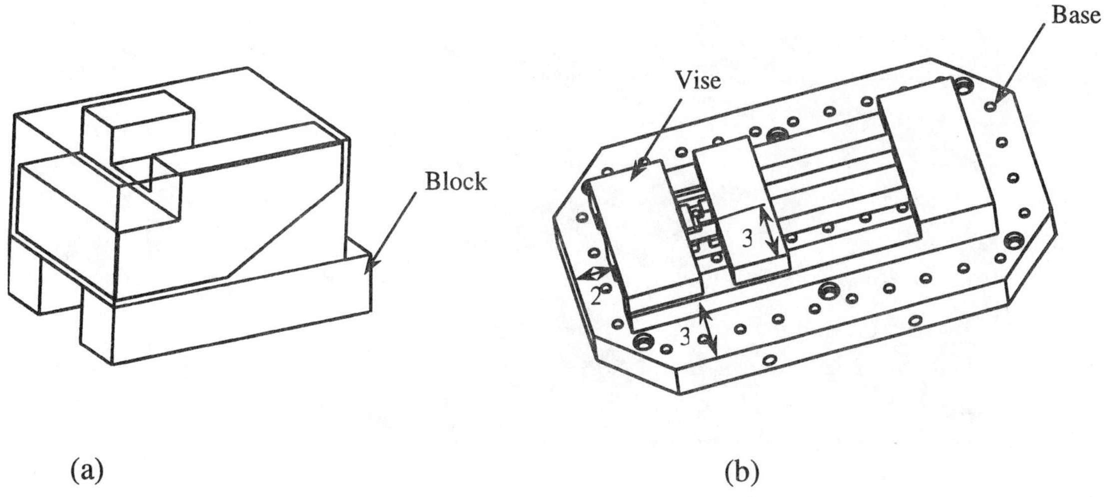

Figure 6-4 Assemble the Workpiece with Fixture

Assemble the manufacturing model with the vise. Choose **Mfg Setup, Fixture**, **Modify**, *fixture*, **Component**, **Assemble,** and enter the name: *vise*. Use the dimensions in Figure 6-4b. Two surfaces of the vise are mated with the workpiece. Use the **Align offset** command to offset the side of the vise from the workpiece by 3. Make sure the distance between the jaws of the vise is the same as the width of the workpiece.

Assemble the manufacturing model with the base. Choose **Mfg Setup, Fixture**, **and Modify**, *fixture*, **Component**, **Assemble,** and enter the name: *base*. Assemble the vise with the base following the dimensions in Figure 6-5b. The top surface of the base is mated with the bottom surface of the vise. The side surfaces of the *base* and *vise* are offset by 2 and 3 inches.

Choose **Mfg Setup**, **Fixture**, and **active** to set the current fixture setup as the active fixture.

Select the Cutting Tools for Machining

The next step is to select and set the tools that will be used in the machining. In Pro/ENGINEER, you can either set your own tools or select tools from the tooling library.

First, we will set our own tool for drilling the hole. Choose **Mfg Setup, Tooling, Create, Drill, Set**. Pro/ENGINEER will display a dialog window that lets you set the parameters of the tools as shown in Figure 6-5a. The first column shows the parameter names. You need to specify the values in the second column as –1. The other values with '-' are optional. Since the diameter of the hole is 10 mm in example 1, we need to change the *LENGTH_UNITS* to *MILLIMETER* by pressing the F4 key and choosing from the list. Set the CUTTER_DIAM to be 10 and the LENGTH as 95. Save and exit from the File Menu. Choose **Show.** The section of the tool is shown in Figure 6-5b.

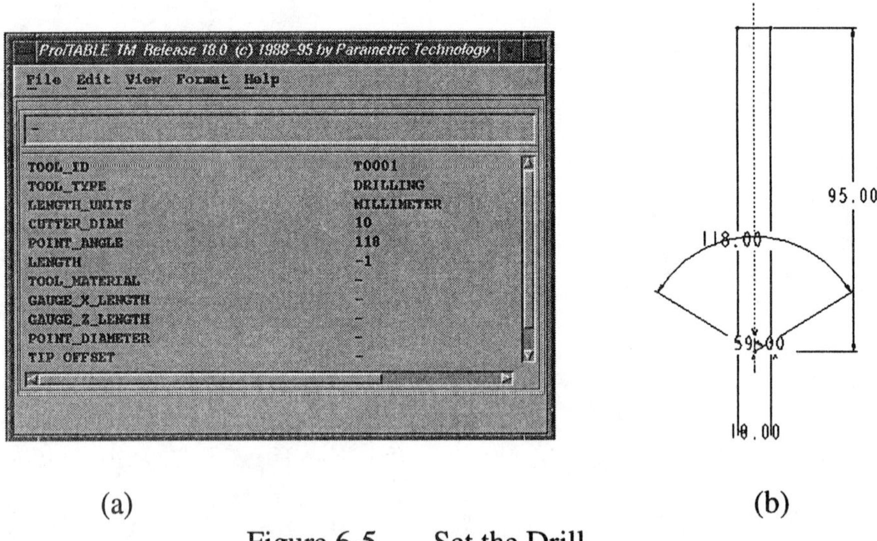

Figure 6-5 Set the Drill

If you have a tooling library, you can also select from the library, in which more than 11,000 standard tools, fixture components and tool holders, are available. By using the library, it will greatly enhance the productivity and accuracy of the processing of NC code generation. The tool library includes three sub-libraries: *Toolings, Holders, and Fixture*.

We will select a milling tool from the library to machine the slot. Choose **Tooling, Create, Retrieve, Tool Library, Pro/Library**, */standard_t, /mills_s*. The CUTTER_DIAM is 0.5 and the LENGTH of the tool is 3.25(Total length). In Pro/ENGINEER, a tool is a part and can be retrieved in the part mode to display the solid model as shown in Figure 6-6a.

Now we will select a tool to face the surface. We will also select a holder for this tool. This is done in assembly mode. Choose **Mode** from the **MAIN** menu, **Assembly**, **Create**, and enter the name for the first tool: mytool1. Then, choose **Component**, **Assemble**. Type '?' after the prompt in message window. Choose **Pro/Library**, */tools_lib, /standard_t, /mills_s, TSMCWS, TW5.* The Cutter diameter of this tool is 1.25" and the diameter of the shank is 1.00". Choose **Component**, **Assemble**, Type '?' after the prompt in message window. Choose Pro/Library, */holder_lib, /bt_flange, /endmill_b, btem45, BTEM45002.* The holder is retrieved in a new window. Choose **Coord Sys** from the **PLACE** menu; select the coordinate system *HOLDER* for the tool and holder. The tool is inserted into the holder as shown in Figure 6-6b. The coordinate system HOLDER of the holder is placed at a depth relative to the nose of the holder. Modify this value to adjust the depth to which the tool enters the holder. The tool is shown in figure x. To add this tool, choose **Tooling, Create, Mill, Retrieve, Tool Library, By Copy, Done**, '?', and select 'mytool1.asm'. You can use **Show** to see a section view of the tool.

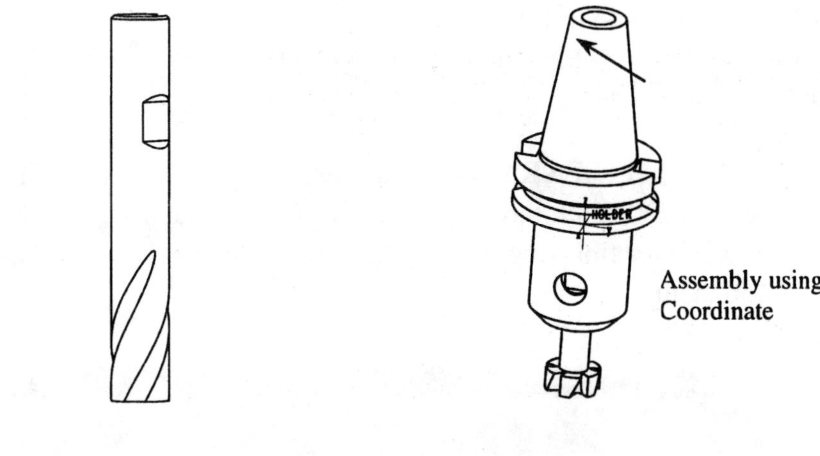

(a) The Second tool (b) The Third Tool with Holder

Figure 6-6 Using Tooling Library

Set up the Machining Operation:

In Pro/MFG, an operation is a series of NC sequences performed at a particular workcell while using a particular coordinate system for NC code generation. An operation is a workpiece feature that contains the following information: Name, workcell (machine) to be used, coordinate system, a set of manufacturing parameters that will be used by NC sequences, and FROM and HOME points.

From the MANUFACTURE menu, choose **Machining, Operation**. Add checkmarks next to **Name**, **Workcell**, **Mach Csys,** and **Activate**. Select **Done Oper**. Give a name for the operation: *oper1*. Choose **Create, Mill, 3 Axis, Done, Done**. The workcell is created. Choose **Create** from the **MACH CSYS** menu and choose the workpiece as the model. Select **Pnt+2Axes, Done**. Follow the illustration in Figure 6-7 to set the Csys at the corner of the workpiece. Pick the corner and two edges. Make sure that the directions of the three axes are following the figure.

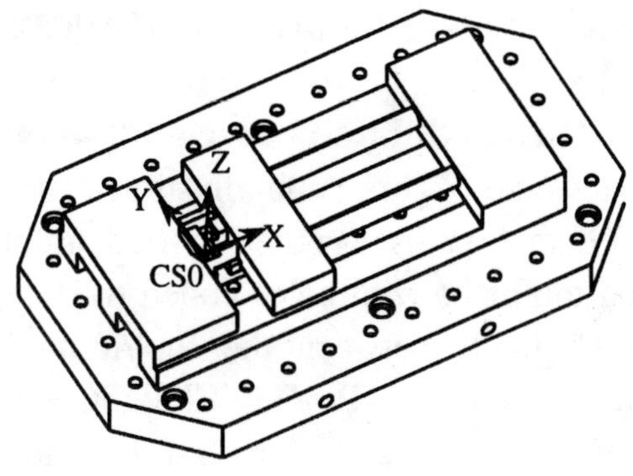

Figure 6-7 Set up Coordinate System for Machining Operation

NC Sequence #1 Face Milling. Mill the bottom surface of the slider basket.

1. Choose **Machining, NC Sequence, Face, Done.**
2. The **SEQ SETUP** menu comes up. Add checkmarks next to **Tool, Parameters, Retract**, and **Surfaces**. Choose **Done**.
3. Set tool to *T001*.
4. Choose **Set** from the **MFG PARAMS** menu. Set up the manufacturing parameters (refer to the table in Figure 6-8). Click the Advanced button on the upper right corner of the table to see more options. After finishing the settings, select **File, Exit** to save all of the settings or **Quit** to abandon them.

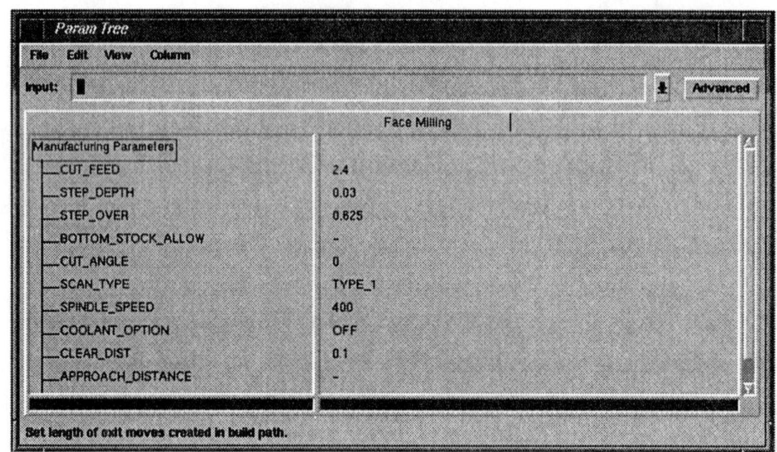

Figure 6-8 Set up Manufacturing Parameters

CUT_FEED:	2.4
STEP_DEPTH:	0.03
STEP_OVER:	1.25/2
SPINDLE_SPEED:	400
CLEAR_DIST:	0.1

5. Specify the retract plane by choosing **From CSYS** and enter 0.1 for Z value.
6. Select the surface to be machined. Choose **Model** from **SURF PICK**. Select the bottom surface of the slider basket. **Done**.
7. To check the tool path, choose **Play Path** or **NC check**. If satisfied, choose **Done Seq**. Figure 6-9a shows the tool path, and Figure 6-9b illustrates the simulation of the face milling process. In Pro/MFG, tool paths can be verified graphically by viewing the models, CL data, and simulated toolpaths with color-coded trajectories of the cutting tool. Interference checks can also be made between the tool/toolholder assembly fixtures, clamps, workpiece, and the design model. Pro/NC-CHECK enhances this function by providing a graphical representation of the tool path for dynamic and exact material removal.

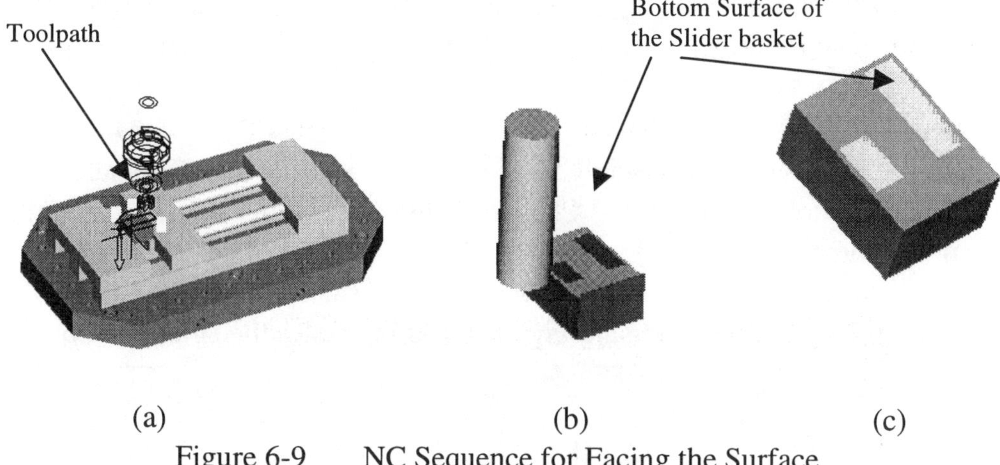

(a) (b) (c)

Figure 6-9 NC Sequence for Facing the Surface

8. Create Material Removal for this NC sequence. This can avoid air machining in future NC sequences. The representation of the workpiece with material removed may also be necessary to design the fixture and document the process plan (Remember that the workpiece is also a part and can be retrieved as mentioned before). Choose **Matrl Removal** from the **MACHINING** menu and **NC Sequence** from the **SELECT FEAT** menu. Select **Face Milling**. Pro/MFG provides **Automatic** and **Construct** methods of generating material removal simulation. For this NC sequence, only the Construct method is applicable --creating material removal as a regular feature. To do this, choose **Cut**, **Extrude** and **Solid**, **Done**, then **One Side, Done**. Select the bottom surface of the slider basket as the sketch plane and sketch a feature to represent the removed material. Regenerate and select the depth option **Up to Surface**. The workpiece after the face milling sequence is shown in Figure 6-9c.
9. Generate NC Codes for this sequence. Choose **CL Data, Output, NC Sequence,** *Face Milling*, **File,** add checkmarks next to **CL File, MCD File** to create an .ncl file and a .tap file. In Pro/MFG, The .ncl file is in a format similar to APT format. The .tap file contains the actual NC code to drive the machine. Enter a filename and choose **Done.** Select a post processor for your specific NC machine. Pro/MFG displays a window and generates the NC code.

NC Sequence #2 Conventional Milling. Mill the corner slot.

1. Choose **Machining**, **NC Sequence, Conventl Srf, Done**.
2. The SEQ SETUP menu comes up. Add checkmarks next to **Tool, Parameters, Retract,** and **Surfaces**. Choose **Done**.
3. Set tool to *RLTF8*.
4. Choose **Set** from **MFG PARAMS** menu. Set up the manufacturing parameters.

CUT_FEED:	2.4
STEP_DEPTH:	0.01
STEP_OVER:	0.5/2
SPINDLE_SPEED:	400
CLEAR_DIST:	0.1

5. Specify the retract plane by choosing **CSYS** and entering 0.1 for the Z value.
6. Select the surface to be machined. Choose **Model** from **SURF PICK**. Pick the bottom surface of the corner slot. **Done**.
7. To check the tool path, choose **Play Path** or **NC check**. If satisfied, choose **Done Seq**. Figure 6-10a shows the tool path, and Figure 6-10b illustrates the simulation of the milling of the corner slot.

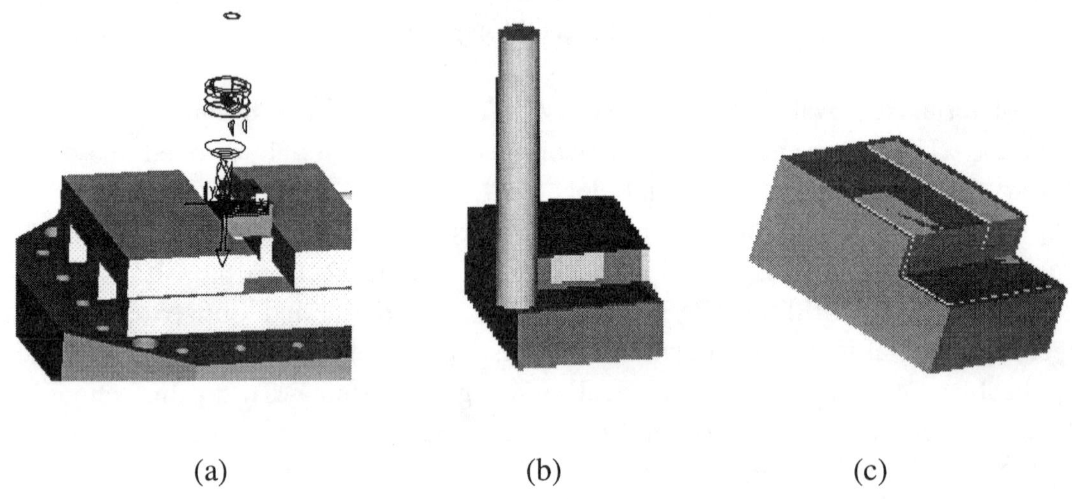

(a) (b) (c)

Figure 6-10 NC Sequence of Milling the Corner Slot

8. Create Material Removal for this NC sequence. Choose **Matrl Removal** from the **MACHINING** menu and **NC Sequence** from the **SELECT FEAT** menu. Select **Conventl Surface**. Construct the material removal feature in the same way as in Sequence #1. The workpiece following this sequence is shown in Figure 6-10c.
9. Generate NC Codes for this sequence.

NC sequence #3 Homemaking. Drill the hole in the middle of the design model.

1. Choose **Machining**, **NC Sequence, Holemaking, Done, Drill Standard, Done**.
2. The SEQ SETUP menu comes up. Add checkmarks next to **Tool, Parameters, Retract,** and **Holes**. Choose **Done**.
3. Set tool to T001.
4. Choose **Set** from MFG PARAMS menu. Set up manufacturing parameters. Specify the retract plane by choosing **CSYS** and enter a 0.1 for the Z value.

CUT_FEED:	1.2

SPINDLE_SPEED: 800
CLEAR_DIST: 0.1

5. Select Hole. Choose **Auto, By Tip, Done** from the **DRILL DEPTH** menu. Select the axis of the hole in the design model.
6. To check the tool path, choose **Play Path** or **NC check**. If satisfied, choose **Done Seq**.

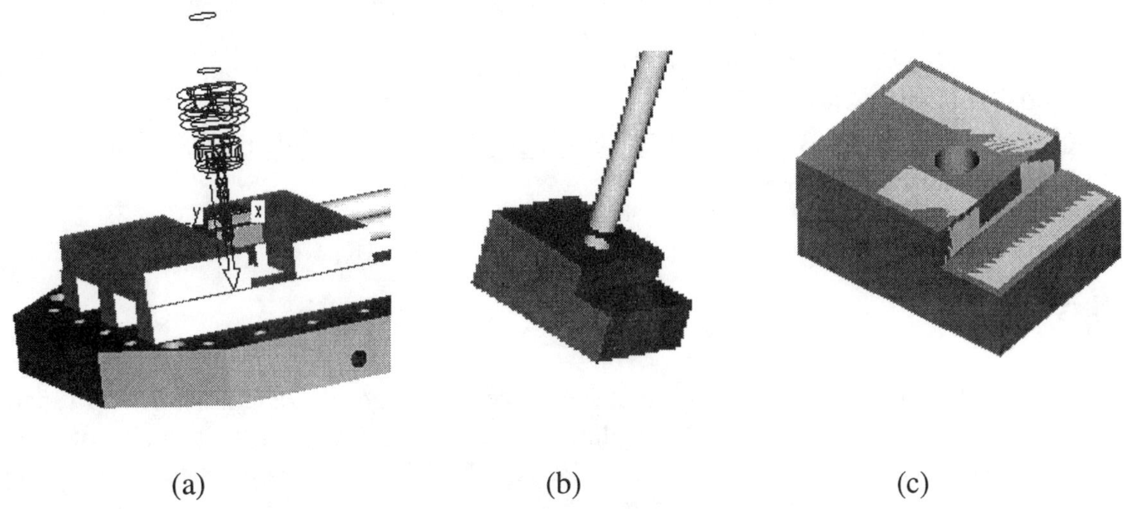

(a) (b) (c)

Figure 6-11 NC Tool Path Sequence of Drilling a Hole

7. Create Material Removal for this NC sequence. Choose **Matrl Removal** from the **MACHINING** menu and **NC Sequence** from the SELECT FEAT menu. Select **Conventl Surface**. **Automatic** can be selected for this sequence. The workpiece following the hole drilling sequence is shown in Figure 6-11c.
8. Generate NC Codes for this sequence.

In this example, three machining operations performed under the Pro/ENGINEER design environment have been demonstrated. Following a similar procedure as described, readers should be able to complete the machining operations that finish the slope on the top surface and the two other slots.

REFERENCES

1. F. L. Amirouche, Computer-aided Design and Manufacturing, Prentice Hall, Englewood Cliffs, New Jersey, 1993.
2. D. D. Bedworth, M. R. Henderson, and P. M. Wolfe, Computer-Integrated Design and Manufacturing, McGraw-Hill, New York, NY, 1991.
3. T. C. Chang, R. A. Wysk, and H. P. Wang, Computer-aided Manufacturing, , Prentice Hall, Englewood Cliffs, New Jersey, 1991.
4. M. P. Groover and E. W. Zimmers, Computer-aided Design and Manufacturing, Englewood Cliffs, NJ, 1984.
5. C. McMahon and J. Browne, CADCAM: from Principles to Practice, Addison Wesley, Workingham, England, 1993.

EXERCISES

1. Use Pro/ENGINEER to design a plate on which the logo of NBC is shown. Prepare a NC program to construct the logo. The plate size is 4 x 6 inches and is made of plastic material. You may design a logo other than NBC's. There will be a contest to select the best logo made and will be on display!

2. Form a team of three and four student. We need the team to build a box. Plastic plates will be used for construction. The box is to be used to collect fees for making prints. Students will drop dimes, nickels, quarters, and, sometimes, dollar bills to the box when prints are made. Design the box first and share the responsibility of preparing NC codes to manufacture the components and finally assemble them together.

3. When machining parts on a CNC milling machine tool, setting the rotation speed of an end mill is critical regarding the tool life. In the following cases, recommend a range for rotation speed that would keep the rate of tool wear at a normal pace. Provide your support evidence for selections.

(1)	tool material:	High speed steel	diameter: 25.4 mm
	workpiece material:	Low carbon steel	coolant: ON
(2)	tool material:	High speed steel	diameter: 100 mm
	workpiece material:	1040 carbon steel	coolant: ON
(3)	drill material:	High speed steel	diameter: 2 mm
	workpiece material:	aluminum	coolant: ON
(4)	tool material:	Carbide Inserts end mill	diameter: 25.4 mm
	workpiece material:	low carbon steel	coolant: OFF
(5)	tool material:	Carbide Inserts face mill	diameter: 80 mm
	workpiece material:	high carbon steell	coolant: OFF

4. The following figure is a representative engineering drawing in the two dimensional space. All the geometrical features are constructed in a 2D space with the third dimension - depth (or height) specified in a format called "NOTES".

DESIGN NOTES:

1. The thickness of the plate is 20 mm.
2. The depth of the pocket is 2 mm.
3. The plate material is aluminum.

You are asked to prepare a NC program written in G-code and M-code. In the process of completing this homework assignment, the following assumptions are made:

(1) The clearance plane is 2 mm above the top surface of the workpiece.

(2) There is only one tool available. It is an end mill with a diameter equal to 12 mm. The machine tool available has to be the machine tool you have used in the lab session.

(3) The tool height compensation information is installed in register H9.

(4) You have the choice of setting x=0 and y=0.

(5) Your program should be prepared in a standard format.

(6) Justify your selections of spindle speed, feedrate and depth of cut.

(7) Show your work on the calculation of tool position compensations.

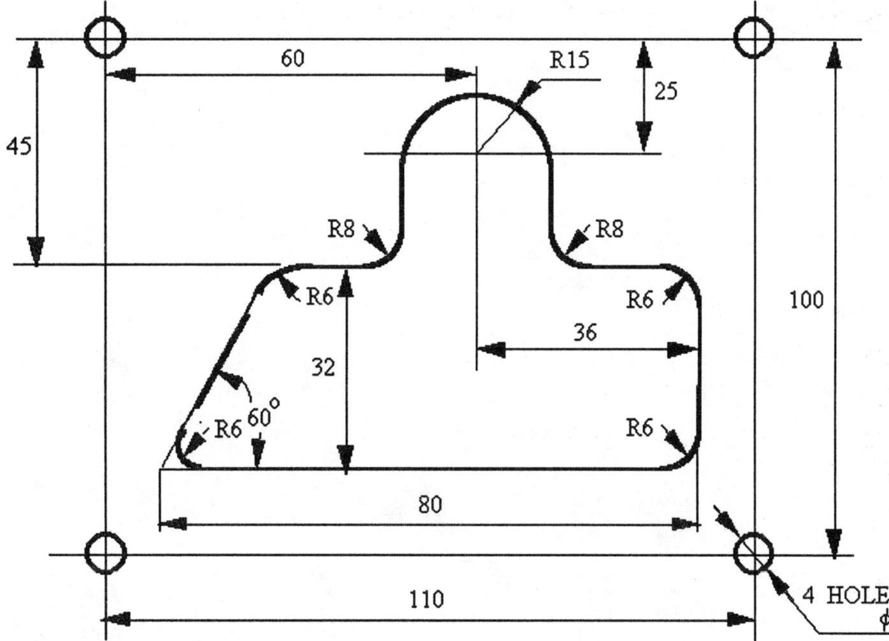

5. Use Pro/ENGINEER to create the part shown in Problem 4. Use Pro/MFG-MILL to generate the tool path and compare the computer generated path code with the manual written code.

APPENDICES

Appendix 1 American Standard Running and Sliding Fits (hole basis)

Limits are in thousandths of an inch.
Limits for hole and shaft are applied algebraically to the basic size to obtain the limits of size for the parts.
Data in bold face are in accordance with ABC agreements.
Symbols H5, g5 etc., are Hole and Shaft designations used in ABC System.

Nominal Size Range Inches	Class RC1			Class RC2			Class RC3			Class RC4		
	Limits of Clearance	Standard Limits		Limits of Clearance	Standard Limits		Limits of Clearance	Standard Limits		Limits of Clearance	Standard Limits	
Over To		Hole H5	Shaft g4		Hole H6	Shaft g5		Hole H7	Shaft F6		Hole H8	Shaft f7
0 - 0.12	0.1 0.45	+0.2 0	-.01 -.025	0.1 0.55	+0.25 0	-0.1 -0.3	0.3 0.95	+0.4 0	-0.3 -.055	0.3 1.3	+0.6 0	-0.3 -.0.7
0.12 - 0.24	0.15 0.5	+0.2 0	-0.15 -0.3	0.15 0.65	+0.3 0	-.15 -.35	0.4 1.12	+0.5 0	-.04 -0.7	0.4 1.6	+0.7 0	-0.4 -0.9
0.24 - 0.40	0.2 0.6	0.25 0	-0.2 -0.35	0.2 0.85	+0.4 0	-0.2 -0.45	0.5 1.5	+0.6 0	-0.5 -0.9	0.5 2.0	+0.9 0	-0.5- 1.1
0.40 - 0.71	0.25 0.75	+0.3 0	-0.25 -.45	0.25 0.95	+0.4 0	-0.25 -0.55	0.6 1.7	+0.7 0	-0.6 -1.0	0.6 2.3	+1.0 0	-0.6 -1.3
0.71 – 1.19	0.3 0.95	+0.4 0	-0.3 -0.55	0.3 1.2	+0.5 0	-0.3 -0.7	0.8 2.1	+0.8 0	-0.8 -1.3	0.8 2.8	+1.2 0	-0.8 -1.6
1.19 – 1.97	0.4 1.1	+0.4 0	-0.4 -0.7	0.4 1.4	+0.6 0	-0.4 -0.8	1.0 2.6	+1.0 0	-1.0 -1.6	1.0 3.6	+1.6 0	-1.0 -2.0
1.97 - 3.15	0.4 1.2	+0.5 0	-0.4 -0.7	0.4 1.6	+0.7 0	-0.4 -0.9	1.2 3.1	+1.2 0	-1.2 -1.9	1.2 4.2	+1.8 0	-1.2 -2.4
3.15 – 4.73	0.5 1.5	+0.6 0	-0.5 -0.9	0.5 2.0	+0.9 0	-0.5 -1.1	1.4 3.7	+1.4 0	-1.4 -2.3	1.4 5.0	+2.2 0	-1.4 -2.8
4.73 – 7.09	0.6 1.8	+0.7 0	-0.6 -1.1	0.6 2.3	+1.0 0	-0.6 -1.3	1.6 4.2	+1.6 0	-1.6 -2.6	1.6 5.7	+2.5 0	-1.6 -3.2
7.09 – 9.85	0.6 2.0	+0.8 0	-0.6 -1.2	0.6 2.6	+1.2 0	-0.6 -1.4	2.0 5.0	+1.8 0	-2.0 -3.2	2.0 6.6	+2.8 0	-2.0 -3.8
9.85 – 12.41	0.8 2.3	+0.9 0	-0.8 -1.4	0.8 2.9	+1.2 0	-0.8 -1.7	2.5 5.7	+2.0 0	-2.5 -3.7	2.5 7.5	+3.0 0	-2.5 -4.5
12.41- 15.75	1.0 2.7	+1.0 0	-1.0 -1.7	1.0 3.4	+1.4 0	-1.0 -2.0	3.0 6.6	+2.2 0	-3.0 -4.4	3.0 8.7	+3.5 0	-3.0 -5.2
15.75 – 19.69	1.2 3.0	+1.0 0	-1.2 -2.0	1.2 3.8	+1.6 0	-1.2 -2.2	4.0 8.1	+1.6 0	-4.0 -5.6	4.0 10.5	+4.0 0	-4.0 -6.5
19.69 – 30.09	1.6 3.7	+1.2 0	-1.6 -2.5	1.6 4.8	+2.0 0	-1.6 -2.8	5.0 10.0	+3.0 0	-5.0 -7.0	5.0 13.0	+5.0 0	-5.0 -8.0
30.09 – 41.49	2.0 4.6	+1.6 0	-2.0 -3.0	2.0 6.1	+2.5 0	-2.0 -3.6	6.0 12.5	+4.0 0	-6.0 -8.5	6.0 16.0	+6.0 0	-6.0 -10.0
41.49 – 56.19	2.5 5.7	+2.0 0	-2.5 -3.7	2.5 7.5	+3.0 0	-2.5 -4.5	8.0 16.0	+5.0 0	-8.0 -11.0	8.0 21.0	+8.0 0	-8.0 -13.0
56.19 – 76.39	3.0 7.1	+2.5 0	-3.0 -4.6	3.0 9.5	+4.0 0	-3.0 -5.5	10.0 20.0	+6.0 0	-10.0 -14.0	10.0 26.0	+10.0 0	-10.0 -16.0
76.39 – 100.9	4.0 9.0	+3.0 0	-4.0 -6.0	4.0 12.0	+5.0 0	-4.0 -7.0	12.0 25.0	+8.0 0	-12.0 -17.0	12.0 32.0	+12.0 0	-12.0 -20.0
100.9 – 131.9	5.0 11.5	+4.0 0	-5.0 -7.5	5.0 15.0	+6.0 0	-5.0 -9.0	16.0 32.0	+10.0 0	-16.0 -22.0	16.0 36.0	+16.0 0	-16.0 -26.0
131.9 – 171.9	6.0 14.0	+5.0 0	-6.0 -9.0	6.0 19.0	+8.0 0	-6.0 -11.0	18.0 38.0	+8.0 0	-18.0 -26.0	18.0 50.0	+20.0 0	-18.0 -30.0
171.9 - 200	8.0 18.0	+6.0 0	-8.0 -12.0	8.0 22.0	+10.0 0	-8.0 -12.0	22.0 48.0	+16.0 0	-22.0 -32.0	22.0 63.0	+25.0 0	-22.0 -38.0

Source: Courtesy of USASI; B4.1 – 1955.

Appendix 1 (continued)

Class RC5			Class RC6			Class RC7			Class RC8			Class RC9			Nominal Size Range Inches
Limits of Clearance	Standard Limits		Limits of Clearance	Standard Limits		Limits of Clearance	Standard Limits		Limits of Clearance	Standard Limits		Limits of Clearance	Standard Limits		
	Hole H8	Shaft e7		Hole H9	Shaft e8		Hole H9	Shaft d8		Hole H10	Shaft c9		Hole H11	Shaft	Over To
0.6 1.6	+0.6 -0	-0.6 -1.0	0.6 2.2	+1.0 -0	-0.6 -1.2	1.0 2.6	+1.0 0	-1.0 -1.6	2.5 5.1	+1.6 0	-2.5 -3.5	4.0 8.1	+2.5 0	-4.0 -5.6	0 - 0.12
0.8 2.0	+0.7 -0	-0.8 -1.3	0.8 2.7	+1.2 -0	-0.8 -1.5	1.2 3.1	+1.2 0	-1.2 -1.9	2.8 5.8	+1.8 0	-2.8 -4.0	4.5 9.0	+3.0 0	-4.5 -6.0	0.12 - 0.24
1.0 2.5	+0.9 -0	-1.0 -1.6	1.0 3.3	+1.4 -0	-1.0 -1.9	1.6 3.9	+1.4 0	-1.6 -2.5	3.0 6.6	+2.2 0	-3.0 -4.4	5.0 10.7	+3.5 0	-5.0 -7.2	0.24 – 0.40
1.2 2.9	+1.0 -0	-1.2 -1.9	1.2 3.8	+1.6 -0	-1.2 -2.2	2.0 4.6	+1.6 0	-2.0 -3.0	3.5 7.9	+2.8 0	-3.5 -5.1	6.0 12.8	+4.0 -0	-6.0 -8.8	0.40 - 0.71
1.6 3.6	+1.2 -0	-1.6 -2.4	1.6 4.8	+2.0 -0	-1.6 -2.8	2.5 5.7	+2.0 0	-2.5 -3.7	4.5 10.0	+3.5 0	-4.5 -6.5	7.0 15.5	+5.0 0	-7.0 -10.5	0.71 - 1.19
2.0 4.6	+1.6 -0	-2.0 -3.0	2.0 6.1	+2.5 -0	-2.0 -3.6	3.0 7.1	+2.5 0	-3.0 -4.6	5.0 11.5	+4.0 0	-5.0 -7.5	8.0 18.0	+6.0 0	-8.0 -12.0	1.19 - 1.97
2.5 5.5	+1.8 -0	-2.5 -3.7	2.5 7.3	+3.0 -0	-2.5 -4.3	4.0 8.8	+3.0 0	-4.0 -5.8	6.0 13.5	+4.5 0	-6.0 -9.0	9.0 20.5	+7.0 0	-9.0 -13.5	1.97 – 3.15
3.0 6.6	+2.2 -0	-3.0 -4.4	3.0 8.7	+3.5 -0	-3.0 -5.2	5.0 10.7	+3.5 0	-5.0 -7.2	7.0 15.5	+5.0 0	-7.0 -10.5	10.0 24.0	+9.0 0	-10.0 -15.0	3.15 – 4.73
3.5 7.6	+2.5 -0	-3.5 -5.1	3.5 10.0	+4.0 -0	-3.5 -6.0	6.0 12.5	+4.0 0	-6.0 -8.5	8.0 18.0	+6.0 0	-8.0 -12.0	12.0 28.0	+10.0 0	-12.0 -18.0	4.73 – 7.09
4.0 8.6	+2.8 -0	-4.0 -5.8	4.0 11.3	+4.5 0	-4.0 -6.8	7.0 14.3	+4.5 0	-7.0 -9.8	10.0 21.5	+7.0 0	-10.0 -14.5	15.0 34.0	+12.0 0	-15.0 -22.0	7.09 – 9.85
5.0 10.0	+3.0 0	-5.0 -7.0	5.0 13.0	+5.0 0	-5.0 -8.0	8.0 16.0	+5.0 0	-8.0 11.0	12.0 25.0	+8.0 0	-12.0 -17.0	18.0 38.0	+12.0 0	-18.0 -26.0	9.85 – 12.41
6.0 11.7	+3.5 0	-6.0 -8.2	6.0 15.5	+6.0 0	-6.0 -9.5	10.0 19.5	+6.0 0	-10.0 13.5	14.0 29.0	+9.0 0	-14.0 -20.0	22.0 45.0	+14.0 0	-22.0 -31.0	12.41 – 15.75
8.0 14.5	+4.0 0	-8.0 -10.5	8.0 18.0	+6.0 0	-8.0 -12.0	12.0 22.0	+6.0 0	-12.0 -16.0	16.0 32.0	+10.0 0	-16.0 -22.0	25.0 51.0	+16.0 0	-25.0 -35.0	15.75 – 19.69
10.0 18.0	+5.0 0	-10.0 -13.0	10.0 23.0	+8.0 0	-10.0 -15.0	16.0 29.0	+8.0 0	-16.0 -21.0	20.0 40.0	+12.0 0	-20.0 -28.0	30.0 62.0	+20.0 0	-30.0 -42.0	19.69 – 30.09
12.0 22.0	+6.0 0	-12.0 -16.0	12.0 28.0	+10.0 0	-12.0 -18.0	20.0 36.0	+10.0 0	-20.0 -26.0	25.0 51.0	+16.0 0	-25.0 -35.0	40.0 81.0	+25.0 0	-40.0 -56.0	30.09 – 41.49
16.0 29.0	+8.0 0	-16.0 -21.0	16.0 36.0	+12.0 0	-16.0 -24.0	25.0 45.0	+12.0 0	-25.0 -33.0	30.0 62.0	+20.0 0	-30.0 -42.0	50.0 100.0	+30.0 0	-50.0 -70.0	41.49 – 56.19
20.0 36.0	+10.0 0	-20.0 -26.0	20.0 46.0	+16.0 0	-20.0 -30.0	30.0 56.0	+16.0 0	-30.0 -40.0	40.0 81.0	+25.0 0	-40.0 -56.0	60.0 125.0	+40.0 0	-60.0 -85.0	56.19 – 76.39
25.0 45.0	+12.0 0	-25.0 -33.0	25.0 57.0	+20.0 0	-25.0 -37.0	40.0 72.0	+20.0 0	-40.0 -52.0	50.0 100	+30.0 0	-50.0 -70.0	80.0 160	+50.0 0	-80.0 -110	76.39 – 100.9

30.0 56.0	+16.0 0	-30.0 -40.0	30.0 71.0	+25.0 0	-30.0 -46.0	50.0 91.0	+25.0 0	-50.0 -66.0	60.0 125	+40.0 0	-60.0 -85.0	100 200	+60.0 0	-100 -140	100.9 – 131.9
35.0 67.0	+20.0 0	-35.0 -47.0	35.0 85.0	+30.0 0	-35.0 -55.0	60.0 110	+30.0 0	-60.0 -80.0	80.0 160	+50.0 0	-80.0 -110	130 260	+80.0 0	-130 -180	131.9 – 171.9
45.0 86.0	+25.0 0	-45.0 -61.0	45.0 110	+40.0 0	-45.0 -70.0	80.0 145	+4.0 0	-80.0 -105	100 200	+60.0 0	-100 -140	150 310	+100 0	-150 -210	171.9 - 200

CLASS RC 9: RUNNING & CLEARANCE FIT

BASIC DIA.		2.0000
HOLE	+7.0 0	+ .0070 0000
SHAFT	- 9.0 -13.5	- .0090 - .0135
MAX CLEAR.		.0205
MIN CLEAR. (ALLOWANCE)		.0090

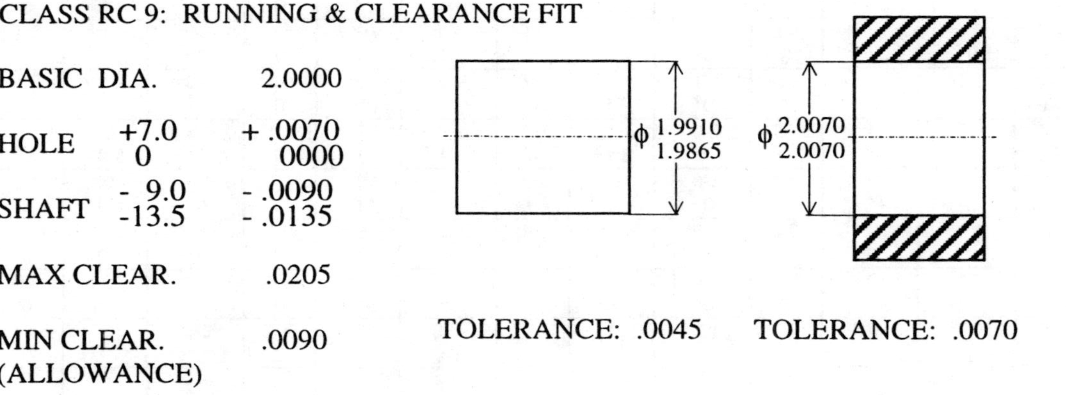

Appendix 2 American Standard Clearance Locational Fits (hole basis)

Limits are in thousandths of an inch.
Limits for hole and shaft are applied algebraically to the basic size to obtain the limits of size for the parts.
Data in bold face are in accordance the ABC agreements.
Symbols H9, f8, are Hole and Shaft designations used in ABC System.

	Class LC1			Class LC2			Class LC3			Class LC4			Class LC5		
Nominal Size Range Inches	Limits of Clearance	Standard Limits		Limits of Clearance	Standard Limits		Limits of Clearance	Standard Limits		Limits of Clearance	Standard Limits		Limits of Clearance	Standard Limits	
Over To		Hole H6	Shaft h5		Hole H7	Shaft h6		Hole H8	Shaft h7		Hole H10	Shaft h9		Hole H7	Shaft g6
0 - 0.12	0 0.45	+0.25 -0	+0 -0.2	0 0.63	+0.4 -0	+0 -0.25	0 1	+0.6 -0	+0 -0.4	0 2.6	+1.6 -0	+0 -1.0	0.1 0.75	+0.4 -0	0.1 -0.35
0.12 - 0.24	0 0.5	+0.3 -0	+0 -0.2	0 0.8	+0.5 -0	+0 -0.3	0 1.2	+0.7 -0	+0 -0.5	0 3.0	+1.8 -0	+0 -1.2	0.15 0.95	+0.5 -0	-0.15 -0.45
0.24 - 0.40	0 0.65	+0.4 -0	+0 -0.25	0 1.0	+0.6 -0	+0 -0.4	0 1.5	+0.9 -0	+0 -0.6	0 3.6	+2.2 -0	+0 -1.4	0.2 1.2	+0.6 -0	-0.2 -0.6
0.40 - 0.71	0 0.7	+0.4 -0	+0 -0.3	0 1.1	+0.7 -0	+0 -0.4	0 1.7	+1.0 -0	+0 -0.7	0 4.4	+2.8 -0	+0 -1.6	0.25 1.35	+0.7 -0	-0.25 -0.65
0.71 - 1.19	0 0.9	+0.5 -0	+0 -0.4	0 1.3	+0.8 -0	+0 -0.5	0 2.0	+1.2 -0	+0 -0.8	0 5.5	+3.5 -0	+0 -2.0	0.3 1.6	+0.8 -0	-0.3 -0.8
1.19 - 1.97	0 1.0	+0.6 -0	+0 -0.4	0 1.6	+1.0 -0	+0 -0.6	0 2.6	+1.6 -0	+0 -1.0	0 6.5	+4.0 -0	+0 -2.5	0.4 2.0	+1.0 -0	-0.4 -1.0
1.97 - 3.15	0 1.2	+0.7 -0	+0 -0.5	0 1.9	+1.2 -0	+0 -0.7	0 3.0	+1.8 -0	+0 -1.2	0 7.5	+4.5 -0	+0 -3.0	0.4 2.3	+1.2 -0	-0.4 -1.1
3.15 - 4.73	0 1.5	+0.9 -0	+0 -0.6	0 2.3	+1.4 -0	+0 -0.9	0 3.6	+2.2 -0	+0 -1.4	0 8.5	+5.0 -0	+0 -3.5	0.5 2.8	+1.4 -0	-0.5 -1.4
4.73 - 7.09	0 1.7	+1.0 -0	+0 -0.7	0 2.6	+1.6 -0	+0 -1.0	0 4.1	+2.5 -0	+0 -1.6	0 10	+6.0 -0	+0 -4.0	0.6 3.2	+1.6 -0	-0.6 -1.6
7.09 - 9.85	0 2.0	+1.2 -0	+0 -0.8	0 3.0	+1.8 -0	+0 -1.2	0 4.6	+2.8 -0	+0 -1.8	0 11.5	+7.0 -0	+0 -4.5	0.6 3.6	+1.8 -0	-0.6 -1.8
9.85 - 12.41	0 2.1	+1.2 -0	+0 -0.9	0 3.2	+2.0 -0	+0 -1.2	0 5.0	+3.0 -0	+0 -2.0	0 13.0	+8.0 -0	+0 -5.0	0.7 3.9	+2.0 -0	-0.7 -1.9
12.41 - 15.75	0 2.4	+1.4 -0	+0 -1.0	0 3.6	+2.2 -0	+0 -1.4	0 5.7	+3.5 -0	+0 -2.2	0 15.0	+9.0 -0	+0 -6.0	0.7 4.3	+2.2 -0	-0.7 -2.1
15.75 - 19.69	0 2.6	+1.6 -0	+0 -1.0	0 4.1	+2.5 -0	+0 -1.6	0 6.5	+4.0 -0	+0 -2.5	0 16.0	+10.0 -0	+0 -6.0	0.8 4.9	+2.5 -0	-0.8 -2.4

Nominal Size Range Inches	Class LC1			Class LC2			Class LC3			Class LC4			Class LC5		
	Limits of Clearance	Standard Limits		Limits of Clearance	Standard Limits		Limits of Clearance	Standard Limits		Limits of Clearance	Standard Limits		Limits of Clearance	Standard Limits	
Over To		Hole H6	Shaft h5		Hole H7	Shaft h6		Hole H8	Shaft h7		Hole H10	Shaft h9		Hole H7	Shaft g6
19.69 - 30.09	0 3.2	+2.0 -0	+0 -1.2	0 5.0	+3.0 -0	+0 -2.0	0 8.0	+5.0 -0	+0 -3.0	0 20.0	+12.0 -0	+0 -8.0	0.9 5.9	+3.0 -0	-0.9 -2.9
30.09 - 41.49	0 4.1	+2.5 -0	+0 -1.6	0 6.5	+4.0 -0	+0 -2.5	0 10.0	+6.0 -0	+0 -4.0	0 26.0	+16.0 -0	+0 -10.0	1.0 7.5	+4.0 -0	-1.0 -3.5
41.49 - 56.19	0 5.0	+3.0 -0	+0 -2.0	0 8.0	+5.0 -0	+0 -3.0	0 13.0	+8.0 -0	+0 -5.0	0 32.0	+20.0 -0	+0 -12.0	1.2 9.2	+5.0 -0	-1.2 -4.2
56.19 - 76.39	0 6.5	+4.0 -0	+0 -2.5	0 10.0	+6.0 -0	+0 -4.0	0 16.0	+10.0 -0	+0 -6.0	0 41.0	+25.0 -0	+0 -16.0	1.2 11.2	+6.0 -0	-1.2 -5.2
76.39 - 100.9	0 8.0	+5.0 -0	+0 -3.0	0 13.0	+8.0 -0	+0 -5.0	0 20.0	+12.0 -0	+0 -8.0	0 50.0	+30.0 -0	+0 -20.0	1.4 14.4	+8.0 -0	-1.4 -6.4
100.9 - 131.9	0 10.0	+6.0 -0	+0 -4.0	0 16.0	+10.0 -0	+0 -6.0	0 26.0	+16.0 -0	+0 -10.0	0 63.0	+40.0 -0	+0 -25.0	1.6 17.6	+10.0 -0	-1.6 -7.6
131.9 - 171.9	0 13.0	+8.0 -0	+0 -5.0	0 20.0	+12.0 -0	+0 -8.0	0 32.0	+20.0 -0	+0 -12.0	0 80.0	+50.0 -0	+0 -30.0	1.8 21.8	+12.0 -0	-1.8 -9.8
171.9 - 200	0 16.0	+10.0 -0	+0 -6.0	0 26.0	+16 -0	+0 -10.0	0 41.0	+25.0 -0	+0 -16.0	0 100. 0	+60.0 -0	+0 -40.0	1.8 27.8	+16.0 -0	-1.8 -11.8

Source: Courtesy of USASI; B4.1-1955

Appendix 2 (continued)

	Class LC6			Class LC7			Class LC8			Class LC9			Class LC10			Class LC11		
Nominal Size Range Inches	Limits of Clearance	Standard Limits		Limits of Clearance	Standard Limits		Limits of Clearance	Standard Limits		Limits of Clearance	Standard Limits		Limits of Clearance	Standard Limits		Limits of Clearance	Standard Limits	
Over To		Hole H9	Shaft f8		Hole H10	Shaft e9		Hole H10	Shaft d9		Hole H11	Shaft e10		Hole H12	Shaft		Hole H13	Shaft
0 - 0.12	0.3 1.9	+1.0 0	-0.3 -0.9	0.6 3.2	+1.6 0	-0.6 -1.6	1.0 3.6	+1.5 0	-1.0 -2.0	2.5 6.6	+2.5 0	-2.5 -4.1	4 12	+4 0	-4 -8	5 17	+6 -0	-5 -11
0.12 - 0.24	0.4 2.3	+1.2 0	-0.4 -1.1	0.8 3.8	+1.8 0	-0.8 -2.0	1.2 4.2	+1.8 0	-1.2 -2.4	2.8 7.6	+3.0 0	-2.8 -4.5	4.5 14.5	+5 0	-4.5 -9.5	6 20	+7 -0	-6 -13
0.24 - 0.40	0.5 2.8	+1.4 0	-0.5 -1.4	1.0 4.6	+2.2 0	-1.0 -2.4	1.6 5.2	+2.2 0	-1.6 -3.0	3.0 8.7	+3.5 0	-3.0 -5.2	5 17	+6 0	-5 -11	7 25	+9 -0	-7 -16
0.40 - 0.71	0.6 3.2	+1.6 0	-0.6 -1.6	1.2 5.6	+2.8 0	-1.2 -2.8	2.0 6.4	+2.8 0	-2.0 -3.6	3.5 10.3	+4.0 0	-3.5 -6.3	6 20	+7 0	-6 -13	8 28	+10 -0	-8 -18
0.71 - 1.19	0.8 4.0	+2.0 0	-0.8 -2.0	1.6 7.1	+3.5 0	-1.6 -3.6	2.5 8.0	+3.5 0	-2.5 -4.5	4.5 13.0	+5.0 0	-4.5 -8.0	7 23	+8 0	-7 -15	10 34	+12 -0	-10 -22
1.19 - 1.97	1.0 5.1	+2.5 0	-1.0 -2.6	2.0 8.5	+4.0 0	-2.0 -4.5	3.0 9.5	+4.0 0	-3.0 -5.5	5.0 15.0	+6.0 0	-5.0 -9.0	8 28	+10 0	-8 -18	12 44	+16 -0	-12 -28
1.97 - 3.15	1.2 6.0	+3.0 0	-1.2 -3.0	2.5 10.0	+4.5 0	-2.5 -5.5	4.0 11.5	+4.5 0	-4.0 -7.0	6.0 17.5	+7.0 0	-6.0 -10.5	10 34	+12 0	-10 -22	14 50	+18 -0	-14 -32
3.15 - 4.73	1.4 7.1	+3.5 0	-1.4 -3.6	3.0 11.5	+5.0 0	-3.0 -6.5	5.0 13.5	+5.0 0	-5.0 -8.5	7.0 21.0	+9.0 0	-7.0 -12.0	11 39	+14 0	-11 -25	16 60	+22 -0	-16 -38
4.73 - 7.09	1.6 8.1	+4.0 0	-1.6 -4.1	3.5 11.5	+6.0 0	-3.5 -7.5	6.0 16.0	+6.0 0	-6.0 -10.0	8.0 24.0	+10.0 0	-8.0 -14.0	12 44	+16 0	-12 -28	18 68	+25 -0	-18 -43
7.09 - 9.85	2.0 9.3	+4.5 0	-2.0 -4.8	3.5 13.5	+7.0 0	-4.0 -8.5	7.0 18.5	+7.0 0	-7.0 -11.5	10.0 29.0	+12.0 0	-10.0 -17.0	16 52	+18 0	-16 -34	22 78	+28 -0	-22 -50
9.85 - 12.41	22 10.2	+5.0 0	-2.2 -5.2	4.0 15.5	+8.0 0	-4.5 -9.5	7.0 20.0	+8.0 0	-7.0 -12.0	12.0 32.0	+12.0 0	-12.0 -20.0	20 60	+20 0	-20 -40	28 88	+30 -0	-28 -58
12.41 - 15.75	2.5 12.0	+6.0 0	-2.5 -6.0	4.5 17.5	+9.0 0	-5.0 -11.0	8.0 23.0	+9.0 0	-8.0 -14.0	14.0 37.0	+14.0 0	-14.0 -23.0	22 66	+22 0	-22 -44	30 100	+35 -0	-30 -65
15.75 - 19.69	2.8 12.8	+6.0 0	-2.8 -6.8	5.0 20.0	+10.0 0	-5.0 -11.0	9.0 25.0	+10.0 0	-9.0 -15.0	16.0 42.0	+16.0 0	-16.0 -26.0	25 75	+25 0	-25 -50	35 115	+40 -0	-35 -75
19.69 - 30.09	3.0 16.0	+8.0 0	-3.0 -8.0	6.0 26.0	+12.0 0	-6.0 -14.0	10.0 30.0	+12.0 0	-10.0 -18.0	18.0 50.0	+20.0 0	-18.0 -30.0	28 88	+30 0	-28 -58	40 140	+50 -0	-40 -90
30.09 - 41.49	3.5 19.5	+10.0 0	-3.5 -9.5	7.0 33.0	+16.0 0	-7.0 -17.0	12.0 38.0	+16.0 0	-12.0 -22.0	20 61	+25.0 0	-20.0 -36.0	30 110	+40 0	-30 -70	45 165	+60 -0	-45 -105
41.49 - 56.19	4.0 24.0	+12.0 0	-4.0 -12.0	8.0 40.0	+20.0 0	-8.0 -20.0	14.0 46.0	+20.0 0	-14.0 -26.0	25 75	+30.0 0	-25.0 -45.0	40 140	+50 0	-40 -90	60 220	+80 -0	-60 -140
56.19 - 76.39	4.5 30.5	+16.0 0	-4.5 -14.5	9.0 50.0	+25.0 0	-9.0 -25.0	16 57	+25.0 0	-16.0 -32.0	30 95	+40.0 0	-30.0 -55.0	50 170	+60 0	-50 -110	70 270	+100 -0	-70 -170

Nominal Size Range Inches	Class LC6			Class LC7			Class LC8			Class LC9			Class LC10			Class LC11		
	Limits of Clearance	Standard Limits		Limits of Clearance	Standard Limits		Limits of Clearance	Standard Limits		Limits of Clearance	Standard Limits		Limits of Clearance	Standard Limits		Limits of Clearance	Standard Limits	
Over To		Hole H9	Shaft f8		Hole H10	Shaft e9		Hole H10	Shaft d9		Hole H11	Shaft e10		Hole H12	Shaft		Hole H13	Shaft
76.39 - 100.9	5.0 37.0	+20.0 0	-5.0 -17.0	10.0 60.0	+30.0 0	-10.0 -30.0	18 68	+30.0 0	-18.0 -38.0	35 115	+50.0 0	-35.0 -65.0	50 210	+80 0	-50 -130	80 330	+125 -0	-80 -205
100.9 - 131.9	6.0 47.0	+25.0 0	-6.0 -22.0	12.0 67.0	+40.0 0	-12.0 -27.0	20 85	+40.0 0	-20.0 -45.0	40 140	+60.0 0	-40.0 -80.0	60 260	+100 0	-60 -160	90 410	+160 -0	-90 -250
131.9 - 171.9	7.0 57.0	+30.0 0	-7.0 -27.0	14.0 94.0	+50.0 0	-14.0 -44.0	25 105	+50.0 0	-25.0 -55.0	50 180	+80.0 0	-50 -100	80 330	+125 0	-80 -205	100 500	+200 -0	-100 -300
171.9 - 200	7.0 72.0	+40.0 0	-7.0 -32.0	14.0 114.0	+60.0 0	-14.0 -54.0	25 125	+60.0 0	-25.0 -65.0	50 210	+100.0 0	-50 -110	90 410	+160 0	-90 -250	125 625	+250 -0	-125 -375

CLASS LC 11: CLEARANCE LOCATIONAL FIT

BASIC DIA.		3.0000
HOLE	+18 0	+ .0180 0000
SHAFT	-14 -32	- .0140 - .0320
MAX CLEAR.		.0160
MIN CLEAR. (ALLOWANCE)		.0600

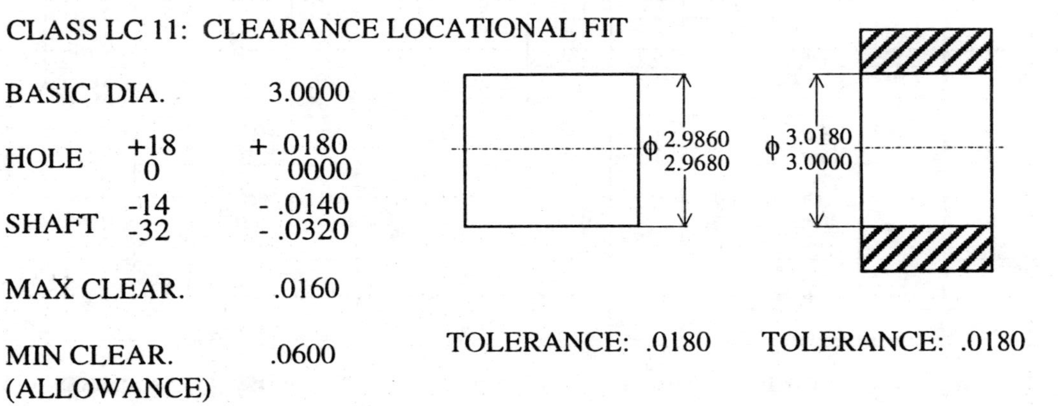

Appendix 3 American Standard Transition Locational Fits (hole basis)

Limits are in thousandths of an inch.
Limits for hole and shaft are applied algebraically to the basic size to obtain the limits of size for the mating parts.
Data in bold face are in accordance with ABC agreements.
"Fit" represents the maximum interference (minus values) and the maximum clearance (plus values).
Symbols H7, js6 etc., are Hole and Shaft designations used in ABC System.

Nominal Size Range Inches	Class LT1			Class LT2			Class LT3			Class LT4			Class LT5			Class LT6		
	Fit	Standard Limits		Fit	Standard Limits		Fit	Standard Limits		Fit	Standard Limits		Fit	Standard Limits		Fit	Standard Limits	
Over To		Hole H7	Shaft js6		Hole H8	Shaft js7		Hole H7	Shaft k6		Hole H8	Shaft k7		Hole H7	Shaft n6		Hole H7	Shaft n7
0 - 0.12	-0.10 +0.50	+0.4 -0	+0.10 -0.10	-0.2 +0.8	+0.6 -0	+0.2 -0.2							-0.5 +0.15	+0.4 -0	+0.5 +0.25	-0.65 +0.15	+0.4 -0	-0.65 +0.25
0.12 - 0.24	-0.15 +0.65	+0.5 -0	+0.15 -0.15	-0.25 +0.95	+0.7 -0	+0.25 -0.25							-0.6 +0.2	+0.5 -0	+0.6 +0.3	-0.8 +0.2	+0.5 -0	+0.8 +0.3
0.24 - 0.40	-0.2 +0.8	+0.6 -0	+0.2 -0.2	-0.3 +1.2	+0.9 -0	+0.3 -0.3	-0.5 +0.5	+0.6 -0	+0.5 +0.1	-0.7 +0.8	+0.9 -0	+0.7 +0.1	-0.8 +0.2	+0.6 -0	+0.8 +0.4	-1.0 +0.2	+0.6 -0	+1.0 +0.4
0.40 - 0.71	-0.2 +0.9	+0.7 -0	+0.2 -0.2	-0.35 +1.35	+1.0 -0	+0.35 -0.35	-0.5 +0.6	+0.7 -0	+0.5 +0.1	-0.8 +0.9	+1.0 -0	+0.8 +0.1	-0.9 +0.2	+0.7 -0	+0.9 +0.5	-1.2 +0.2	+0.7 -0	+1.2 +0.5
0.71 - 1.19	-0.25 +1.05	+0.8 -0	+0.25 -0.25	-0.4 +1.6	+1.2 -0	+0.4 -0.4	-0.6 +0.7	+0.8 -0	+0.6 +0.1	-0.9 +1.1	+1.2 -0	+0.9 +0.1	-1.1 +0.2	+0.8 -0	+1.1 +0.6	-1.4 +0.2	+0.8 -0	+1.4 +0.6
1.19 - 1.97	-0.3 +1.3	+1.0 -0	+0.3 -0.3	-0.5 +2.1	+1.6 -0	+0.5 -0.5	-0.7 +0.9	+1.0 -0	+0.7 +0.1	-1.1 +1.5	+1.6 -0	+1.1 +0.1	-1.3 +0.3	+1.0 -0	+1.3 +0.7	-1.7 +0.3	+1.0 -0	+1.7 +0.7
1.97 – 3.15	-0.3 +1.5	+1.2 -0	+0.3 -0.3	-0.6 +2.4	+1.8 -0	+0.6 -0.6	-0.8 +1.1	+1.2 -0	+0.8 +0.1	-1.3 +1.7	+1.8 -0	+1.3 +0.1	-1.5 +0.4	+1.2 -0	+1.5 +0.8	-2.0 +0.4	+1.2 -0	+2.0 +0.8
3.15 – 4.73	-0.4 +1.8	+1.4 -0	+0.4 -0.4	-0.7 +2.9	+2.2 -0	+0.7 -0.7	-1.0 +1.3	+1.4 -0	+1.0 +0.1	-1.5 +2.1	+2.2 -0	+1.5 +0.1	-1.9 +0.4	+1.4 -0	+1.9 +1.0	-2.4 +0.4	+1.4 -0	+2.4 +1.0
4.73 – 7.09	-0.5 +2.1	+1.6 -0	+0.5 -0.5	-0.8 +3.3	+2.5 -0	+0.8 -0.8	-1.1 +1.5	+1.6 -0	+1.1 +0.1	-1.7 +2.4	+2.5 -0	+1.7 +0.1	-2.2 +0.4	+1.6 -0	+2.2 +1.2	-2.8 +0.4	+1.6 -0	+2.8 +1.2
7.09 – 9.85	-0.6 +2.4	+1.8 -0	+0.6 -0.6	-0.9 +3.7	+2.8 -0	+0.9 -0.9	-1.4 +1.6	+1.8 -0	+1.4 +0.2	-2.0 +2.6	+2.8 -0	+2.0 +0.2	-2.6 +0.4	+1.8 -0	+2.6 +1.4	-3.2 +0.4	+1.8 -0	+3.2 +1.4
9.85 – 12.41	-0.6 +2.6	+2.0 -0	+0.6 -0.6	-1.0 +4.0	+3.0 -0	+1.0 -1.0	-1.4 +1.8	+2.0 -0	+1.4 +0.2	-2.2 +2.8	+3.0 -0	+2.2 +0.2	-2.6 +0.6	+2.0 -0	+2.6 +1.4	-3.4 +0.6	+2.0 -0	+3.4 +1.4
12.41 – 15.75	-0.7 +2.9	+2.2 -0	+0.7 -0.7	-1.0 +4.5	+3.5 -0	+1.0 -1.0	-1.6 +2.0	+2.2 -0	+1.6 +0.2	-2.4 +3.3	+3.5 -0	+2.4 +0.2	-3.0 +0.6	+2.2 -0	+3.0 +1.6	-3.8 +0.6	+2.2 -0	+3.8 +1.6
15.75 – 19.69	-0.8 +3.3	+2.5 -0	+0.8 -0.8	-1.2 +5.2	+4.0 -0	+1.2 -1.2	-1.8 +2.3	+2.5 -0	+1.8 +0.2	-2.7 +3.8	+4.0 -0	+2.7 +0.2	-3.4 +0.7	+2.5 -0	+3.4 +1.8	-4.3 +0.7	+2.5 -0	+4.3 +1.8

Source: Courtesy of ANSI; B4.1-1955.

Appendix 4 American Standard Interference Locational Fits (hole basis)

Limits are in thousandths of an inch.
Limits for hole and shaft are applied algebraically to the basic size to obtain the limits of size for the parts.
Data in bold face are in accordance with ABC agreements.
Symbols H7, p6 etc., are Hole and Shaft designations used in ABC System.

Nominal Size Range Inches	Class LN1			Class LN2			Class LN3		
	Limits of Interference	Standard Limits		Limits of Interference	Standard Limits		Limits of Interference	Standard Limits	
Over To		Hole H6	Shaft n5		Hole H7	Shaft p6		Hole H7	Shaft r6
0 - 0.12	0 0.45	+0.25 -0	+0.45 +0.25	0 0.65	+0.4 -0	0.65 +0.4	0.1 0.75	+0.4 -0	+0.75 +0.5
0.12 – 0.24	0 0.5	+0.3 -0	+0.5 +0.3	0 0.8	+0.5 -0	+0.8 +0.5	0.1 0.9	+0.5 0	+0.9 +0.6
0.24 – 0.40	0 0.65	+0.4 -0	+0.65 +0.4	0 1.0	+0.6 -0	+1.0 +0.6	0.2 1.2	+0.6 -0	+1.2 +0.8
0.40 – 0.71	0 0.8	+0.4 -0	+0.8 +0.4	0 1.1	+0.7 -0	+1.1 +0.7	0.3 1.4	+0.7 -0	+1.4 +1.0
0.71 – 1.19	0 1.0	+0.5 -0	+1.0 +0.5	0 1.3	+0.8 -0	+1.3 +0.8	0.4 1.7	+0.8 -0	+1.7 +1.2
1.19 – 1.97	0 1.1	+0.6 -0	+1.1 +0.6	0 1.6	+1.0 -0	+1.6 +1.0	0.4 2.0	+1.0 -0	+2.0 +1.4
1.97 - 3.15	0.1 1.3	+0.7 -0	+1.3 +0.7	0.2 2.1	+1.2 -0	+2.1 +1.4	0.4 2.3	+1.2 -0	+2.3 +1.6
3.15 - 4.73	0.1 1.6	+0.9 -0	+1.6 +1.0	0.2 2.5	+1.4 -0	+2.5 +1.6	0.6 2.9	+1.4 -0	+2.9 +2.0
4.73 - 7.09	0.2 1.9	+1.0 -0	+1.9 +1.2	0.2 2.8	+1.6 -0	+2.8 +1.8	0.9 3.5	+1.6 -0	+3.5 +2.5
7.09 - 9.85	0.2 2.2	+1.2 -0	+2.2 +1.4	0.2 3.2	+1.8 -0	+3.2 +2.0	1.2 4.2	+1.8 -0	+4.2 +3.0
9.85 - 12.41	0.2 2.3	+1.2 -0	+2.3 +1.4	0.2 3.4	+2.0 -0	+3.4 +2.2	1.5 4.7	+2.0 -0	+4.7 +3.5
12.41 - 15.75	0.2 2.6	+1.4 -0	+2.6 +1.6	0.3 3.9	+2.2 -0	+3.9 +2.5	2.3 5.9	+2.2 -0	+5.9 +4.5
15.75 - 19.69	0.2 2.8	+1.6 -0	+2.8 +1.8	0.3 4.4	+2.5 -0	+4.4 +2.8	2.5 6.6	+2.5 -0	+6.6 +5.0
19.69 - 30.09		+2.0 -0		0.5 5.5	+3 -0	+5.5 +3.5	4 9	+3 -0	+9 +7
30.09 - 41.49		+2.5 -0		0.5 7.0	+4 -0	+7.0 +4.5	5 11.5	+4 -0	+11.5 +9
41.49 - 56.19		+3.0 -0		1 9	+3 -0	+9 +6	7 15	+5 -0	+15 +12
56.19 - 76.39		+4.0 -0		1 11	+6 -0	+11 +7	10 20	+6 -0	+20 +16
76.39 - 100.9		+5.0 -0		1 14	+8 -0	+14 +9	12 25	+8 -0	+25 +20
100.9 - 131.9		+6.0 -0		2 18	+10 -0	+18 +12	15 31	+10 -0	+31 +25
131.9 - 171.9		+8.0 -0		4 24	+12 -0	+24 +16	18 38	+12 -0	+38 +30
171.9 - 200		+10.0 -0		4 30	+16 -0	+30 +20	24 50	+16 -0	+50 +40

Source: Courtesy of ANSI; B4. 1-1955

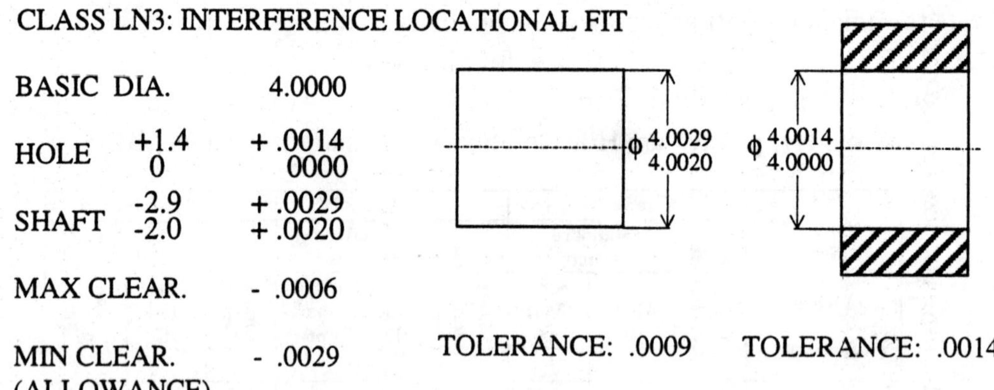
CLASS LN3: INTERFERENCE LOCATIONAL FIT
BASIC DIA. 4.0000
HOLE +1.4 0 +.0014 0000
SHAFT -2.9 -2.0 +.0029 +.0020
MAX CLEAR. - .0006
MIN CLEAR. (ALLOWANCE) - .0029
ϕ 4.0029 4.0020
ϕ 4.0014 4.0000
TOLERANCE: .0009
TOLERANCE: .0014

Appendix 5 American Standard Force and Shrink Fits (hole basis)

Limits are in thousandths of an inch.
Limits for hole and shaft are applied algebraically to the basic size to obtain the limits of size for the parts.
Data in bold face are in accordance with ABC agreements.
Symbols H7, s6 etc., are Hole and Shaft designations used in ABC System.

Nominal Size Range Inches	Class FN 1			Class FN 2			Class FN 3			Class FN 4			Class FN 5		
	Limits of Interference	Standard Limits		Limits of Interference	Standard Limits		Limits of Interference	Standard Limits		Limits of Interference	Standard Limits		Limits of Interference	Standard Limits	
Over - To		Hole H6	Shaft		Hole H7	Shaft s6		Hole H7	Shaft k6		Hole H7	Shaft u6		Hole H8	Shaft s7
0 - 0.12	0.05 0.5	+0.25 -0	+0.5 +0.3	0.2 0.85	+0.4 -0	+0.85 +0.6				0.3 0.95	+0.4 -0	0.95 +0.7	0.3 1.3	+0.6 -0	+0.3 +0.9
0.12 - 0.24	0.1 0.6	+0.3 -0	+0.6 +0.4	0.2 1.0	+0.5 -0	+1.0 +0.7				0.4 1.2	+0.5 -0	+1.2 +0.9	0.5 1.7	+0.7 -0	+1.7 +1.2
0.24 - 0.40	0.1 0.75	+0.4 -0	+0.75 +0.5	0.4 1.4	+0.6 -0	+1.4 +1.0				0.6 1.6	+0.6 -0	+1.6 +1.2	0.5 2.0	+0.9 -0	+2.0 +1.4
0.40 - 0.56	0.1 0.8	-0.4 -0	+0.8 +0.5	0.5 1.6	+0.7 -0	+1.6 +1.2				0.7 1.8	+0.7 -0	+1.8 +1.4	0.6 2.3	+1.0 -0	+2.3 +1.6
0.56 - 0.71	0.2 0.9	+0.4 -0	+0.9 +0.6	0.5 1.6	+0.7 -0	+1.6 +1.2				0.8 2.1	+0.8 -0	+2.1 +1.6	1.0 3.0	+1.2 -0	+3.0 +2.2
0.71 - 0.95	0.2 1.1	+0.5 -0	+1.1 +0.7	0.6 1.9	+0.8 -0	+1.9 +1.4				0.8 2.1	+0.8 -0	+2.1 +1.6	1.0 3.0	+1.2 -0	+3.0 +2.2
0.95 - 1.19	0.3 1.2	+0.5 -0	+1.2 +0.8	0.6 1.9	+0.8 -0	+1.9 +1.4	0.8 2.1	+0.8 -0	+2.1 +1.6	1.0 2.3	+0.8 -0	+2.3 +1.8	1.3 3.3	+1.2 -0	+3.3 +2.5
1.19 - 1.58	0.3 1.3	+0.6 -0	+1.3 +0.9	0.8 2.4	+1.0 -0	+2.4 +1.8	1.0 2.6	+1.0 -0	+2.6 +2.0	1.5 3.1	+1.0 -0	+3.1 +2.5	1.4 4.0	+1.6 -0	+4.0 +3.0
1.58 - 1.97	0.4 1.4	+0.6 -0	+1.4 +1.0	0.8 2.4	+1.0 -0	+2.4 +1.8	1.2 2.8	+1.0 -0	+2.8 +2.2	1.8 3.4	+1.0 -0	+3.4 +2.8	2.4 5.0	+1.6 -0	+5.0 +4.0
1.97 - 2.56	0.6 1.8	+0.7 -0	+1.8 +1.3	0.8 2.7	+1.2 -0	+2.7 +2.0	1.3 3.2	+1.2 -0	+3.2 +2.5	2.3 4.2	+1.2 -0	+4.2 +3.5	3.2 6.2	+1.8 -0	+6.2 +5.0
2.56 - 3.15	0.7 1.9	+0.7 -0	+1.9 +1.4	1.0 2.9	+1.2 -0	+2.9 +2.2	1.8 3.7	+1.2 -0	+3.7 +3.0	2.8 4.7	+1.2 -0	+4.7 +4.0	4.2 7.2	+1.8 -0	+7.2 +6.0
3.15 - 3.94	0.9 2.4	+0.9 -0	+2.4 +1.8	1.4 3.7	+1.4 -0	+3.7 +2.8	2.1 4.4	+1.4 -0	+4.4 +3.5	3.6 5.9	+1.4 -0	+5.9 +5.0	4.8 8.4	+2.2 -0	+8.4 +7.0
3.94 - 4.73	1.1 2.6	+0.9 -0	+2.6 +2.0	1.6 3.9	+1.4 -0	+3.9 +3.0	2.6 4.9	+1.4 -0	+4.9 +4.0	4.6 6.9	+1.4 -0	+6.9 +6.0	5.8 9.4	+2.2 -0	+9.4 +8.0

4.73 - 5.52	1.2 2.9	+1.0 -0	+2.9 +2.2	1.9 4.5	+1.6 -0	+4.5 +3.5	3.4 6.0	+1.6 -0	+6.0 +5.0	5.4 8.0	+1.6 -0	+8.0 +7.0	7.5 11.6	+2.5 -0	+11.6 +10.0
5.52 - 6.30	1.5 3.2	+1.0 -0	+3.2 +2.5	2.4 5.0	+1.6 -0	+5.0 +4.0	3.4 6.0	+1.6 -0	+6.0 +5.0	5.4 8.0	+1.6 -0	+8.0 +7.0	9.5 13.6	+2.5 -0	+13.6 +12.0
6.30 - 7.09	1.8 3.5	+1.0 -0	+3.5 +2.8	2.9 5.5	+1.6 -0	+5.5 +4.5	4.4 7.0	+1.6 -0	+7.0 +6.0	6.4 9.0	+1.6 -0	+9.0 +8.0	9.5 13.6	+2.5 -0	+13.6 +12.0
7.09 - 7.88	1.8 3.8	+1.2 -0	+3.8 +3.0	3.2 6.2	+1.8 -0	+6.2 +5.0	5.2 8.2	+1.8 -0	+8.2 +7.0	7.2 10.2	+1.8 -0	+10.2 +9.0	11.2 15.8	+2.8 -0	+15.8 +14.0
7.88 - 8.86	2.3 4.3	+1.2 -0	+4.3 +3.5	3.2 6.2	+1.8 -0	+6.2 +5.0	5.2 8.2	+1.8 -0	+8.2 +7.0	8.2 11.2	+1.8 -0	+11.2 +10.0	13.2 17.8	+2.8 -0	+17.8 +16.0
8.86 - 9.85	2.3 4.3	+1.2 -0	+4.3 +3.5	4.2 7.2	+1.8 -0	+7.2 +6.0	6.2 9.2	+1.8 -0	+9.2 +8.0	10.2 13.2	+1.8 -0	+13.2 +12.0	13.2 17.8	+2.8 -0	+17.8 +16.0
9.85 - 11.03	2.8 4.9	+1.2 -0	+4.9 +4.0	4.0 7.2	+2.0 -0	+7.2 +6.0	7.0 10.2	+2.0 -0	+10.2 +9.0	10.0 13.2	+2.0 -0	+13.2 +12.0	15.0 20.0	+3.0 -0	+20.0 +18.0
11.03 - 12.41	2.8 4.9	+1.2 -0	+4.9 +4.0	5.0 8.2	+2.0 -0	+8.2 +7.0	7.0 10.2	+2.0 -0	+10.2 +9.0	12.0 15.2	+2.0 -0	+15.2 +14.0	17.0 22.0	+3.0 -0	+22.0 +20.0
12.41 - 13.98	3.1 5.5	+1.4 -0	+5.5 +4.5	5.8 9.4	+2.2 -0	+9.4 +8.0	7.8 11.4	+2.2 -0	+11.4 +10.0	13.8 17.4	+2.2 -0	+17.4 +16.0	18.5 24.2	+3.5 -0	+24.2 +22.0
13.98 - 15.75	3.6 6.1	+1.4 -0	+6.1 +5.0	5.8 9.4	+2.2 -0	+9.4 +8.0	9.8 13.4	+2.2 -0	+13.4 +12.0	15.8 19.4	+2.2 -0	+19.4 +18.0	21.5 27.2	+3.5 -0	+27.2 +25.0
15.75 - 17.72	4.4 7.0	+1.6 -0	+7.0 +6.0	6.5 10.6	+2.5 -0	+10.6 +9.0	9.5 13.6	+2.5 -0	+13.6 +12.0	17.5 21.6	+2.5 -0	+21.6 +20.0	24.0 30.5	+4.0 -0	+30.5 +28.0
17.72 - 19.69	4.4 7.0	+1.6 -0	+7.0 +6.0	7.5 11.6	+2.5 -0	+11.6 +10.0	11.5 15.6	+2.5 -0	+15.6 +14.0	19.5 23.6	+2.5 -0	+23.6 +22.0	26.0 32.5	+4.0 -0	+32.5 +30.0

Source: Courtesy of ANSI; B4.1-1955.

Appendix 6 Preferred Hole Basis Clearance Fits --- Cylindrical Fits (ANSI B4.2)

Dimensions in mm.

BASIC SIZE		LOOSE RUNNING			FREE RUNNING			CLOSE RUNNING			SLIDING			LOCATIONAL CLEARANCE		
		Hole H11	Shaft c11	Fit	Hole H9	Shaft d9	Fit	Hole H8	Shaft f7	Fit	Hole H7	Shaft g6	Fit	Hole H7	Shaft h6	Fit
1.	MAX	1.060	0.904	0.180	1.025	0.980	0.070	1.014	0.994	0.030	1.010	0.998	0.018	1.010	1.000	0.016
	MIN	1.000	0.880	0.060	1.000	0.955	0.020	1.000	0.984	0.006	1.000	0.992	0.002	1.000	0.994	0.000
1.2	MAX	1.260	1.140	0.180	1.225	1.180	0.070	1.214	1.194	0.030	1.210	1.196	0.018	1.210	1.200	0.016
	MIN	1.200	1.080	0.060	1.200	1.155	0.020	1.200	1.184	0.006	1.200	1.192	0.002	1.200	1.194	0.000
1.6	MAX	1.560	1.540	0.180	1.525	1.580	0.070	1.514	1.594	0.030	1.510	1.596	0.018	1.510	1.600	0.016
	MIN	1.500	1.480	0.060	1.500	1.555	0.020	1.500	1.584	0.006	1.500	1.592	0.002	1.500	1.594	0.000
2.	MAX	2.060	1.940	0.180	2.025	1.980	0.070	2.014	1.994	0.030	2.010	1.996	0.018	2.010	2.000	0.016
	MIN	2.000	1.880	0.060	2.000	1.955	0.020	2.000	1.984	0.006	2.000	1.992	0.002	2.000	1.994	0.000
2.5	MAX	2.560	2.440	0.180	2.525	2.480	0.070	2.514	2.494	0.030	2.510	2.496	0.018	2.510	2.500	0.016
	MIN	2.500	2.380	0.060	2.500	2.455	0.020	2.500	2.484	0.006	2.500	2.492	0.002	2.500	2.494	0.000
3.	MAX	3.060	2.940	0.180	3.025	2.980	0.070	3.014	2.994	0.030	3.010	2.996	0.018	3.010	3.000	0.016
	MIN	3.000	2.880	0.060	3.000	2.955	0.020	3.000	2.984	0.006	3.000	2.992	0.002	3.000	2.994	0.000
4.	MAX	4.075	3.930	0.220	4.030	3.970	0.090	4.018	3.990	0.040	4.012	3.996	0.024	4.012	4.000	0.020
	MIN	4.000	3.855	0.070	4.000	3.940	0.030	4.000	3.978	0.010	4.000	3.988	0.004	4.000	3.992	0.000
5.	MAX	5.075	4.930	0.220	5.030	4.970	0.090	5.018	4.990	0.040	5.012	4.996	0.024	5.012	5.000	0.020
	MIN	5.000	4.855	0.070	5.000	4.940	0.030	5.000	4.978	0.010	5.000	4.988	0.004	5.000	4.992	0.000
6.	MAX	6.075	5.930	0.220	6.030	5.970	0.090	6.018	5.990	0.040	6.012	5.996	0.024	6.012	6.000	0.020
	MIN	6.000	5.855	0.070	6.000	5.940	0.030	6.000	5.978	0.010	6.000	5.988	0.004	6.000	5.992	0.000
8	MAX	8.090	7.920	0.260	8.036	7.960	0.112	8.022	7.967	0.050	8.015	7.995	0.029	8.015	8.000	0.024
	MIN	8.000	7.830	0.080	8.000	7.924	0.040	8.000	7.972	0.013	8.000	7.986	0.005	8.000	7.991	0.000
10	MAX	10.090	9.920	0.260	10.036	9.960	0.112	10.022	9.967	0.050	10.015	9.995	0.029	10.015	10.000	0.024
	MIN	10.000	9.830	0.080	10.000	9.924	0.040	10.000	9.972	0.013	10.000	9.986	0.005	10.000	9.991	0.000
12	MAX	12.110	11.905	0.315	12.043	11.950	0.136	12.027	11.984	0.061	12.018	11.994	0.035	12.018	12.000	0.029
	MIN	12.000	11.795	0.095	12.000	11.907	0.050	12.000	11.966	0.016	12.000	11.983	0.006	12.000	11.989	0.000
16	MAX	16.110	15.905	0.315	16.043	15.950	0.136	16.027	15.984	0.061	16.018	15.994	0.035	16.018	16.000	0.029
	MIN	16.000	15.795	0.095	16.000	15.907	0.050	16.000	15.966	0.016	16.000	15.983	0.006	16.000	15.989	0.000
20	MAX	20.130	19.890	0.370	20.052	19.935	0.169	20.033	19.980	0.074	20.021	19.993	0.041	20.021	20.000	0.034
	MIN	20.000	19.760	0.110	20.000	19.883	0.065	20.000	19.959	0.020	20.000	19.980	0.007	20.000	19.987	0.000
25	MAX	25.130	24.890	0.370	25.052	24.935	0.169	25.033	24.980	0.074	25.021	24.993	0.041	25.021	25.000	0.034
	MIN	25.000	24.760	0.110	25.000	24.883	0.065	25.000	24.959	0.020	25.000	24.980	0.007	25.000	24.987	0.000
30	MAX	30.130	29.890	0.370	30.052	29.935	0.169	30.033	29.980	0.074	30.021	29.993	0.041	30.021	30.000	0.034
	MIN	30.000	29.760	0.110	30.000	29.883	0.065	30.000	29.959	0.020	30.000	29.980	0.007	30.000	29.987	0.000

Source: American National Standard Preferred Metric Limits and Fits, ANSI B4.2 – 1978.

Appendix 6 (continued)

Dimensions in mm.

BASIC SIZE		LOOSE RUNNING			FREE RUNNING			CLOSE RUNNING			SLIDING			LOCATIONAL CLEARANCE		
		Hole H11	Shaft c11	Fit	Hole H9	Shaft d9	Fit	Hole H8	Shaft f7	Fit	Hole H7	Shaft g6	Fit	Hole H7	Shaft h6	Fit
40	MAX	40.160	39.880	0.440	40.062	39.920	0.204	40.039	39.975	0.089	40.025	39.991	0.050	40.025	40.000	0.041
	MIN	40.000	39.720	0.120	40.000	39.858	0.080	40.000	39.950	0.025	40.000	39.975	0.009	40.000	39.984	0.000
50	MAX	50.160	49.870	0.450	50.062	49.920	0.204	50.039	49.975	0.089	50.025	49.991	0.050	50.025	50.000	0.041
	MIN	50.000	49.710	0.130	50.000	49.858	0.080	50.000	49.950	0.025	50.000	49.975	0.009	50.000	49.984	0.000
60	MAX	60.190	59.860	0.520	60.074	59.900	0.248	60.046	59.970	0.106	60.030	59.990	0.059	60.030	60.000	0.049
	MIN	60.000	59.670	0.140	60.000	59.826	0.100	60.000	59.940	0.030	60.000	59.971	0.010	60.000	59.961	0.000
80	MAX	80.190	79.850	0.530	80.074	79.900	0.248	80.046	79.970	0.106	80.030	79.990	0.059	80.030	80.000	0.049
	MIN	80.000	79.660	0.150	80.000	79.826	0.100	80.000	79.940	0.030	80.000	79.971	0.010	80.000	79.961	0.000
100	MAX	100.220	99.830	0.610	100.067	99.880	0.294	100.054	99.964	0.125	100.035	99.986	0.069	100.035	100.000	0.057
	MIN	100.000	99.610	0.170	100.000	99.793	0.120	100.000	99.929	0.036	100.000	99.966	0.012	100.000	99.978	0.000
120	MAX	120.220	119.820	0.620	120.067	119.880	0.294	120.054	119.964	0.125	120.035	119.986	0.069	120.035	120.000	0.057
	MIN	120.000	119.500	0.180	120.000	119.793	0.120	120.000	119.929	0.036	120.000	119.966	0.012	120.000	119.978	0.000
160	MAX	160.250	159.790	0.710	160.100	159.855	0.345	160.063	159.957	0.146	160.040	159.986	0.079	160.040	160.000	0.065
	MIN	160.000	159.540	0.210	160.000	159.755	0.145	160.000	159.917	0.043	160.000	159.961	0.014	160.000	159.975	0.000
200	MAX	200.290	199.760	0.820	200.115	199.830	0.400	200.072	199.950	0.168	200.046	199.985	0.090	200.046	200.000	0.075
	MIN	200.000	199.470	0.240	200.000	199.715	0.170	200.000	199.904	0.050	200.000	199.958	0.015	200.000	199.971	0.000
250	MAX	250.290	249.720	0.860	250.115	249.830	0.400	250.072	249.950	0.168	250.046	249.985	0.090	250.046	250.000	0.075
	MIN	250.000	249.430	0.280	250.000	249.715	0.170	250.000	249.904	0.050	250.000	249.958	0.015	250.000	249.971	0.000
300	MAX	300.320	299.570	0.970	300.130	299.810	0.450	300.061	299.944	0.189	300.052	299.963	0.101	300.052	300.000	0.084
	MIN	300.000	299.350	0.330	300.000	299.680	0.190	300.000	299.892	0.056	300.000	299.951	0.017	300.000	299.968	0.000
400	MAX	400.360	399.600	1.120	400.140	399.790	0.490	400.089	399.938	0.206	400.057	399.962	0.111	400.057	400.000	0.083
	MIN	400.000	399.240	0.400	400.000	399.650	0.210	400.000	399.881	0.062	400.000	399.946	0.018	400.000	399.964	0.000
500	MAX	500.400	499.520	1.280	500.155	499.770	0.540	500.097	499.932	0.228	500.063	499.980	0.123	500.063	500.000	0.103
	MIN	500.000	499.120	0.480	500.000	499.615	0.230	500.000	499.869	0.068	500.000	499.940	0.020	500.000	499.960	0.000

METRIC H11/c11: LOOSE RUNNING FIT

BASIC DIA.	40 mm
HOLE	40.160 40.000
SHAFT	39.880 39.720
MAX CLEAR.	0.440

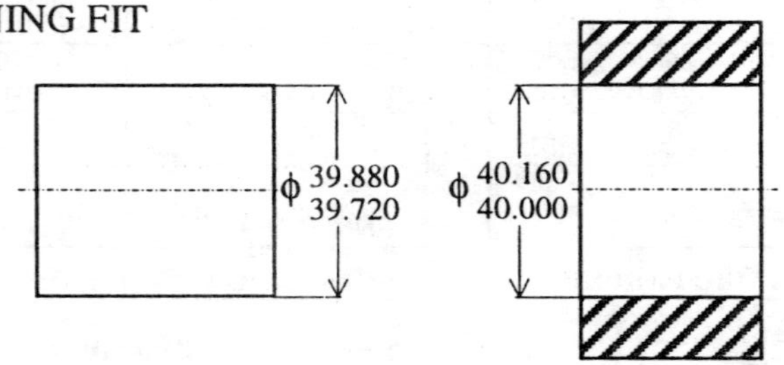

TOLERANCE: 0.160 TOLERANCE: 0.160

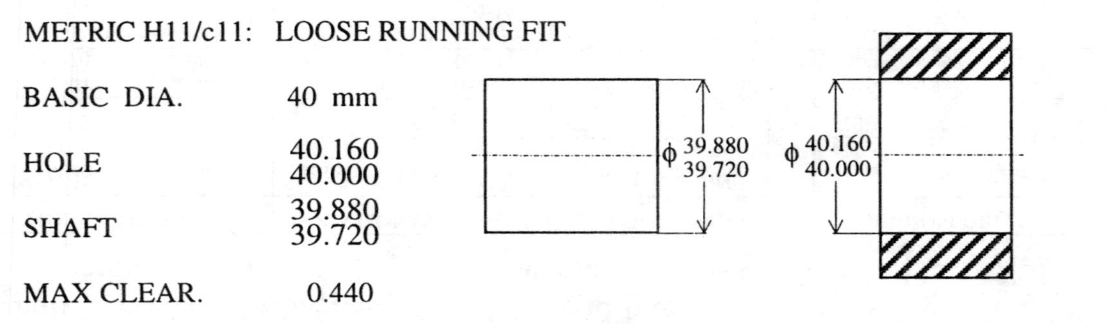

Appendix 7 Preferred Hole Basis Transition and Interference Fits --- Cylindrical Fits (ANSI B4.2)

Dimensions in mm.

BASIC SIZE		LOCATIONAL TRANSH.			LOCATIONAL TRANSH			LOCATIONAL INTERF.			MEDIUM DRIVE			FORCE		
		Hole H7	Shaft k6	Fit	Hole H7	Shaft n6	Fit	Hole H7	Shaft p6	Fit	Hole H7	Shaft s6	Fit	Hole H7	Shaft u6	Fit
1	MAX	1.010	1.006	0.010	1.010	1.010	0.006	1.010	1.012	0.004	1.010	1.020	-0.004	1.010	1.024	-0.008
	MIN	1.000	1.000	-0.006	1.000	1.004	-0.010	1.000	1.006	-0.012	1.000	1.014	-0.020	1.000	1.018	-0.024
1.2	MAX	1.210	1.206	0.010	1.210	1.210	0.006	1.210	1.212	0.004	1.210	1.220	-0.004	1.210	1.224	-0.008
	MIN	1.200	1.200	-0.006	1.200	1.204	-0.010	1.200	1.206	-0.012	1.200	1.214	-0.020	1.200	1.218	-0.024
1.6	MAX	1.610	1.606	0.010	1.610	1.610	0.006	1.610	1.612	0.004	1.610	1.620	-0.004	1.610	1.624	-0.008
	MIN	1.600	1.600	-0.006	1.600	1.604	-0.010	1.600	1.606	-0.012	1.600	1.614	-0.020	1.600	1.618	-0.024
2	MAX	2.010	2.006	0.010	2.010	2.010	0.006	2.010	2.010	0.004	2.010	2.020	-0.004	2.010	2.024	-0.008
	MIN	2.000	2.000	-0.006	2.000	2.004	-0.010	2.000	2.006	-0.012	2.000	2.014	-0.020	2.000	2.018	-0.024
2.5	MAX	2.510	2.506	0.010	2.510	2.510	0.006	2.510	2.512	0.004	2.510	2.520	-0.004	2.510	2.524	-0.008
	MIN	2.500	2.500	-0.006	2.500	2.504	-0.010	2.500	2.506	-0.012	2.500	2.514	-0.020	2.500	2.518	-0.024
3	MAX	3.010	3.006	0.010	3.010	3.010	0.006	3.010	3.012	0.004	3.010	3.020	-0.004	3.010	3.024	-0.008
	MIN	3.000	3.000	-0.006	3.000	3.004	-0.010	3.000	3.006	-0.012	3.000	3.014	-0.020	3.000	3.018	-0.024
4	MAX	4.012	4.009	0.011	4.012	4.016	0.004	4.012	4.020	0.000	4.012	4.027	-0.007	4.012	4.031	-0.011
	MIN	4.000	4.001	-0.009	4.000	4.008	-0.016	4.000	4.012	-0.020	4.000	4.019	-0.027	4.000	4.023	-0.031
5	MAX	5.012	5.009	0.011	5.012	5.016	0.004	5.012	5.020	0.000	5.012	5.027	-0.007	5.012	5.031	-0.011
	MIN	5.000	5.001	-0.009	5.000	5.008	-0.016	5.000	5.012	-0.020	5.000	5.019	-0.027	5.000	5.023	-0.031
6	MAX	6.012	6.009	0.011	6.012	6.016	0.004	6.012	6.020	0.000	6.012	6.027	-0.007	6.012	6.031	-0.011
	MIN	6.000	6.001	-0.009	6.000	6.008	-0.016	6.000	6.012	-0.020	6.000	6.019	-0.027	6.000	6.023	-0.031
8	MAX	8.015	8.010	0.014	8.015	8.019	0.005	8.015	8.024	0.000	8.015	8.032	-0.008	8.015	8.037	-0.013
	MIN	8.000	8.001	-0.010	8.000	8.010	-0.019	8.000	8.015	-0.024	8.000	8.023	-0.032	8.000	8.028	-0.037
10	MAX	10.015	10.010	0.014	10.015	10.019	0.005	10.015	10.024	0.000	10.015	10.032	-0.008	10.015	10.037	-0.013
	MIN	10.000	10.001	-0.010	10.000	10.010	-0.019	10.000	10.015	-0.024	10.000	10.023	-0.032	10.000	10.028	-0.037
12	MAX	12.018	12.012	0.017	12.018	12.023	0.006	12.018	12.029	0.000	12.018	12.039	-0.010	12.018	12.044	-0.015
	MIN	12.000	12.001	-0.012	12.000	12.012	-0.023	12.000	12.018	-0.029	12.000	12.028	-0.039	12.000	12.033	-0.044
16	MAX	16.018	16.012	0.017	16.018	16.023	0.006	16.018	16.029	0.000	16.018	16.039	-0.010	16.018	16.044	-0.015
	MIN	16.000	16.001	-0.012	16.000	16.012	-0.023	16.000	16.018	-0.029	16.000	16.028	-0.039	16.000	16.033	-0.044
20	MAX	20.021	20.015	0.019	20.021	20.028	0.006	20.021	20.035	- 0.001	20.021	20.048	-0.014	20.021	20.054	-0.020
	MIN	20.000	20.002	-0.015	20.000	20.015	-0.028	20.000	20.022	-0.035	20.000	20.035	-0.048	20.000	20.041	-0.054
25	MAX	25.021	25.015	0.019	25.021	25.028	0.006	25.021	25.035	- 0.001	25.021	25.048	-0.014	25.021	25.061	-0.027
	MIN	25.000	25.002	-0.015	25.000	25.015	-0.028	25.000	25.022	-0.035	25.000	25.035	-0.048	25.000	25.048	-0.061
30	MAX	30.021	30.015	0.019	30.021	30.028	0.006	30.021	30.035	- 0.001	30.021	30.048	-0.014	30.021	30.061	-0.027
	MIN	30.000	30.002	-0.015	30.000	30.015	-0.028	30.000	30.022	-0.035	30.000	30.035	-0.048	30.000	30.048	-0.061

Source: American National Standard Preferred Metric Limits and Fits, ANSI B4.2 – 1978

Appendix 7 (Continue)

Dimensions in mm.

BASIC SIZE		LOCATIONAL TRANSH. Hole H7	Shaft k6	Fit	LOCATIONAL TRANSH Hole H7	Shaft n6	Fit	LOCATIONAL INTERF. Hole H7	Shaft p6	Fit	MEDIUM DRIVE Hole H7	Shaft s6	Fit	FORCE Hole H7	Shaft u6	Fit
40	MAX	40.025	40.018	0.023	40.025	40.033	0.008	40.025	40.042	-0.001	40.025	40.059	-0.018	40.025	40.076	-0.035
	MIN	40.000	40.002	-0.018	40.000	40.017	-0.033	40.000	40.026	-0.042	40.000	40.043	-0.059	40.000	40.060	-0.076
50	MAX	50.025	50.018	0.023	50.025	50.033	0.008	50.025	50.042	-0.001	50.025	50.059	-0.018	50.025	50.086	-0.045
	MIN	50.000	50.002	-0.018	50.000	50.017	-0.033	50.000	50.026	-0.042	50.000	50.043	-0.059	50.000	50.070	-0.086
60	MAX	60.030	60.021	0.028	60.030	60.039	0.010	60.030	60.051	-0.002	60.030	60.072	-0.023	60.030	60.106	-0.057
	MIN	60.000	60.002	-0.021	60.000	60.020	-0.039	60.000	60.032	-0.051	60.000	60.053	-0.072	60.000	60.087	-0.106
80	MAX	80.030	80.021	0.028	80.030	80.039	0.010	80.030	80.051	-0.002	80.030	80.078	-0.029	80.030	80.121	-0.072
	MIN	80.000	80.002	-0.021	80.000	80.020	-0.039	80.000	80.032	-0.051	80.000	80.059	-0.078	80.000	80.102	-0.121
100	MAX	100.035	100.025	0.032	100.035	100.045	0.012	100.035	100.059	-0.002	100.035	100.093	-0.036	100.035	100.146	-0.089
	MIN	100.000	100.003	-0.025	100.000	100.023	-0.045	100.000	100.037	-0.059	100.000	100.071	-0.093	100.000	100.124	-0.146
120	MAX	120.035	120.025	0.032	120.035	120.045	0.012	120.035	120.059	-0.002	120.035	120.101	-0.044	120.035	120.166	-0.109
	MIN	120.000	120.003	-0.025	120.000	120.023	-0.045	120.000	120.037	-0.059	120.000	120.079	-0.101	120.000	120.144	-0.166
160	MAX	160.040	160.028	0.037	160.040	160.052	0.013	160.040	160.068	-0.003	160.040	160.125	-0.060	160.040	160.215	-0.150
	MIN	160.000	160.003	-0.028	160.000	160.027	-0.052	160.000	160.043	-0.068	160.000	160.100	-0.125	160.000	160.190	-0.215
200	MAX	200.046	200.033	0.042	200.046	200.060	0.015	200.046	200.079	-0.004	200.046	200.151	-0.076	200.046	200.265	-0.190
	MIN	200.000	200.004	-0.033	200.000	200.031	-0.060	200.000	200.050	-0.079	200.000	200.122	-0.151	200.000	200.236	-0.265
250	MAX	250.046	250.033	0.042	250.046	250.060	0.015	250.046	250.079	-0.004	250.046	250.169	-0.094	250.046	250.313	-0.238
	MIN	250.000	250.004	-0.033	250.000	250.031	-0.060	250.000	250.050	-0.079	250.000	250.140	-0.169	250.000	250.284	-0.313
300	MAX	300.052	300..036	0.048	300.052	300.066	0.018	300.052	300.088	-0.004	300.052	300.202	-0.118	300.052	300.382	-0.298
	MIN	300.000	300.004	0.036	300.000	300.034	-0.066	300.000	300.056	-0.088	300.000	300.170	-0.202	300.000	300.350	-0.382
400	MAX	400.057	400.040	0.053	400.057	400.073	0.020	400.057	400.098	-0.005	400.057	400.244	-0.151	400.057	400.471	-0.378
	MIN	400.000	400.004	-0.040	400.000	400.037	-0.073	400.000	400.062	-0.098	400.000	400.208	-0.244	400.000	400.435	-0.471
500	MAX	500.063	500.045	0.058	500.063	500.080	0.023	500.063	500.108	-0.005	500.063	500.292	-0.189	500.063	500.580	-0.477
	MIN	500.000	500.005	-0.045	500.000	500.040	-0.080	500.000	500.068	-0.108	500.000	500.252	-0.292	500.000	500.540	-0.580

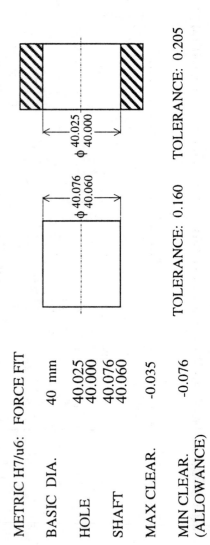

METRIC H7/u6: FORCE FIT

BASIC DIA.	40 mm
HOLE	40.025 40.000
SHAFT	40.076 40.060
MAX CLEAR.	-0.035
MIN CLEAR. (ALLOWANCE)	-0.076

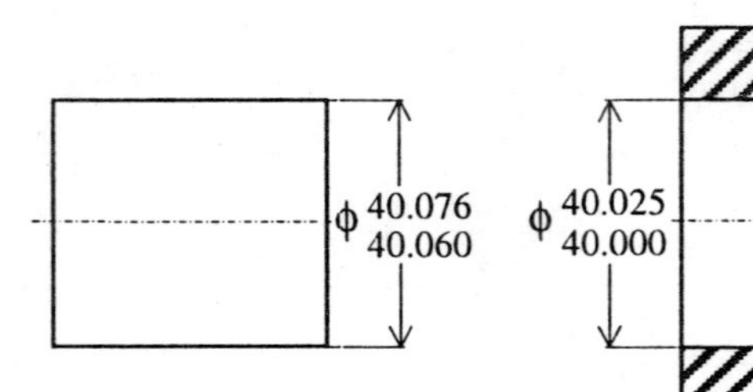

TOLERANCE: 0.160 TOLERANCE: 0.205

Appendix 8 Preferred Shaft Basis Clearance Fits --- Cylindrical Fits (ANSI B4.2)

Dimensions in mm.

BASIC SIZE		LOOSE RUNNING Hole C11	Shaft h11	Fit	FREE RUNNING Hole D9	Shaft h9	Fit	CLOSE RUNNING Hole F8	Shaft h7	Fit	SLIDING Hole G7	Shaft h6	Fit	LOCATIONAL CLEARANCE Hole H7	Shaft h6	Fit
1	MAX	1.120	1.000	0.180	1.045	1.000	0.070	1.020	1.000	0.030	1.012	1.000	0.018	1.010	1.000	0.016
	MIN	1.060	0.940	0.060	1.020	0.975	0.020	1.006	0.990	0.006	1.002	0.994	0.002	1.000	0.994	0.000
1.2	MAX	1.320	1.200	0.180	1.245	1.200	0.070	1.220	1.200	0.030	1.212	1.200	0.018	1.210	1.200	0.016
	MIN	1.260	1.140	0.060	1.220	1.175	0.020	1.206	1.190	0.006	1.202	1.194	0.002	1.200	1.194	0.000
1.6	MAX	1.720	1.600	0.180	1.645	1.60 0	0.070	1.620	1.600	0.030	1.612	1.600	0.018	1.610	1.500	0.016
	MIN	1.660	1.540	0.060	1.620	1.575	0.020	1.606	1.590	0.006	1.602	1.594	0.002	1.600	1.594	0.000
2	MAX	2.120	2.000	0.180	2.045	2.000	0.070	2.020	2.000	0.030	2.012	2.000	0.018	2.010	2.000	0.016
	MIN	2.060	1.940	0.060	2.020	1.975	0.020	2.006	1.990	0.006	2.002	1.994	0.002	2.000	1.994	0.000
2.5	MAX	2.620	2.500	0.180	2.545	2.500	0.070	2.520	2.500	0.030	2.512	2.500	0.018	2.510	2.500	0.016
	MIN	2.580	2.440	0.060	2.520	2.475	0.020	2.506	2.490	0.006	2.502	2.494	0.002	2.500	2.494	0.000
3	MAX	3.120	3.000	0.180	3.045	3.000	0.070	3.020	3.000	0.030	3.012	3.000	0.018	3.010	3.000	0.016
	MIN	3.060	2.940	0.060	3.020	2.975	0.020	3.006	2.990	0.006	3.002	2.994	0.002	3.000	2.994	0.000
4	MAX	4.145	4.000	0.220	4.060	4.000	0.090	4.028	4.000	0.040	4.016	4.000	0.024	4.012	4.000	0.020
	MIN	4.070	3.925	0.070	4.030	3.970	0.030	4.010	3.968	0.010	4.004	3.992	0.004	4.000	3.992	0.000
5	MAX	5.145	5.000	0.220	5.060	5.000	0.090	5.028	5.000	0.040	5.016	5.000	0.024	5.012	5.000	0.020
	MIN	5.070	4.925	0.070	5.030	4.970	0.030	5.010	4.968	0.010	5.004	4.992	0.004	5.000	4.992	0.000
6	MAX	6.145	6.000	0.220	6.060	6.000	0.090	6.028	6.000	0.040	6.016	6.000	0.024	6.012	6.000	0.020
	MIN	6.070	5.925	0.070	6.030	5.970	0.030	6.010	5.968	0.010	6.004	5.992	0.004	6.000	5.992	0.000
8	MAX	8.170	8.000	0.260	8.076	8.000	0.112	8.035	8.000	0.050	8.020	8.000	0.029	8.015	8.000	0.024
	MIN	8.080	7.910	0.080	8.040	7.964	0.040	8.013	7.965	0.013	8.005	7.991	0.005	8.000	7.991	0.000
10	MAX	10.170	10.000	0.260	10.076	10.000	0.112	10.035	10.000	0.050	10.020	10.000	0.029	10.015	10.000	0.024
	MIN	10.080	9.910	0.080	10.040	9.964	0.040	10.013	9.985	0.013	10.005	9.991	0.005	10.000	9.991	0.000
12	MAX	12.205	12.000	0.315	12.093	12.000	0.136	12.043	12.000	0.061	12.024	12.000	0.035	12.018	12.000	0.029
	MIN	12.095	11.890	0.095	12.050	11.957	0.050	12.016	11.982	0.016	12.006	11.989	0.006	12.000	11.989	0.000
16	MAX	16.205	16.000	0.315	16.093	16.000	0.136	16.043	16.000	0.061	16.024	16.000	0.035	16.018	16.000	0.029
	MIN	16.095	15.890	0.095	16.050	15.957	0.050	16.016	15.982	0.016	16.006	5.969	0.006	16.000	15.969	0.000
20	MAX	20.240	20.000	0.370	20.117	20.000	0.136	20.053	20.000	0.074	20.028	20.000	0.041	20.021	20.000	0.034
	MIN	20.110	19.870	0.110	20.065	19.948	0.050	20.020	19.979	0.020	20.007	19.967	0.007	20.000	19.987	0.000

25	MAX	25.240	25.000	0.370	25.117	25.000	0.169	25.053	25.000	0.074	25.028	25.000	0.041	25.021	25.000	0.034
	MIN	25.110	24.870	0.110	25.065	24.948	0.065	25.020	24.979	0.020	25.007	24.967	0.007	25.000	24.987	0.000
30	MAX	30.240	30.000	0.370	30.117	30.000	0.169	30.053	30.000	0.074	30.028	30.000	0.041	30.021	30.000	0.034
	MIN	30.110	29.870	0.110	30.065	29.948	0.065	30.020	29.979	0.020	30.007	29.987	0.007	30.000	29.987	0.000

Source: American National Standard Preferred Metric Limits and Fits, ANSI B4.2 – 1978

Appendix 8 (continued)

Dimensions in mm.

BASIC SIZE		LOOSE RUNNING Hole C1	Shaft h11	Fit	FREE RUNNING Hole D9	Shaft h9	Fit	CLOSE RUNNING Hole F8	Shaft h7	Fit	SLIDING Hole G7	Shaft h6	Fit	LOCATIONAL CLEARANCE Hole H7	Shaft h6	Fit
40	MAX	40.280	40.000	0.440	40.142	40.000	0.204	40.064	40.000	0.089	40.034	40.000	0.050	40.025	40.000	0.041
	MIN	40.120	39.540	0.120	40.080	39.938	0.080	40.025	39.975	0.025	40.009	39.984	0.009	40.000	39.964	0.000
50	MAX	50.290	50.000	0.450	50.142	50.000	0.204	50.064	50.000	0.089	50.034	50.000	0.050	50.025	50.000	0.041
	MIN	50.130	49.840	0.130	50.080	49.938	0.080	50.025	49.975	0.025	50.009	19.984	0.009	50.000	49.964	0.000
60	MAX	60.330	80.000	0.520	60.174	60.000	0.248	60.076	50.000	0.106	60.040	60.000	0.059	60.030	60.000	0.049
	MIN	60.140	59.810	0.140	60.100	59.926	0.100	60.030	59.970	0.030	60.010	59.981	0.010	60.000	59.961	0.000
80	MAX	80.340	80.000	0.530	80.174	80.000	0.248	80.076	80.000	0.106	80.040	80.000	0.059	80.030	80.000	0.049
	MIN	80.150	79.810	0.150	80.100	79.926	0.100	80.030	79.970	0.030	80.010	79.981	0.010	80.000	79.961	0.000
100	MAX	100.390	100.000	0.610	100.207	100.000	0.294	100.090	100.000	0.125	100.047	100.000	0.069	100.035	100.000	0.057
	MIN	100.170	99.780	0.170	100.120	99.913	0.120	100.036	99.965	0.036	100.012	99.978	0.012	100.000	99.978	0.000
120	MAX	120.400	120.000	0.620	120.207	120.000	0.294	120.090	120.000	0.125	120.047	120.000	0.069	120.035	120.000	0.057
	MIN	120.180	119.780	0.180	120.120	119.913	0.120	120.036	119.965	0.036	120.012	119.978	0.012	120.000	119.978	0.000
160	MAX	160.460	160.000	0.710	160.245	160.000	0.345	160.106	160.000	0.146	160.054	160.000	0.079	160.040	160.000	0.065
	MIN	160.210	159.750	0.210	160.145	159.900	0.145	160.043	159.960	0.043	160.014	159.975	0.014	160.000	159.975	0.000
200	MAX	200.530	200.000	0.820	200.285	200.000	0.400	200.122	200.000	0.168	200.061	200.000	0.090	200.046	200.000	0.075
	MIN	200.240	199.710	0.240	200.170	199.585	0.170	200.050	199.954	0.050	200.015	199.971	0.015	200.000	199.971	0.000
250	MAX	250.570	250.000	0.860	250.285	250.000	0.400	250.122	250.000	0.168	250.061	250.000	0.090	250.048	250.000	0.075
	MIN	250.280	249.710	0.280	250.170	249.885	0.170	250.050	249.954	0.050	250.015	249.971	0.015	250.000	249.971	0.000
300	MAX	300.650	300.000	0.970	300.320	300.000	0.450	300.137	300.000	0.189	300.089	300.000	0.101	300.052	300.000	0.084
	MIN	300.330	299.680	0.330	300.190	299.870	0.190	300.056	299.948	0.058	300.017	299.968	0.017	300.000	299.968	0.000
400	MAX	400.780	400.000	1.120	400.350	400.000	0.490	400.151	400.000	0.208	400.075	400.000	0.111	400.570	400.000	0.093
	MIN	400.400	399.640	0.400	400.210	399.860	0.210	400.062	399.943	0.062	400.018	399.964	0.016	400.000	399.964	0.000
500	MAX	500.860	500.000	1.280	500.385	500.000	0.540	500.165	500.000	0.228	500.063	500.000	0.123	500.063	500.000	0.103
	MIN	500.480	499.600	0.480	500.230	499.845	0.230	500.058	499.937	0.068	500.020	499.960	0.020	500.000	499.960	0.000

METRIC C 11/h 11: LOOSE RUNNING FIT

BASIC DIA.	40 mm	
HOLE	40.280 40.120	
SHAFT	40.000 39.840	
MAX CLEAR.	0.440	
MIN CLEAR. (ALLOWANCE)	0.120	TOLERANCE: 0.160 TOLERANCE: 0.160

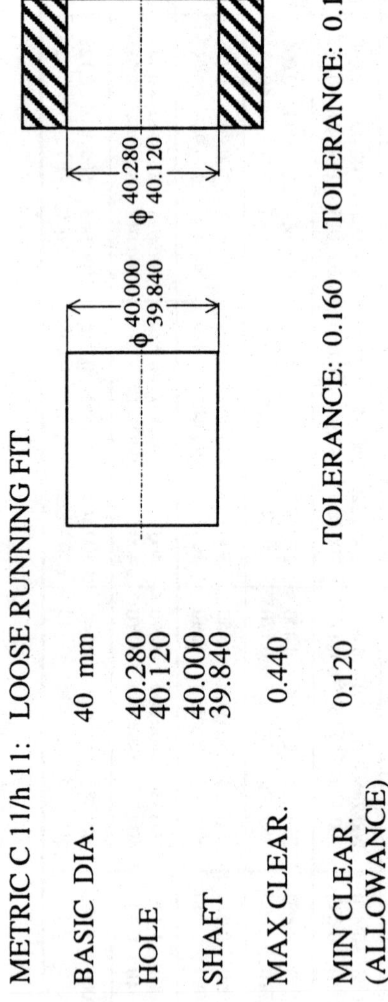

Appendix 9 Preferred Shaft Basis Clearance Fits --- Cylindrical Fits (ANSI B4.2)

Dimensions in mm.

BASIC SIZE		LOCATIONAL TRANSH.			LOCATIONAL TRANSH			LOCATIONAL INTERF.			MEDIUM DRIVE			FORCE		
		Hole K7	Shaft h6	Fit	Hole N7	Shaft h6	Fit	Hole F7	Shaft h6	Fit	Hole S7	Shaft h6	Fit	Hole U7	Shaft h6	Fit
1	MAX	1.000	1.000	0.006	0.996	1.000	0.002	0.994	1.000	0.000	0.986	1.000	-0.006	0.982	1.000	-0.012
	MIN	0.990	0.994	-0.010	0.986	0.994	-0.014	0.984	0.994	-0.016	0.976	0.994	-0.024	0.972	0.994	-0.028
1.2	MAX	1.200	1.200	0.006	1.196	1.200	0.002	1.194	1.200	0.000	1.186	1.200	-0.006	1.182	1.200	-0.012
	MIN	1.190	1.194	-0.010	1.186	1.194	-0.014	1.184	1.194	-0.016	1.176	1.194	-0.024	1.172	1.194	-0.028
1.6	MAX	1.500	1.600	0.006	1.596	1.600	0.002	1.594	1.600	0.000	1.586	1.600	-0.006	1.582	1.600	-0.012
	MIN	1.590	1.594	-0.010	1.586	1.594	-0.014	1.584	1.594	-0.016	1.576	1.594	-0.024	1.572	1.594	-0.028
2.0	MAX	2.000	2.000	0.006	1.996	2.000	0.002	1.994	2.000	0.000	1.986	2.000	-0.006	1.982	2.000	-0.012
	MIN	1.990	1.994	-0.010	1.986	1.994	-0.014	1.984	1.994	-0.016	1.976	1.994	-0.024	1.972	1.994	-0.028
2.5	MAX	2.500	2.500	0.006	2.496	2.500	0.002	2.494	2.500	0.000	2.486	2.500	-0.006	2.482	2.500	-0.012
	MIN	2.490	2.494	-0.010	2.486	2.494	-0.014	2.484	2.494	-0.016	2.476	2.494	-0.024	2.472	2.494	-0.028
3.0	MAX	3.000	3.000	0.006	2.996	3.000	0.002	2.994	3.000	0.000	2.986	3.000	-0.006	2.982	3.000	-0.012
	MIN	2.990	2.994	-0.010	2.986	2.994	-0.014	2.984	2.994	-0.016	2.976	2.994	-0.024	2.972	2.994	-0.028
4.0	MAX	4.003	4.000	0.011	3.996	4.000	0.004	3.992	4.000	0.000	3.985	4.000	-0.007	3.981	4.000	-0.011
	MIN	3.991	3.992	-0.009	3.984	3.992	-0.016	3.980	3.992	-0.020	3.973	3.992	-0.027	3.969	3.992	-0.031
5.0	MAX	5.003	5.000	0.011	4.996	5.000	0.004	4.992	5.000	0.000	4.985	5.000	-0.007	4.981	5.000	-0.011
	MIN	4.991	4.992	-0.009	4.984	4.992	-0.016	4.980	4.992	-0.020	4.973	4.992	-0.027	4.969	4.992	-0.031
6.0	MAX	6.003	6.000	0.011	5.996	6.000	0.004	5.992	6.000	0.000	5.985	6.000	-0.007	5.981	6.000	-0.011
	MIN	5.991	5.992	-0.009	5.984	5.992	-0.016	5.980	5.992	-0.020	5.973	5.992	-0.027	5.969	5.992	-0.031
8.0	MAX	8.005	8.000	0.014	7.996	8.000	0.005	7.991	8.000	0.000	7.983	8.000	-0.008	7.976	8.000	-0.015
	MIN	7.990	7.991	-0.010	7.981	7.991	-0.019	7.976	7.991	-0.024	7.968	7.991	-0.032	7.963	7.991	-0.037
10.0	MAX	10.005	10.000	0.014	9.995	10.000	0.005	9.991	10.000	0.000	9.983	10.000	-0.006	9.978	10.000	-0.013
	MIN	9.990	9.991	-0.010	9.981	9.991	-0.019	9.976	9.991	-0.024	9.968	9.991	-0.032	9.963	9.991	-0.037
12.0	MAX	12.006	12.000	0.017	11.995	12.000	0.006	11.989	12.000	0.000	11.979	12.000	-0.010	11.974	12.000	-0.015
	MIN	11.988	11.989	-0.012	11.977	11.989	-0.023	11.971	11.989	-0.029	11.961	11.989	-0.039	11.956	11.989	-0.044
16.0	MAX	16.006	16.000	0.017	15.995	16.000	0.006	15.989	16.000	0.000	15.979	16.000	-0.010	15.974	16.000	-0.015
	MIN	15.988	15.989	-0.012	15.977	15.989	-0.023	15.971	15.989	-0.029	15.961	15.989	-0.039	15.956	15.989	-0.044
20.0	MAX	20.006	20.000	0.019	19.993	20.000	0.006	19.986	20.000	-0.001	19.973	20.000	-0.014	19.967	20.000	-0.020
	MIN	19.985	19.987	-0.015	19.972	19.987	-0.028	19.965	19.987	-0.035	19.952	19.987	-0.048	19.946	19.987	-0.054
25.0	MAX	25.006	25.000	0.019	24.993	25.000	0.006	24.986	25.000	-0.001	24.973	25.500	-0.014	24.960	25.000	-0.027
	MIN	24.985	24.987	-0.015	24.972	24.987	-0.028	24.965	24.987	-0.035	24.952	24.987	-0.048	24.939	24.987	-0.061
30.0	MAX	30.006	30.000	0.019	29.993	30.000	0.006	29.986	30.000	-0.001	29.973	30.000	-0.014	29.960	30.000	-0.027
	MIN	29.985	29.987	-0.015	29.972	29.987	-0.028	29.965	29.987	-0.035	29.952	29.987	-0.048	29.939	29.987	-0.061

Source: American National Standard Preferred Metric Limits and Fits, ANSI B4.2-1978

Appendix 9 (continued)

Dimensions in mm.

BASIC SIZE		LOCATIONAL TRANSH. Hole K7	Shaft h6	Fit	LOCATIONAL TRANSH Hole N7	Shaft h6	Fit	LOCATIONAL INTERF. Hole F7	Shaft h6	Fit	MEDIUM DRIVE Hole S7	Shaft h6	Fit	FORCE Hole U7	Shaft h6	Fit
40	MAX	40.007	40.000	0.023	39.992	40.000	0.008	39.983	40.000	-0.001	39.968	40.000	-0.018	39.949	40.000	-0.035
	MIN	39.982	39.984	-0.018	39.967	39.984	-0.033	39.958	39.984	-0.042	39.941	39.984	-0.059	39.924	39.984	-0.076
50	MAX	50.007	50.000	0.023	49.992	50.000	0.008	49.983	50.000	-0.001	49.968	50.000	-0.018	49.939	50.000	-0.045
	MIN	49.982	49.984	-0.018	49.967	49.984	-0.033	49.958	49.984	-0.042	49.941	49.984	-0.058	49.914	49.984	-0.086
60	MAX	60.009	60.000	0.028	59.991	60.000	0.010	59.979	60.000	-0.002	59.958	60.000	-0.023	59.924	60.000	-0.057
	MIN	59.979	59.981	-0.021	59.961	59.981	-0.039	59.949	59.981	-0.051	59.928	59.981	-0.072	59.894	59.981	-0.108
80	MAX	80.009	80.000	0.028	79.991	80.000	0.010	79.979	80.000	-0.002	79.952	80.000	-0.029	79.909	80.000	-0.072
	MIN	79.979	79.981	-0.021	79.961	79.981	-0.039	79.949	79.981	-0.051	79.922	79.981	-0.078	79.879	79.981	-0.121
100	MAX	100.010	100.000	0.032	99.990	100.000	0.012	99.976	100.000	-0.002	99.942	100.000	-0.036	99.889	100.000	-0.089
	MIN	99.975	99.978	-0.025	99.955	99.978	-0.045	99.941	99.978	-0.305	99.907	99.978	-0.093	99.854	99.978	-0.146
120	MAX	120.010	120.000	0.032	119.990	120.000	0.012	119.976	120.000	-0.002	119.934	120.000	-0.044	119.869	120.000	-0.109
	MIN	119.975	119.978	-0.025	119.955	119.978	-0.045	119.941	119.978	-0.059	119.899	119.978	-0.101	119.834	119.978	-0.166
160	MAX	160.012	160.000	0.037	159.988	160.000	0.013	159.972	160.000	-0.003	159.915	160.000	-0.060	159.825	160.000	-0.150
	MIN	159.972	159.975	-0.028	159.948	159.975	-0.052	159.932	159.975	-0.064	159.875	159.975	-0.125	159.785	159.975	-0.215
200	MAX	200.013	200.000	0.042	199.986	200.000	0.015	199.967	200.000	-0.004	199.895	200.000	-0.076	199.781	200.000	-0.190
	MIN	199.967	199.971	-0.033	199.940	199.971	-0.060	199.921	199.971	-0.079	199.849	199.971	-0.151	199.735	199.971	-0.265
250	MAX	250.013	250.000	0.042	249.986	250.000	0.015	249.967	250.000	-0.004	249.877	250.000	-0.094	249.733	250.000	-0.236
	MIN	249.967	249.971	-0.033	249.940	249.971	-0.060	249.921	249.971	-0.079	249.831	249.971	-0.169	249.687	249.971	-0.313
300	MAX	300.016	300.000	0.048	299.986	300.000	0.018	299.964	300.000	-0.004	299.850	300.000	-0.118	299.670	300.000	-0.292
	MIN	299.964	299.968	-0.036	299.934	299.968	-0.066	299.912	299.968	-0.088	299.798	299.968	-0.202	299.618	299.968	-0.382
400	MAX	400.017	400.000	0.053	399.984	400.000	0.020	399.959	400.000	-0.005	399.813	400.000	-0.151	399.588	400.000	-0.378
	MIN	399.960	399.964	-0.040	399.927	399.964	-0.073	399.902	399.964	-0.098	399.756	399.964	-0.244	399.529	399.964	-0.471
500	MAX	500.018	500.000	0.058	499.983	500.000	0.023	499.955	500.000	-0.005	499.771	500.000	-0.189	499.483	500.000	-0.477
	MIN	499.955	499.960	-0.045	499.920	499.960	-0.080	499.892	499.960	-0.108	499.708	499.960	-0.292	499.420	499.960	-0.580

METRIC K 7/h 6: LOCATIONAL TRANSITION FIT

BASIC DIA.	40 mm
HOLE	40.007 39.982
SHAFT	40.000 39.984
MAX CLEAR.	+0.023
MIN CLEAR. (ALLOWANCE)	-0.018

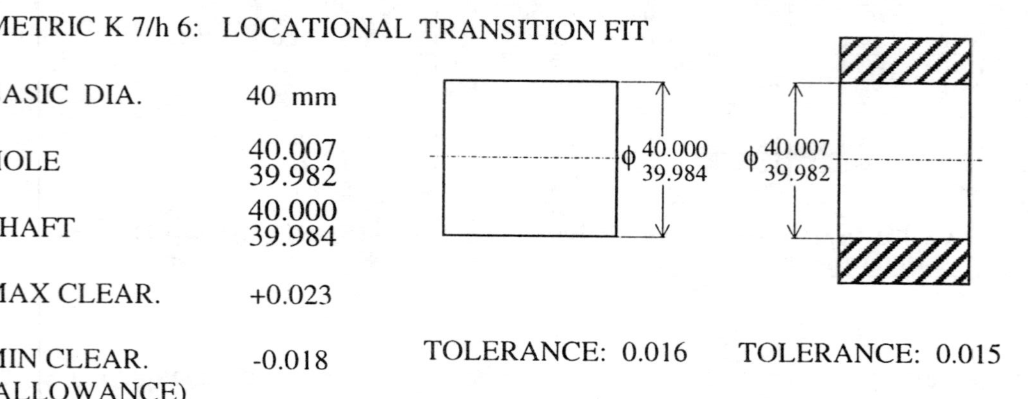

Appendix 10 The International Tolerance Grades (ANSI B4.2)

Basic Size		Tolerance grades																	
Over	Up to and Including	IT01	IT0	IT1	IT2	IT3	IT4	IT5	IT6	IT7	IT8	IT9	IT10	IT11	IT12	IT13	IT14	IT15	IT16
0	3	0.0003	0.0005	0.0008	0.0012	0.002	0.004	0.008	0.010	0.010	0.014	0.026	0.040	0.060	0.100	0.140	0.250	0.400	0.500
3	6	0.0004	0.0006	0.001	0.0015	0.0025	0.004	0.006	0.008	0.012	0.018	0.030	0.048	0.075	0.120	0.180	0.300	0.480	0.750
6	10	0.0004	0.0008	0.001	0.0015	0.0025	0.004	0.006	0.009	0.015	0.022	0.038	0.058	0.090	0.150	0.220	0.360	0.580	0.900
10	18	0.0005	0.0008	0.0012	0.002	0.003	0.006	0.008	0.011	0.018	0.027	0.043	0.070	0.110	0.180	0.270	0.430	0.700	1.100
18	30	0.0006	0.001	0.0016	0.0026	0.004	0.006	0.008	0.013	0.021	0.033	0.052	0.064	0.130	0.210	0.330	0.520	0.840	1.300
30	50	0.0008	0.001	0.0018	0.0026	0.004	0.007	0.011	0.016	0.025	0.039	0.062	0.100	0.160	0.250	0.390	0.620	1.000	1.600
50	80	0.0008	0.0012	0.002	0.003	0.005	0.008	0.013	0.019	0.030	0.048	0.074	0.120	0.190	0.300	0.460	0.740	1.200	1.900
80	120	0.001	0.0015	0.0025	0.004	0.008	0.0010	0.016	0.022	0.036	0.054	0.087	0.140	0.220	0.350	0.540	0.870	1.400	2.200
120	180	0.0012	0.002	0.0035	0.005	0.008	0.012	0.018	0.025	0.040	0.063	0.100	0.160	0.250	0.400	0.630	1.000	1.600	2.500
180	250	0.002	0.003	0.0045	0.007	0.010	0.014	0.020	0.029	0.048	0.072	0.115	0.185	0.290	0.460	0.720	1.150	1.850	2.900
250	315	0.0025	0.004	0.006	0.008	0.012	0.016	0.023	0.032	0.062	0.061	0.130	0.210	0.320	0.520	0.810	1.300	2.100	3.200
315	400	0.003	0.005	0.007	0.009	0.013	0.018	0.025	0.038	0.067	0.089	0.140	0.230	0.360	0.570	0.890	1.400	2.300	3.600
400	500	0.004	0.006	0.008	0.010	0.015	0.020	0.027	0.040	0.063	0.097	0.156	0.250	0.400	0.630	0.970	1.550	2.500	4.000
500	630	0.046	0.006	0.009	0.011	0.016	0.022	0.030	0.044	0.070	0.110	0.175	0.280	0.440	0.700	1.100	1.750	2.800	4.400
630	800	0.005	0.007	0.010	0.013	0.018	0.025	0.035	0.050	0.080	0.125	0.200	0.320	0.500	0.800	1.250	2.000	3.200	5.000
800	1000	0.0055	0.008	0.011	0.015	0.021	0.029	0.040	0.056	0.090	0.140	0.230	0.360	0.560	0.900	1.400	2.300	3.600	5.600
1000	1250	0.0065	0.009	0.013	0.018	0.024	0.034	0.046	0.066	0.105	0.165	0.260	0.420	0.660	1.050	1.860	2.600	4.200	6.600
1260	1800	0.008	0.011	0.015	0.021	0.029	0.040	0.064	0.078	0.125	0.196	0.310	0.500	0.780	1.250	1.950	3.100	5.000	7.800
1800	2000	0.009	0.013	0.018	0.025	0.035	0.048	0.065	0.092	0.150	0.230	0.370	0.600	0.920	1.500	2.300	3.700	6.000	9.200
2000	2500	0.011	0.015	0.022	0.030	0.041	0.067	0.077	0.110	0.175	0.280	0.440	0.700	1.100	1.750	2.800	4.400	7.000	11.000
2500	3150	0.013	0.018	0.026	0.036	0.050	0.069	0.093	0.135	0.210	0.330	0.540	0.860	1.360	2.100	3.300	5.400	8.600	13.500

Formulas: IT17 = IT12 x 10 and IT18 = IT13 x 10, etc.

Appendix 11 Cumulative Standard Normal Distribution Table N(0, 1)

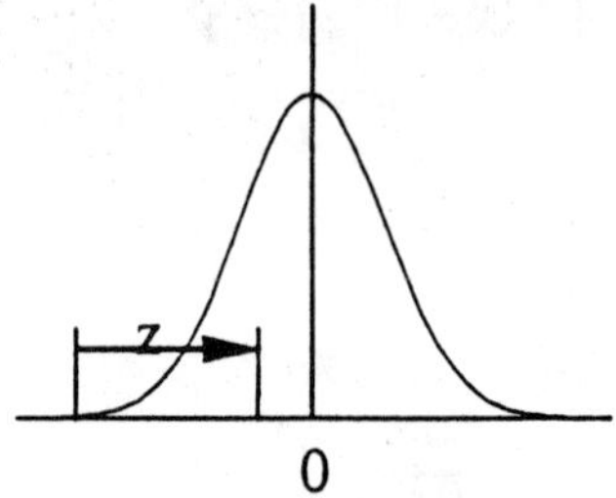

	0.00	0.01	0.02	0.03	0.04	0.05	0.06	0.07	0.08	0.09
0.0	0.5000	0.5040	0.5080	0.5120	0.5160	0.5199	0.5239	0.5279	0.5319	0.5359
0.1	0.5398	0.5438	0.5478	0.5517	0.5557	0.5596	0.5636	0.5675	0.5714	0.5753
0.2	0.5793	0.5832	0.5871	0.5910	0.5948	0.5987	0.6026	0.6064	0.6103	0.6141
0.3	0.6179	0.6217	0.6255	0.6293	0.6331	0.6368	0.6406	0.6443	0.6480	0.6517
0.4	0.6554	0.6591	0.6628	0.6664	0.6700	0.6736	0.6772	0.6808	0.6844	0.6879
0.5	0.6915	0.6950	0.6985	0.7019	0.7054	0.7088	0.7123	0.7157	0.7190	0.7224
0.6	0.7257	0.7291	0.7324	0.7357	0.7389	0.7422	0.7454	0.7486	0.7517	0.7549
0.7	0.7580	0.7611	0.7642	0.7673	0.7704	0.7734	0.7764	0.7794	0.7823	0.7852
0.8	0.7881	0.7910	0.7939	0.7967	0.7995	0.8023	0.8051	0.8078	0.8106	0.8133
0.9	0.8159	0.8186	0.8212	0.8238	0.8264	0.8289	0.8315	0.8340	0.8365	0.8389
1.0	0.8413	0.8438	0.8461	0.8485	0.8508	0.8531	0.8554	0.8577	0.8599	0.8621
1.1	0.8643	0.8665	0.8686	0.8708	0.8729	0.8749	0.8770	0.8790	0.8810	0.8830
1.2	0.8849	0.8869	0.8888	0.8907	0.8925	0.8944	0.8962	0.8980	0.8997	0.9015
1.3	0.9032	0.9049	0.9066	0.9082	0.9099	0.9115	0.9131	0.9147	0.9162	0.9177
1.4	0.9192	0.9207	0.9222	0.9236	0.9251	0.9265	0.9279	0.9292	0.9306	0.9319
1.5	0.9332	0.9345	0.9357	0.9370	0.9382	0.9394	0.9406	0.9418	0.9429	0.9441
1.6	0.9452	0.9463	0.9474	0.9484	0.9495	0.9505	0.9515	0.9525	0.9535	0.9545
1.7	0.9554	0.9564	0.9573	0.9582	0.9591	0.9599	0.9608	0.9616	0.9625	0.9633
1.8	0.9641	0.9649	0.9656	0.9664	0.9671	0.9678	0.9686	0.9693	0.9699	0.9706
1.9	0.9713	0.9719	0.9726	0.9732	0.9738	0.9744	0.9750	0.9756	0.9761	0.9767
2.0	0.9772	0.9778	0.9783	0.9788	0.9793	0.9798	0.9803	0.9808	0.9812	0.9817
2.1	0.9821	0.9826	0.9830	0.9834	0.9838	0.9842	0.9846	0.9850	0.9854	0.9857
2.2	0.9861	0.9864	0.9868	0.9871	0.9875	0.9878	0.9881	0.9884	0.9887	0.9890
2.3	0.9893	0.9896	0.9898	0.9901	0.9904	0.9906	0.9909	0.9911	0.9913	0.9916
2.4	0.9918	0.9920	0.9922	0.9925	0.9927	0.9929	0.9931	0.9932	0.9934	0.9936
2.5	0.9938	0.9940	0.9941	0.9943	0.9945	0.9946	0.9948	0.9949	0.9951	0.9952
2.6	0.9953	0.9955	0.9956	0.9957	0.9959	0.9960	0.9961	0.9962	0.9963	0.9964
2.7	0.9965	0.9966	0.9967	0.9968	0.9969	0.9970	0.9971	0.9972	0.9973	0.9974
2.8	0.9974	0.9975	0.9976	0.9977	0.9977	0.9978	0.9979	0.9979	0.9980	0.9981
2.9	0.9981	0.9982	0.9982	0.9983	0.9984	0.9984	0.9985	0.9985	0.9986	0.9986
3.0	0.9987	0.9987	0.9987	0.9988	0.9988	0.9989	0.9989	0.9989	0.9990	0.9990
3.1	0.9990	0.9991	0.9991	0.9991	0.9992	0.9992	0.9992	0.9992	0.9993	0.9993
3.2	0.9993	0.9993	0.9994	0.9994	0.9994	0.9994	0.9994	0.9995	0.9995	0.9995
3.3	0.9995	0.9995	0.9995	0.9996	0.9996	0.9996	0.9996	0.9996	0.9996	0.9997
3.4	0.9997	0.9997	0.9997	0.9997	0.9997	0.9997	0.9997	0.9997	0.9997	0.9998
3.5	0.9998	0.9998	0.9998	0.9998	0.9998	0.9998	0.9998	0.9998	0.9998	0.9998
3.6	0.9998	0.9998	0.9999	0.9999	0.9999	0.9999	0.9999	0.9999	0.9999	0.9999
3.7	0.9999	0.9999	0.9999	0.9999	0.9999	0.9999	0.9999	0.9999	0.9999	0.9999
3.8	0.9999	0.9999	0.9999	0.9999	0.9999	0.9999	0.9999	0.9999	0.9999	0.9999
3.9	1.0000	1.0000	1.0000	1.0000	1.0000	1.0000	1.0000	1.0000	1.0000	1.0000

INDEX

T

V

W

X

Z